Ocean Modelling in Support of Operational Ocean and Coastal Services

Ocean Modelling in Support of Operational Ocean and Coastal Services

Editor

Marcos G. Sotillo

MDPI • Basel • Beijing • Wuhan • Barcelona • Belgrade • Manchester • Tokyo • Cluj • Tianjin

Editor
Marcos G. Sotillo
Nologin Oceanic Weather Systems,
Madrid, Spain

Editorial Office
MDPI
St. Alban-Anlage 66
4052 Basel, Switzerland

This is a reprint of articles from the Special Issue published online in the open access journal *Journal of Marine Science and Engineering* (ISSN 2077-1312) (available at: https://www.mdpi.com/journal/jmse/special_issues/ocean_support_services).

For citation purposes, cite each article independently as indicated on the article page online and as indicated below:

LastName, A.A.; LastName, B.B.; LastName, C.C. Article Title. *Journal Name* **Year**, *Volume Number*, Page Range.

ISBN 978-3-0365-6261-2 (Hbk)
ISBN 978-3-0365-6262-9 (PDF)

Cover image courtesy of Marcos García Sotillo.

Contents

About the Editor

Marcos G. Sotillo

Marcos G. Sotillo, currently Technical Director of Nologin Oceanic Weather systems (NOW), is a senior scientist (Ph.D. in physics) with more than 25 years of experience as an ocean and atmospheric modeler. He is the developer of and is responsible for different Operational Oceanographical Forecast services at regional and coastal/local scales (built in the context of several projects and initiatives developed along his professional career in centers such as Puertos del Estado and Mercator Ocean International). He has authors papers published in more than 45 scientific peer-reviewed journals (first author of more than 20) and has participated in more than 25 R&D projects and contracts at the national and European levels (in several as the IP, coordinator or responsible). In the context of the Copernicus Marine Service, he is currently the leader element (deputy) of the regional service for the European Atlantic façade (the IBI-MFC Iberia–Biscay–Ireland Monitoring and Forecasting Centre) and is the former Copernicus Marine Product Quality Coordinator.

Preface to "Ocean Modelling in Support of Operational Ocean and Coastal Services"

Operational oceanography is maturing rapidly. Its capabilities are being noticeably enhanced in response to a growing demand for regularly updated ocean information. Today, several core forecasting and monitoring services, such as the Copernicus Marine ones focused on global and regional scales, are well-established. The sustained availability of oceanography products has favored the proliferation of specific downstream services devoted to coastal monitoring and forecasting. Ocean models are a key component of these operational oceanographic systems (especially in a context marked by extensive application of dynamical downscaling approaches), and progress in ocean modeling is certainly a driver for the evolution of these services.

The goal of this Special Issue is to publish research papers on ocean modeling that advance modeling applications and that support existing operational oceanographic services. Eleven high-quality papers are compiled in this Special Issue. They are mainly focused on combining ocean models with observational products, including data assimilation and coupling topics.

This Special Issue is addressed to an audience with interests on physical oceanography and especially on its operational applications. There is a focus on the numerical modeling needed for better forecasting in marine environments using seamless modeling approaches to simulate global to coastal processes. Specifically, the following thematic groups or topics are highlighted:

- Ocean model improvements to support a better (or new) operational forecasting;
- Increasing interactions between systems (including contributions to the river hydrological system), improving data forcing, and promoting coupling approaches;
- Data assimilation progress (towards better analysis products);
- Enhancing product quality assessment and demonstrating downscaling's added value.

Finally, I would like to take this opportunity to acknowledge to the Editorial Team of JMSE, and especially to Ms. Cynthia Li (MDPI editor manager), who assisted during the Special Issue's editing process, as well as to all the talented authors contributing with their research papers and to the hardworking and professional reviewers involved in the peer-review process.

Marcos G. Sotillo
Editor

Journal of
Marine Science and Engineering

MDPI

Editorial

Ocean Modelling in Support of Operational Ocean and Coastal Services

Marcos G. Sotillo

Nologin Ocean Weather Systems-NOW Systems/Nologin Consulting, 28046 Madrid, Spain;
marcos.sotillo@nologin.es

Operational oceanography is maturing rapidly. Its related capabilities are being noticeably enhanced in response to a growing demand for regularly updated ocean information. Today, several core forecasting and monitoring services, focused on global and regional scales, are well established. The sustained availability of their products has favored the proliferation of downstream services devoted to coastal monitoring and forecasting. Ocean models (together with proliferation in the use of data-assimilation schemes to generate analysis) are key components in operational oceanographic systems, as ocean modeling drives the evolution of services (especially in a context marked by the extensive application of dynamic downscaling approaches for coastal applications).

The goal of this Special Issue was to publish research focused on ocean modeling to benefit model applications and support existing operational oceanographic services. Likewise, research works focused on improving combinations of ocean models with observational products, including data assimilation, were also considered, as well as papers addressing model product quality assessments and evaluations of operational service capabilities to simulate outstanding marine processes/features and extreme events.

The book "Ocean Modelling in Support of Operational Ocean and Coastal Services" includes eleven contributions [1–11] to this Special Issue published during 2021–2022. The overall objective of this Special Issue is to draw an updated picture of the heterogeneous challenges that are being addressed in operational oceanographic regional and coastal services, gathering novel and cutting-edge techniques and methods to advance the state of the art of forecast and analysis products. The scientific collection presented in this Special Issue will be valuable for both scientists/technical developers and end-users, since each paper within is motivated by an applied and pragmatic spirit, with the objective of providing better and more fit-for-purpose ocean model solutions (from a widespread set of applications) that meet a growing number of end-user requirements and needs.

Considering the eleven different contributions presented in this Special Issue, it is possible to highlight five major goals or thematic groups that seem to articulate them:

- Ocean model improvements to support better (or new) operational forecasting [2,5,7,9,10];
- Increasing interactions between systems (including contributions to the river hydrological system), improving data forcing and promoting coupling approaches [3,6,7,9–11];
- Data assimilation progress (towards better analysis products) [7,8,10];
- Enhancing product quality assessment and demonstrating downscaling's added value [2,8–10].
- New approaches to improve/optimize specific components in operational services [1,4,11].

A brief overview of all the contributions follows, emphasizing the main investigation topic and the outcome of the analysis.

Some of the contributions are related to new services or developments achieved in the context of the Copernicus Marine Regional Monitoring and Forecasting Services.

Ciliberti et al. [7] describe the latest developments in the context of the Copernicus Marine Black Sea Monitoring and Forecasting Center to improve the skill of their ocean

Citation: Sotillo, M.G. Ocean Modelling in Support of Operational Ocean and Coastal Services. *J. Mar. Sci. Eng.* **2022**, *10*, 1482. https://doi.org/10.3390/jmse10101482

Received: 9 October 2022
Accepted: 9 October 2022
Published: 12 October 2022

analyses, forecasts, and re-analyses (for hydrodynamics, biogeochemistry, and waves). Model solutions are generated through a set of dedicated online coupled systems which involve physics, biogeochemistry, and waves, together with the atmosphere. A special focus on the physical component of this regional Black Sea Copernicus Marine service is provided by Ciliberti et al. [8], including insight for the ocean model and data assimilation scheme used, as well as details on the operational quality assessment framework. On the European–Atlantic side, Toledano et al. [10] show the most recent advances of the new Copernicus Marine regional IBI wave forecast service, emphasizing the impacts that both a new altimetric wave data assimilation scheme and current–wave coupling have on operational wave products. The demonstrated benefits, related to the herein proposed upgrades, supported the IBI-MFC decision to evolve its operational wave system, using (since the March 2020 Copernicus Marine Release) the resulting wave model set-up, with data assimilation and current–wave coupling for operational purposes.

Some other contributions are focused on new services implemented at national or local coastal levels, with modelling applications covering sites of specific interest. Thus, García-León et al. [9] show the most recent developments applied to the SAMOA forecast service deployed for 31 Spanish ports. Research was focused on upgrading the ROMS model set-up, together with atmospheric forcing and the testing of new methodologies to nest the SAMOA system into the Copernicus Marine IBI-MFC regional solution (with emphasis on its 3D hourly dataset). Model sensitivity tests were extensively assessed using the available in situ and remote sensed (including RadarHF) observations, demonstrating the added value of the downscaled coastal solutions. Ilicak et al. [5] simulate the Turkish Strait System, connecting the Black Sea and the Mediterranean using a model of the Turkish strait system using a high-resolution unstructured grid ocean circulation model, with realistic atmospheric forcing and lateral open-boundary conditions from the Copernicus Marine system. Nagy et al. [2] illustrate the main achievements associated with the development and validation of a new operational model system for an area of the northeast Atlantic that encompasses all of Ireland's territorial waters. The system, based on an ROMS application, uses operationally available atmospheric and boundary forcing, tidal forcing, and a climatological river contribution. Some insights on the skill of the systems are provided, compared with in situ and satellite observations, and specific studies on the simulation of the Irish coastal current are provided.

New approaches to improve/optimize specific components of operational services are also included in the Special Issue: Yang et al. [4] propose a barotropic solver for high-resolution ocean general circulation models that helps to ensure scalability and the optimization of performance. The new algorithm proposed was tested on the NEMO ocean model, allowing good scalability and significant computational execution time reductions. Yuan et al. [1] aim to use singular spectrum analysis (SSA) to reduce the noise level in Jason-1 altimeter waveforms in order to obtain SSA-denoised waveforms, and thus improve the accuracy of a mean sea surface height (MSSH) model. This methodology can help to reduce the contamination of altimeter waveforms (usually due to nonmarine surfaces or inhomogeneous sea-state conditions). Model accuracy enhancements over the China Sea are obtained using SSA-denoised waveforms, compared with those for raw waveforms.

Related to the increase in interactions between systems, improving uses of data forcing, and promoting coupling approaches, Causio et al. [6] study wave–current interaction for the first time in the Black Sea, implementing a coupled numerical system based on NEMO and WaveWatchIII, evaluating how waves impact surface ocean dynamics, and how ocean currents may impact sea state. Different physical processes are considered in the study (i.e., Stokes–Coriolis force, sea-state-dependent momentum flux, wave-induced vertical mixing, Doppler shift effect, and stability parameter for the computation of effective wind speed).

Finally, with respect to the inclusion of a contribution regarding the river hydrological system, Campuzano et al. [11] aim to explore the potential impacts on the ocean system that can be obtained by using sophisticated land boundary conditions based on the capacities of state-of-the-art hydrologic models combined with observation networks. Together with the assessment of the source of river-flow data, this work also explores the use of estuarine proxies based on simple modelling grids. The estuarine proxies enable the incorporation of the mixing processes that take place in estuaries into the land fluxes, and obtain plume momentum. The watershed, estuarine proxies, and ocean were modelled using the MOHID water modelling system and evaluated in Western Iberia waters (illustrating the sea surface salinity extension of the Western Iberia Buoyant Plume during an extreme event). Sotillo et al. [3] analyses the river freshwater contribution of the European Atlantic margin and its influence on the sea salinity field by means of model sensitivity tests. The paper assesses the potential impacts that newly available river discharge datasets can have on regional ocean model solutions delivered by the Copernicus Marine IBI-MFC regional service when used as part of the freshwater land forcing. Usages of these data result in a better capture of salinity variability and a more realistic simulation of baroclinic frontal structures linked to coastal and river freshwater buoyancy plumes; major impacts are found in areas with bigger river discharge contributions (i.e., the French shelf or the northwestern Iberian coast).

Funding: This research received no external funding.

Conflicts of Interest: The authors declare no conflict of interest.

References

1. Yuan, J.; Guo, J.; Niu, Y.; Zhu, C.; Li, Z.; Liu, X. Denoising Effect of Jason-1 Altimeter Waveforms with Singular Spectrum Analysis: A Case Study of Modelling Mean Sea Surface Height over South China Sea. *J. Mar. Sci. Eng.* **2020**, *8*, 426. [CrossRef]
2. Nagy, H.; Lyons, K.; Nolan, G.; Cure, M.; Dabrowski, T. A Regional Operational Model for the North East Atlantic: Model Configuration and Validation. *J. Mar. Sci. Eng.* **2020**, *8*, 673. [CrossRef]
3. Sotillo, M.G.; Campuzano, F.; Guihou, K.; Lorente, P.; Olmedo, E.; Matulka, A.; Santos, F.; Amo-Baladrón, M.A.; Novellino, A. River Freshwater Contribution in Operational Ocean Models along the European Atlantic Façade: Impact of a New River Discharge Forcing Data on the CMEMS IBI Regional Model Solution. *J. Mar. Sci. Eng.* **2021**, *9*, 401. [CrossRef]
4. Yang, X.; Zhou, S.; Zhou, S.; Song, Z.; Liu, W. A Barotropic Solver for High-Resolution Ocean General Circulation Models. *J. Mar. Sci. Eng.* **2021**, *9*, 421. [CrossRef]
5. Ilicak, M.; Federico, I.; Barletta, I.; Mutlu, S.; Karan, H.; Ciliberti, S.A.; Clementi, E.; Coppini, G.; Pinardi, N. Modeling of the Turkish Strait System Using a High Resolution Unstructured Grid Ocean Circulation Model. *J. Mar. Sci. Eng.* **2021**, *9*, 769. [CrossRef]
6. Causio, S.; Ciliberti, S.A.; Clementi, E.; Coppini, G.; Lionello, P. A Modelling Approach for the Assessment of Wave-Currents Interaction in the Black Sea. *J. Mar. Sci. Eng.* **2021**, *9*, 893. [CrossRef]
7. Ciliberti, S.A.; Grégoire, M.; Staneva, J.; Palazov, A.; Coppini, G.; Lecci, R.; Peneva, E.; Matreata, M.; Marinova, V.; Masina, S.; et al. Monitoring and Forecasting the Ocean State and Biogeochemical Processes in the Black Sea: Recent Developments in the Copernicus Marine Service. *J. Mar. Sci. Eng.* **2021**, *9*, 1146. [CrossRef]
8. Ciliberti, S.A.; Jansen, E.; Coppini, G.; Peneva, E.; Azevedo, D.; Causio, S.; Stefanizzi, L.; Creti', S.; Lecci, R.; Lima, L.; et al. The Black Sea Physics Analysis and Forecasting System within the Framework of the Copernicus Marine Service. *J. Mar. Sci. Eng.* **2022**, *10*, 48. [CrossRef]
9. García-León, M.; Sotillo, M.G.; Mestres, M.; Espino, M.; Fanjul, E.Á. Improving Operational Ocean Models for the Spanish Port Authorities: Assessment of the SAMOA Coastal Forecasting Service Upgrades. *J. Mar. Sci. Eng.* **2022**, *10*, 149. [CrossRef]
10. Toledano, C.; Ghantous, M.; Lorente, P.; Dalphinet, A.; Aouf, L.; Sotillo, M.G. Impacts of an Altimetric Wave Data Assimilation Scheme and Currents-Wave Coupling in an Operational Wave System: The New Copernicus Marine IBI Wave Forecast Service. *J. Mar. Sci. Eng.* **2022**, *10*, 457. [CrossRef]
11. Campuzano, F.; Santos, F.; Simionesei, L.; Oliveira, A.R.; Olmedo, E.; Turiel, A.; Fernandes, R.; Brito, D.; Alba, M.; Novellino, A.; et al. Framework for Improving Land Boundary Conditions in Ocean Regional Products. *J. Mar. Sci. Eng.* **2022**, *10*, 852. [CrossRef]

Journal of
Marine Science and Engineering

MDPI

Article

The Black Sea Physics Analysis and Forecasting System within the Framework of the Copernicus Marine Service

Stefania A. Ciliberti [1,*], Eric Jansen [1], Giovanni Coppini [1], Elisaveta Peneva [2], Diana Azevedo [1], Salvatore Causio [1], Laura Stefanizzi [1], Sergio Creti' [1], Rita Lecci [1], Leonardo Lima [3], Mehmet Ilicak [4], Nadia Pinardi [1,5] and Atanas Palazov [6]

1 Ocean Predictions and Applications Division, Fondazione Centro Euro-Mediterraneo sui Cambiamenti Climatici, 73100 Lecce, Italy; eric.jansen@cmcc.it (E.J.); giovanni.coppini@cmcc.it (G.C.); diana.azevedo@cmcc.it (D.A.); salvatore.causio@cmcc.it (S.C.); laura.stefanizzi@cmcc.it (L.S.); sergio.creti@cmcc.it (S.C.); rita.lecci@cmcc.it (R.L.); nadia.pinardi@unibo.it (N.P.)
2 Department of Meteorology and Geophysics, Faculty of Physics, Sofia University "St. Kliment Ohridski", 1164 Sofia, Bulgaria; elfa@phys.uni-sofia.bg
3 Ocean Modelling and Data Assimilation Division, Fondazione Centro Euro-Mediterraneo sui Cambiamenti Climatici, 40127 Bologna, Italy; leonardo.lima@cmcc.it
4 Eurasia Institute of Earth Sciences, Istanbul Technical University, Istanbul 34467, Turkey; milicak@itu.edu.tr
5 Department of Physics and Astronomy "Augusto Righi", University of Bologna Alma Mater Studiorum, 40126 Bologna, Italy
6 Institute of Oceanology, Bulgarian Academy of Science, 9000 Varna, Bulgaria; palazov@io-bas.bg
* Correspondence: stefania.ciliberti@cmcc.it

Abstract: This work describes the design, implementation and validation of the Black Sea physics analysis and forecasting system, developed by the Black Sea Physics production unit within the Black Sea Monitoring and Forecasting Center as part of the Copernicus Marine Environment and Monitoring Service. The system provides analyses and forecasts of the temperature, salinity, sea surface height, mixed layer depth and currents for the whole Black Sea basin, excluding the Azov Sea, and has been operational since 2016. The system is composed of the NEMO (v 3.4) numerical model and an OceanVar scheme, which brings together real time observations (in-situ temperature and salinity profiles, sea level anomaly and sea surface temperature satellite data). An operational quality assessment framework is used to evaluate the accuracy of the products which set the basic standards for the future upgrades, highlighting the strengths and weaknesses of the model and the observing system in the Black Sea.

Keywords: Black Sea; operational oceanography; numerical modelling; data assimilation; forecasting; validation

Citation: Ciliberti, S.A.; Jansen, E.; Coppini, G.; Peneva, E.; Azevedo, D.; Causio, S.; Stefanizzi, L.; Creti', S.; Lecci, R.; Lima, L.; et al. The Black Sea Physics Analysis and Forecasting System within the Framework of the Copernicus Marine Service. *J. Mar. Sci. Eng.* **2022**, *10*, 48. https://doi.org/10.3390/jmse10010048

Academic Editor: Carlos Guedes Soares

Received: 14 December 2021
Accepted: 30 December 2021
Published: 2 January 2022

1. Introduction

Operational forecasting is now a reality in most ocean areas around the world. It follows the example of weather forecasting [1] and provides products that are crucial for the sustainable development of activities at sea and along the coasts.

The operational quality of the products has steadily improved since the late 1990s, when only a few centers around the world were engaged in ocean analysis and forecasting [2]. The operational system assessment allows Earth System Science to estimate the quality and fitness of the numerical ocean model for purpose of the observing system.

Operational forecasting in the Black Sea region is part of the data production architecture of the Copernicus Marine Environment and Monitoring Service (CMEMS, https://marine.copernicus.eu/ accessed on 29 December 2021, [3]). The Black Sea Monitoring and Forecasting Center (BS-MFC) has been operative since 2016. The center provides regular, systematic and operational marine information, such as numerical analyses, forecasts and reanalyses for the main physical parameters (temperature, salinity, sea level, currents), together with biogeochemical and wave products.

The Black Sea operational analysis and Forecasting System (BSFS), within BS-MFC, is devoted to providing near real-time information of the physical ocean state in the basin. It is implemented in the Black Sea region, covering the area shown in Figure 1.

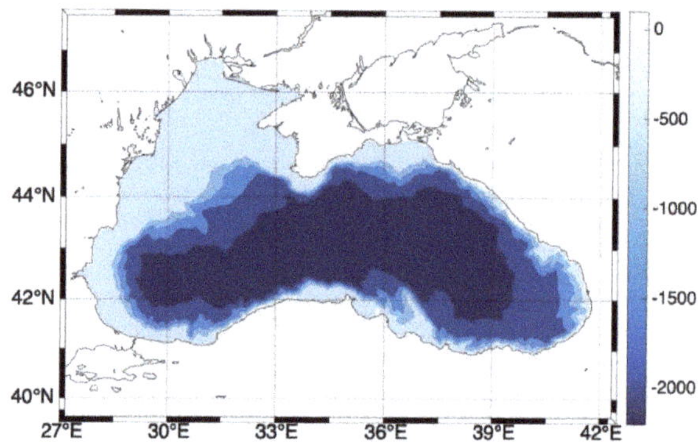

Figure 1. BSFS spatial domain and bathymetry (expressed in meters, from 0 to −2200 m).

The system is based on three major components: collection of upstream data, including atmospheric forcing and observations, the numerical ocean model, and a variational data assimilation scheme. Together with the Mediterranean Sea forecasting system, which began operation in 1998 [4,5], the Black Sea one completed the operational capacity in the Southern Regional European Seas [6]. These operational products also reconstruct the past state of the Black Sea, thus providing the optimal data set to study the ocean climate variability and in general the Black Sea general circulation. We are currently at the third generation of forecasting systems, and this paper analyzes the accuracy of the first ocean forecasting system for the Black Sea physics.

This study presents the numerical setup, operational implementation and product quality assessment for the period January 2018 to December 2020, using observations from in-situ (temperature and salinity profiles) and satellite (sea surface temperature and sea level anomaly) platforms, provided by CMEMS Thematic Assembly Centers (TAC) [3].

The paper is organized as follows: Section 2 presents the system, including the ocean model, the data assimilation method used for the forecasting cycle and the processing chain; Section 3 describes the operational products and discusses the product quality; Section 4 presents the conclusions and future evolutions of the BSFS.

2. System Description

2.1. Ocean Numerical Model

BSFS is based on a free surface implementation of the NEMO hydrodynamical model (v3.4, [7]) for the Black Sea region. The horizontal resolution is approximately 3 km (1/27° in zonal and 1/36° in meridional directions), which conforms to the mesoscale eddy-resolving scale (Rossby radius of deformation in the Black Sea is ~20 km, [8]). On the other hand, 31 unevenly distributed z-levels are used along the vertical direction, with an initial layer at about 2.5 m in depth. BSFS bathymetry is derived from GEBCO at 1′ resolution [9]; a local refinement of the coastline, using a high-resolution NOAA dataset [10], is performed to represent the main coastal features of the basin.

The model is forced by momentum, water and heat fluxes interactively computed by bulk formulae implemented first for the Mediterranean Sea [11] and adapted for the Black Sea, as described in Appendix A. The analysis and forecast atmospheric fields are provided by the European Centre for Medium Range Weather Forecast (ECMWF) at 0.125° horizontal resolution, at 3–6 h frequency.

The momentum equation is written in flux form and is solved with a leapfrog time stepping scheme. The free surface equation uses a linearized form (i.e., the barotropic velocity field is defined with an integral from the flat surface to the bottom) and is integrated implicitly [12] with the time step of the total velocity field equal to 150 s. The advection scheme for the temperature and salinity tracers is the total variance dissipation (TVD) scheme. Horizontal eddy diffusivity is applied for tracers using a Laplacian operator, with a coefficient of 50 m^2/s, while a bi-Laplacian viscosity is applied for momentum, with a coefficient of 1.0×10^8 m^4/s. The vertical diffusivity and viscosity are computed from a turbulent kinetic energy (TKE) closure model, with the parameters set as in [7]. Vertical eddy viscosity and diffusivity coefficients are selected with values of 1.2×10^{-5} m^2/s and 1.2×10^{-6} m^2/s, respectively. The model also implements free-slip lateral boundary conditions and a classical quadratic bottom friction scheme, with a drag coefficient of 1.0×10^{-3} m^2/s^2.

The Bosporus Strait connects the Black Sea with the Sea of Marmara: to model it, a two-layer, narrow strait water mass exchange which maintains a relatively steady state water and salt balance in the Black Sea [13–18] is used, with a closed boundary condition. The excess precipitation and river runoff over evaporation is removed using the outflow from the strait, thus leading to a zero balance in the Black Sea [19]. Considering the horizontal average of the free surface equation [20], the net transport at the Bosporus Strait is given by:

$$\frac{1}{A}Tr_B = -\frac{\partial}{\partial t}\langle\eta\rangle - \langle E - P - R\delta(x - x_i)\rangle \tag{1}$$

where Tr_B is the net transport at the Bosporus Strait, and A is the basin surface area, η is the sea surface height, E, P, R are, respectively, evaporation, precipitation and runoff, the Dirac δ is different from zero at 72 river mouths and the triangular brackets mean horizonal basin average. First, we redefine the transport at the Bosporus Strait in terms of a "discharge" R_B :

$$Tr_B = L_{i,B}H_{i,B}\, R_B \tag{2}$$

Thus, the discharge at the Bosporus Strait is given by:

$$R_B = -\frac{A}{L_{i,B}H_{i,B}}\left[\frac{\partial}{\partial t}\langle\eta\rangle + \langle E - P - R\delta(x - x_i)\rangle\right] \tag{3}$$

where $H_{i,B}$ and $L_{i,B}$ are depth and width in one-grid-sea-point, respectively.

We calculated the Bosporus discharge in (3) from a 10-year simulation by computing the mean free surface tendency and the mean water flux. The values of the Bosporus discharge are stored as monthly mean values and set as vertical velocity boundary conditions as done for the rivers, except with the negative sign, indicating a discharge out of the basin or a "negative river". This parametrization is quite robust for decadal long simulations that do not consider climate change trends in sea level and water fluxes.

With regard to the 72 real river runoff contributions, we use monthly mean discharge data from the SESAME dataset [21] for all rivers, including the Danube, the Dniepr, the Dniester, the Rioni, the Kizil Irmak, and the Sakarya.

2.2. Data Assimilation Scheme

The ocean model is coupled with a data assimilation system in order to produce analyses for optimal initial conditions of the forecasts. The DA ocean state vector x contains the temperature T, salinity S and sea level values at each model grid location. For a model setup with n vertical levels, this vector is defined as:

$$x = (T_0, \ldots, T_n, S_0, \ldots, S_n, \eta) \tag{4}$$

Given an observation vector y_0 and the background model state x_b it is possible to define the innovation, i.e., the difference between the observations and their respective model predictions, denoted $\boldsymbol{d}$:

$$\boldsymbol{d} = H(x_b) - \boldsymbol{y}_0 \tag{5}$$

with $H(x_b)$ being the observation operator that projects the model state onto the observation space. The aim of the DA system is to find a correction to the model state:

$$\delta x = x - x_b \tag{6}$$

This correction needs to minimize the analysis error while taking into account the error covariances of the model state vector and the observations. There are several methods to calculate such corrections; however, they can be broadly categorized as variational methods or Kalman filters. In BSFS, a 3D-variational method is used [22,23]. In this method, the corrections are derived by iteratively minimizing a cost function J, defined as:

$$J = \frac{1}{2}\delta x^T \mathbf{B}^{-1} \delta x + \frac{1}{2}(\mathbf{H}\delta x - \boldsymbol{d})^T \mathbf{R}^{-1}(\mathbf{H}\delta x - \boldsymbol{d}) \tag{7}$$

Here the matrices **B** and **R** are the error covariance matrices of, respectively, the background state and the observations. For a number of state variables n_b, the model covariance matrix **B** is of size n_b^2. In addition, as the cost function contains $\mathbf{B}^{-1}$, the covariance matrix needs to be inverted. It is clear that for large numbers of state variables, this is a computationally costly calculation, which can be avoided if additional constraints are imposed on **B**, in particular if **B** is of the form:

$$\mathbf{B} = \mathbf{V}\mathbf{V}^T \tag{8}$$

with **V** an arbitrary matrix. In this case, Equation (7) can be written as:

$$J = \frac{1}{2}v^T v + \frac{1}{2}(\mathbf{H}\mathbf{V}v - \boldsymbol{d})^T \mathbf{R}^{-1}(\mathbf{H}\mathbf{V}v - \boldsymbol{d}) \tag{9}$$

introducing a coordinate transformation:

$$\delta x = \mathbf{V}v \tag{10}$$

$$v = \mathbf{V}^{-1}\delta x \tag{11}$$

The minimization of the cost function can now be performed in terms of the new control variable v, without the need to calculate $\mathbf{B}^{-1}$. It is sufficient to perform the minimization in terms of v before transforming back to the model state space increment δx with (10).

As the covariance is by definition a positive definite matrix, the matrix **V** exists and could be found by performing a Cholesky decomposition on **B**. However, since the model error covariance **B** is usually derived from model state variable anomalies in a long model run, **V** is simply the anomaly matrix and **B** is already calculated according to (8).

One of the main features of OceanVar is the covariance decomposition. The matrix **V** of (8) is expressed as a product of different components:

$$\mathbf{V} = \mathbf{V}_\mathrm{H}\mathbf{V}_\eta\mathbf{V}_\mathrm{V} \tag{12}$$

where $\mathbf{V}_\mathrm{H}$ is the horizontal and $\mathbf{V}_\mathrm{V}$ is the vertical component of the covariance. The horizontal component provides the correlation between neighboring grid points at each model level. This is implemented by means of a recursive filter with a radius of approximately 25 km in open sea and a gradual falloff near the coast. The vertical component is calculated from a long model simulation and decomposed using empirical orthogonal functions (EOF). This EOF decomposition significantly reduces the number of variables, $n_v \ll n_x$, and reduces spurious correlations due to the finite dataset used to estimate the covariance. V_η is the dynamic height operator, which uses the local hydrostatic adjustment scheme [23] to transform the sea level anomaly innovations into increments for T and S in regions deeper than 1000 m.

For the BSFS system with 31 vertical levels, the vertical covariance of T and S is represented by 15 EOF. The EOF are calculated separately for each month and each grid location to adequately capture the variability of the covariance in time and in space. The DA system is used to assimilate in-situ observations of temperature and salinity from available ARGO profilers in the Black Sea, and satellite observations of sea level anomaly (SLA) and of sea surface temperature (SST). For the latter, the foundation temperature is used and assimilated only at nighttime (as in e.g., [24]). This approach limits possible biases due to the diurnal cycle of SST.

2.3. Operational Chain

BSFS version here considered is EAS3 (European Analysis and forecasting System, version 3, referred also in the next sections) for the Black Sea. It implements technical interfaces with upstream data, used for model runs—including ECMWF atmospheric forcing, CMEMS in-situ and satellite observations from corresponding TACs—as well as with the CMEMS Dissemination Unit for the delivery of analysis and forecast products to users. Observations used by BSFS for assimilation and validation are summarized in Table 1.

Table 1. List of observations used by BSFS in the assimilation and verification steps.

Product Reference	Platforms/Satellite	Upstream Reference	Usage
Temperature and Salinity profiles	ARGO	INSITU_BS_NRT_ OBSERVATIONS_013_034	Validation, Assimilation
Sea Surface Temperature (SST)	SLSTR and AVHRR on Sentinel-3A and 3B, and NOAA, VIIRS, MetOp-B, MODIS AQUA, TERRA and SEVIRI on board of MSG satellite	SST_BS_SST_L4_NRT_ OBSERVATIONS_010_006	Assimilation
		SST_BS_SST_L3S_NRT_ OBSERVATIONS_010_013	Validation
Along track Sea Level Anomaly (SLA)	Altika Cryosat-2 H2B Jason-2 Jason-3 Sentinel-3A Sentinel-3B	SEALEVEL_EUR_PHY_L3_NRT_ OBSERVATIONS_008_059 from CMEMS	Validation, Assimilation

The atmospheric forcing (previous day analyses and 10 days forecast starting at 00:00 UTC) is downloaded from ECMWF through Aeronautica Militare Italiana. As soon as it is available, the BSFS processing system starts (at around 7:00 UTC). The atmospheric forcing availability is a major source of delay for the oceanic forecast.

Every day the BSFS (Figure 1) produces ten days of forecast (J to J+9), one day of simulation (J−), and three days of analysis (J−4 to J−1). Once a week (on Tuesday), 14 days of analysis are produced, from J−15 to J−1, with the assimilation of all available satellite and in-situ data collected over the past weeks. Analysis produced at the weekly cycle represents the best estimation of the Black Sea state because all in-situ and satellite altimetry high quality processing data is used up to J−3. Analysis and simulation runs are forced by ECMWF atmospheric analysis fields at 6 h frequency; the forecast cycle is forced with ECMWF atmospheric forecast fields at 3 h frequency for the first 3 days and 6 h frequency for the remaining 7 days.

At the end of the forecasting cycle, post-processing involves the preparation of all the BSFS files in a format that is compliant with the CMEMS and CF standards, delivering to the CMEMS DU, and archiving of the BSFS native products at CMCC's supercomputing facilities. Every day, the forecast is released to CMEMS within less than 3 h (target delivery time: 12 UTC).

The BSFS product catalogue includes daily and hourly mean datasets, centered at midday of each J, for the Black Sea essential variables: 3D temperature (*T*), salinity (*S*), zonal and meridional velocity components (*U* and *V*, respectively) and 2D sea surface height (*SSH*), bottom temperature (*BottomT*) and mixed layer depth (*MLD*).

The BSFS operational chain is very similar to that of the Mediterranean Forecasting System (MFS, [25]) in terms of general setup (technical interface, analysis-simulation-forecast chain). Unlike MFS, it implements a more frequent analysis cycle to provide higher quality Black Sea products using daily observations.

3. Quality Assessment of the Operational System

The aim of the product quality assessment is to monitor the analysis quality and forecast accuracy of BSFS products using quasi-independent validation assessment. Operationally, a regional website provides daily bulletins (http://bsfs.cmcc.it/, accessed on 29 December 2021) and skill scores (http://oceanlab.cmcc.it/bsfs-evaluation/, accessed on 29 December 2021). Daily bulletins consist of a collection of interactive 2D maps for visualizing the essential variables of the Black Sea (as provided in the CMEMS catalogue). Weekly skill scores are provided through the evaluation section: during the assimilation weekly cycle, as described in Section 2.3, the difference between the model analysis and the observations at the time and location of the observations (i.e., misfits) is stored to compute statistics. Two main statistical metrics for the assessment of the analysis quality are computed using the misfits: the first is the bias, given by:

$$\text{bias}(\phi) = \frac{1}{N}\sum\left(\phi_i^M - \phi_i^O\right) \tag{13}$$

and the Root Mean Square Error (RMSE) given by:

$$\text{RMSE}(\phi) = \sqrt{\frac{1}{N}\sum_{i=1}^{N}(\phi_i^M - \phi_i^O)^2} \tag{14}$$

where N is the number of data used in the evaluation, ϕ^M is the model analysis field, and ϕ^O the observation.

On the other hand, to assess the quality of the forecast, we evaluated the differences of the forecast fields with respect to the analysis considered to be the best estimate of the truth. Murphy (1993, [26]), revisited by [2], defined the "forecast goodness" methodology and here we use his "quality measure" methodology that compares the forecast with the analysis and the "persistence" or the best estimate at initial time. If the difference between forecast and analysis is better than the difference between forecast and persistence, then the forecast is valuable. For a modern oceanographic forecasting system, [27] defined a metrics that we will partially follow here by computing:

- The difference between the analysis and the forecast (*AF*):

$$AF(t) = \sqrt{\frac{\sum_1^T(\phi^{AN}(t) - \phi^{FC}(t))^2}{T}} \tag{15}$$

where ϕ^{AN} and ϕ^{FC} is the temperature (salinity) daily mean forecast and analysis, respectively, at each forecast day $t = day1, day2, \ldots, day10$ (ID day as in Figure 2); T corresponds to the time period covered by the evaluation (here we consider a time period of 1 month). The metrics is normalized horizontally by averaging over the area and at a specific selected depth.

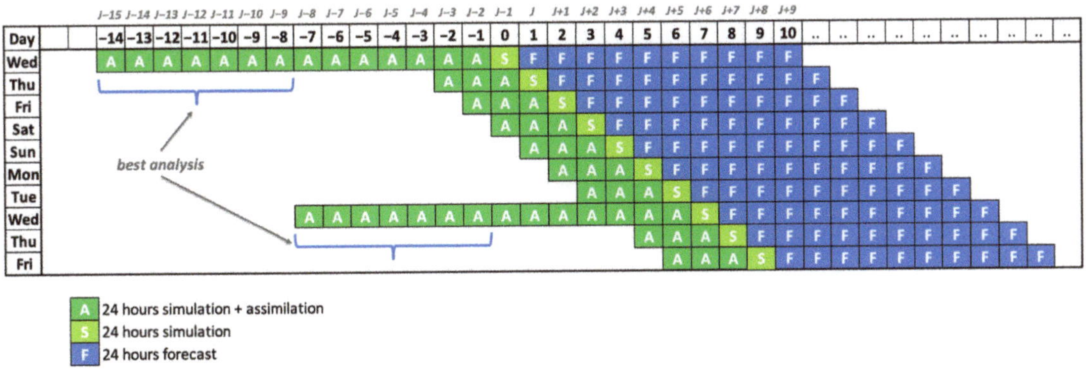

Figure 2. BSFS processing chain.

- The difference between the forecast and the persistence (*PF*):

$$PF(t) = \langle\sqrt{\frac{\sum_1^N(\phi^{FC}(t) - \phi^{AN}(t = day0))^2}{N}}\rangle \quad (16)$$

Here, $\phi^{AN}(t = day0)$ is the persistence (ID day as in Figure 2), e.g., the initial condition for the forecast. The latter is the last time step of a simulation started from an assimilated initial condition at day J−2.

In the following subsections, we present BSFS analysis statistics for the period 2018–2019 for SST and SLA, as well as temperature and salinity at given layers: 2–5 m, 5–10 m, 10–20 m, 20–30 m, 30–50 m, 50–75 m, 75–100 m, 100–200 m, 200–500 m, and 500–1000 m. Forecast assessment is performed for 2 significant months—February and August 2020: metrics have been computed at selected depth, e.g., 2.5, 30, 150 m.

3.1. Sea Surface Temperature

The BSFS analysis sea surface temperature is assessed by comparing analysis model fields against SST satellite data remapped over the Black Sea basin at 1/16° spatial resolution and representative of night SST values and delivered by CMEMS (Table 1).

Figure 3 shows the time series from 2018 to 2019 of the difference between BSFS and satellite SST (BIAS, top) and RMSE (bottom). The numerical SST is slightly warmer than the measured one (+0.1 °C), and the error is about 0.5 °C.

3.2. Sea Surface Height

Figure 4 shows the RMSE for SLA in the operational period (green dotted line) and the number of observational data used, given by the sum of the available along track sea level data provided by Altika, Cryosat-2, Jason-2 and 3, Sentinel 3A and 3B. The mean value of RMSE misfits for SLA is around 2.3 cm. A decrease in the error is also shown once the number of observations increases (June 2018–December 2018; March 2019–October 2019 thanks to inclusion of Sentinel-3B data). The accuracy is comparable to the Mediterranean Sea [25], where the average error over the same period is about 3.35 cm (http://oceanlab.cmcc.it/mfs-evaluation/ accessed on 29 December 2021).

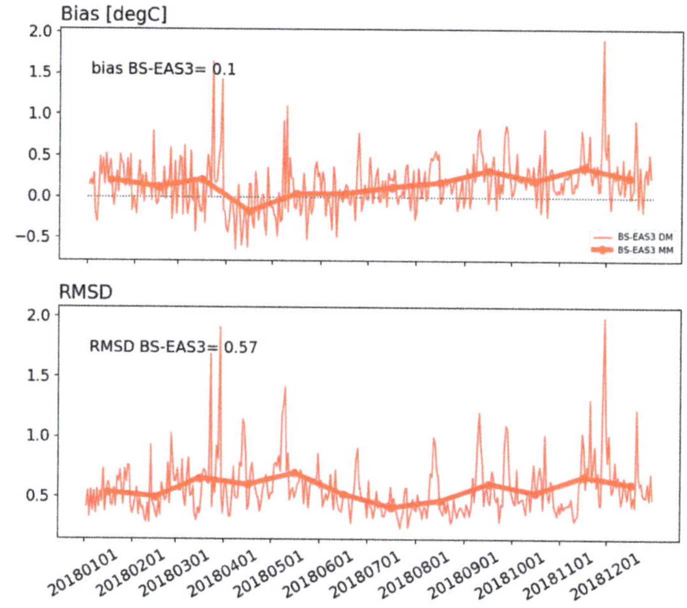

(**a**)

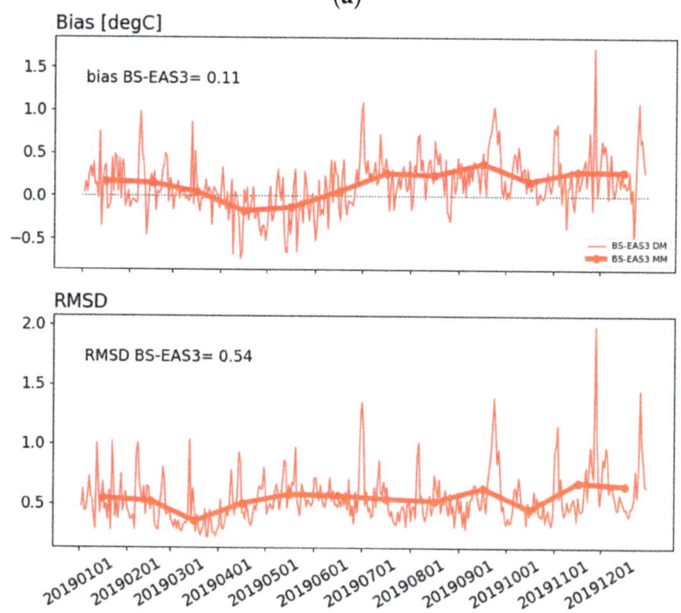

(**b**)

Figure 3. SST bias (top) and RMSE (bottom) timeseries, expressed in °C, at daily (thinner line) and monthly (solid line) frequency in 2018 (**a**) and 2019 (**b**).

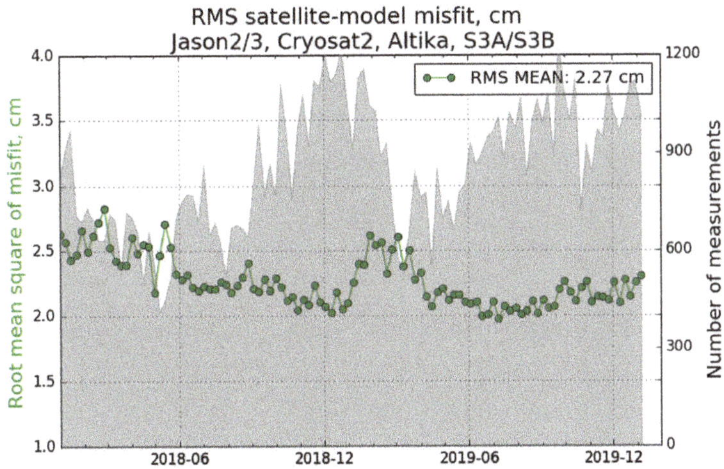

Figure 4. RMSE misfit of SLA, expressed in cm, computed using available observations from Altika, Cryosat-2, Jason-2 and 3, Sentinel-3A and 3B during 2018–2019.

3.3. *Temperature and Salinity*

3.3.1. Analysis Quality

Water column properties given by BSFS are assessed after 3D temperature and salinity are validated against all available observations.

Table 2 shows the basin scale RMSE misfits at specific layers (m) for 2018–2019. In the 2–5 m surface layer, the average RMSE for temperature at the whole basin is about 0.4 °C (Figure 5(a1)) and 0.2 PSU (Figure 5(b1)). However, the error in temperature starts to increase in the sub-surface, from 5 m to 30 m, from 0.5 °C to 2 °C in the summer and below 0.5 °C in the winter (Figure 5(a2–a4)).

Table 2. Temperature and salinity RMSE misfits over the period 2018–2019.

Layer (m)	T RMSE Misfit (°C)	S RMSD Misfit (PSU)
2–5	0.42	0.21
5–10	0.65	0.19
10–20	0.97	0.17
20–30	0.80	0.17
30–50	0.47	0.20
50–75	0.24	0.27
75–100	0.14	0.26
100–200	0.07	0.13
200–500	0.02	0.05

The maximum error for temperature occurs in the 10–20 m layer with a value up to 2.5 °C computed for August–October in both 2018 and 2019 (Figure 5(a3)).

Regarding salinity, the error in the subsurface, on average, is around 0.2 PSU (Figure 5(b2–b4)), which increases from 30 m up to 100 m up to a maximum value of 0.4 PSU (Figure 5(b5–b7)). In the intermediate layers up to 1000 m, the error decreases to quasi-zero for both temperature (Figure 5(a8)) and salinity (Figure 5(b8)).

The increased error in the thermocline (10–30 m layer) and in the halocline (50–100 m layer) in the summer is likely due to a deficiency in the vertical discretization (we use only 31 z-levels) and in the vertical mixing parameterization.

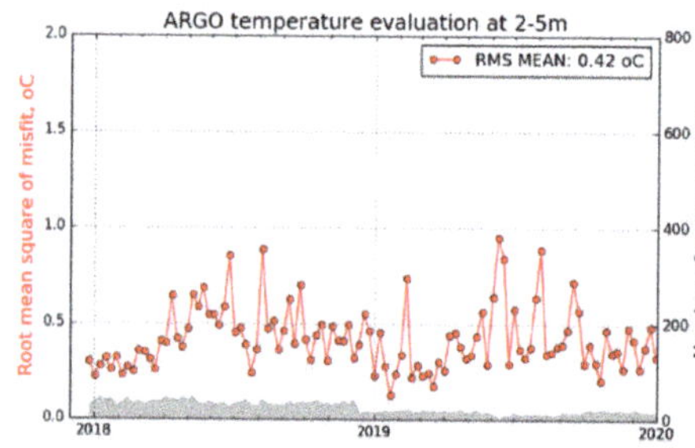

(a1)

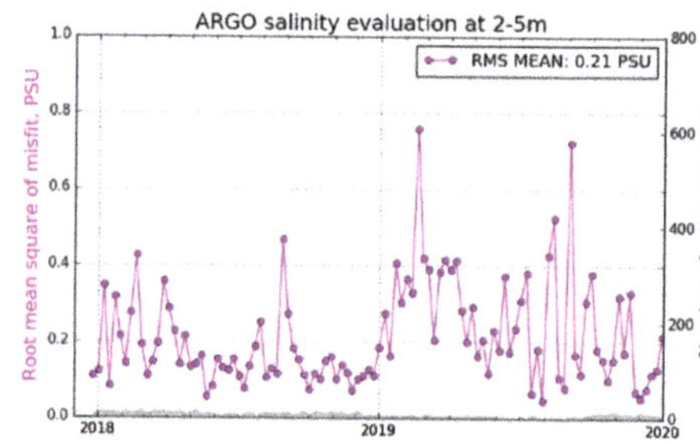

(b1)

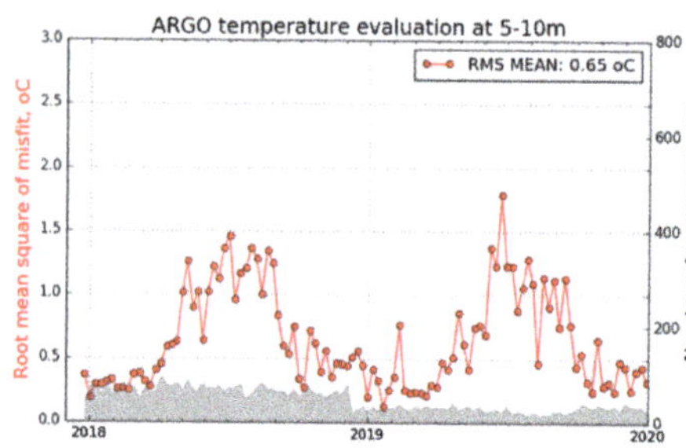

(a2)

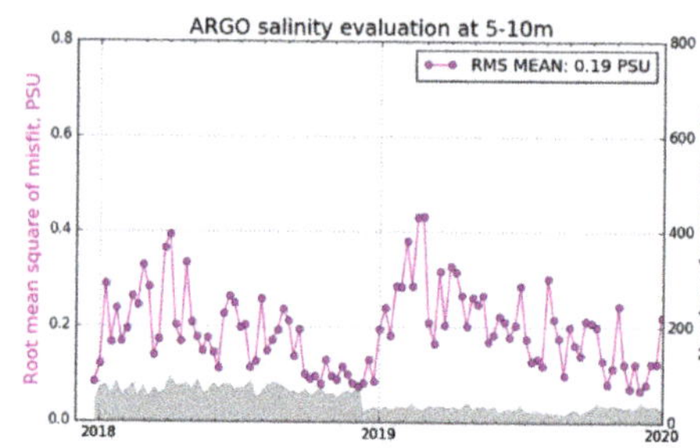

(b2)

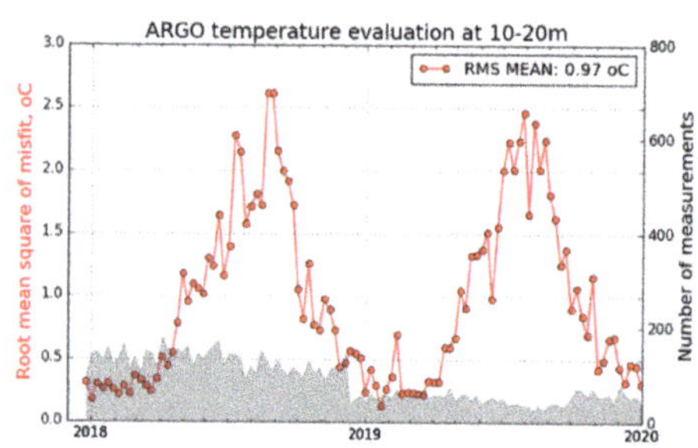

(a3)

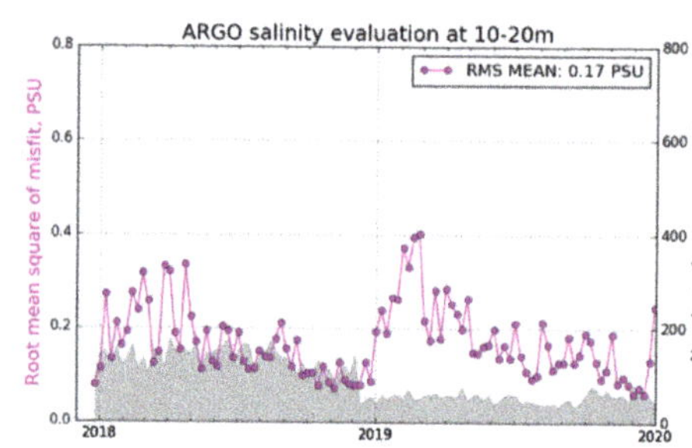

(b3)

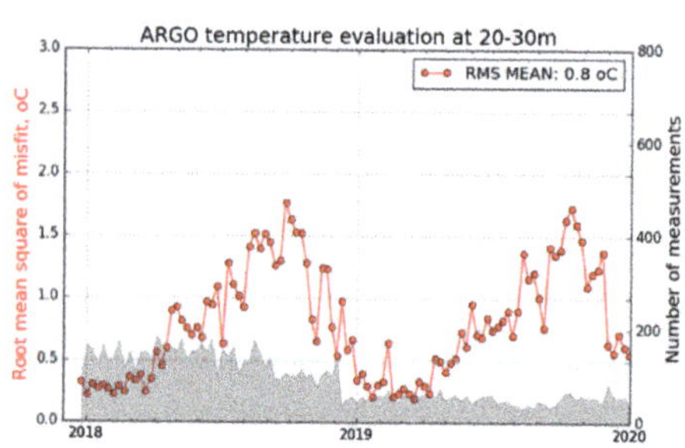

(a4)

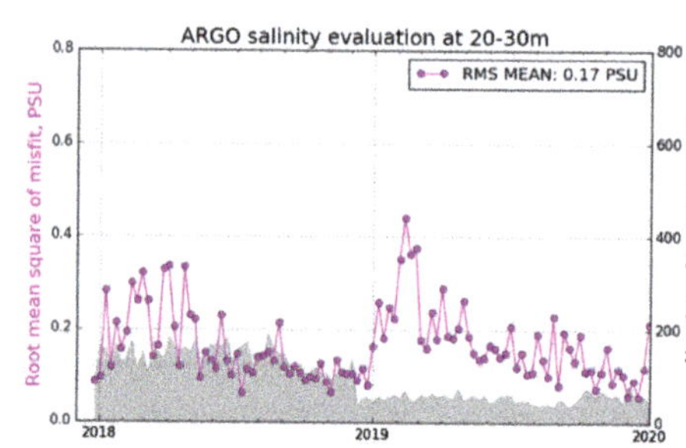

(b4)

Figure 5. *Cont.*

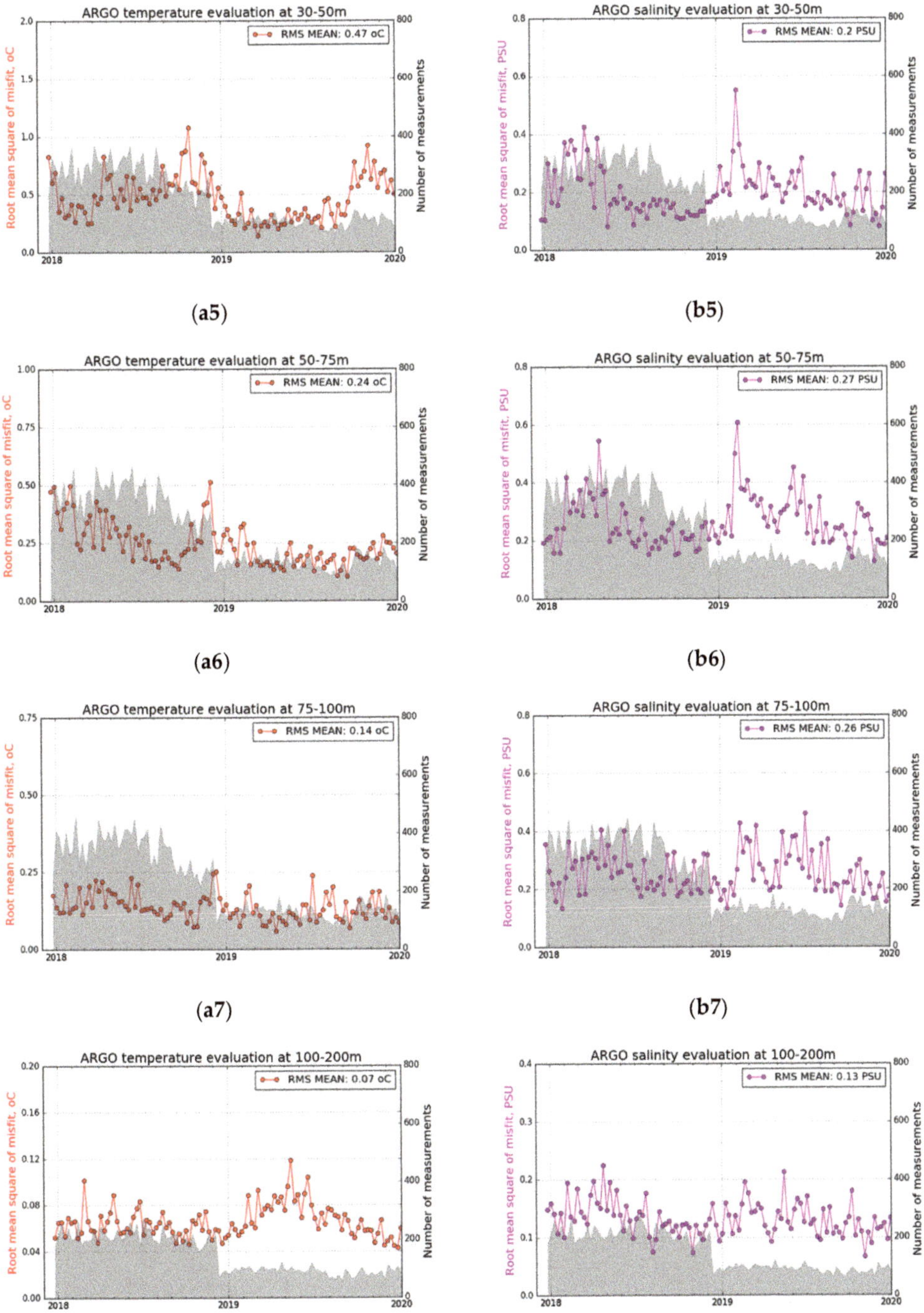

Figure 5. RMSE timeseries between BSFS analysis fields and ARGO observations for temperature (**a1–a8**) (expressed in °C) and salinity (expressed in PSU) (**b1–b8**). The reference layer as considered for the evaluation is written on the top of each panel.

The errors are due to (a) the Bosporus Strait being represented as closed boundary, (b) mixing processes that were not completely resolved since effects induced by waves and tides are not accounted for, and (c) simplified representation of the heat, momentum and water fluxes through bulk formulation, whose effects are particularly important especially during the summer. These are significant challenges to be considered in the development of the next generation of the Black Sea physical system.

3.3.2. Forecast Assessment

In Figures 6 and 7 we present the BSFS forecast quality assessment for temperature and salinity considering 2 reference months, February 2020 and August 2020, respectively. Metrics are given at different depths—2.5, 30 and 150 m. If the forecast is skillful, *AF* score should be better than *PF* score.

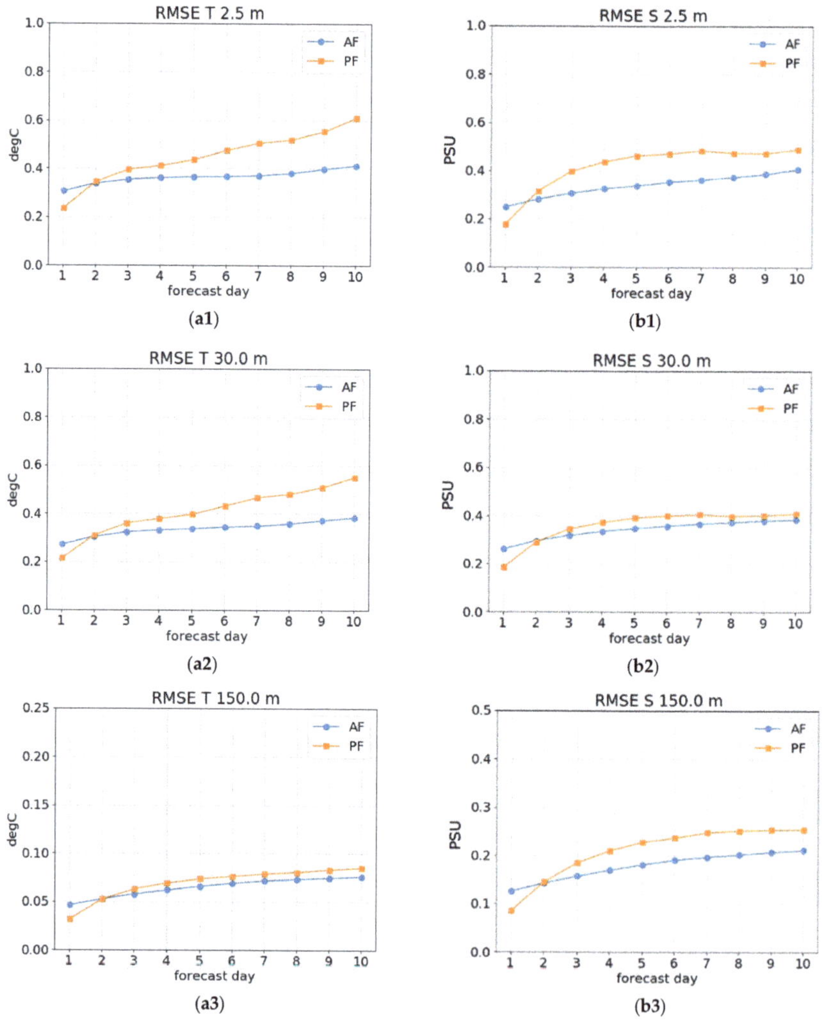

Figure 6. RMSE of *AF* (Analysis-Forecast, in blue), *PF* (Forecast-Persistence, in orange) for temperature (on the left, (**a1–a3**), expressed in °C) and salinity (on the right, (**b1–b3**), expressed in PSU) at 2.5 m (1), 30.0 m (2) and 150 m (3), computed for February 2020.

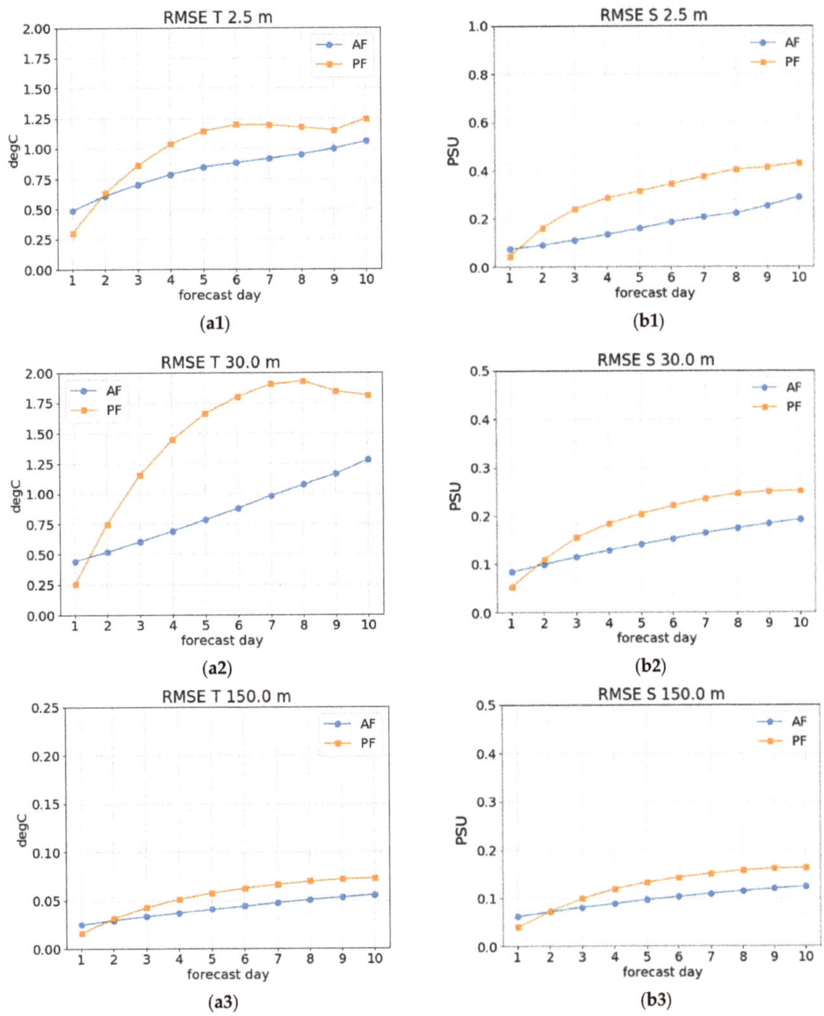

Figure 7. RMS of *AF* (Analysis-Forecast, in blue), *PF* (Forecast-Persistence, in orange) for temperature (on the left, (**a1–a3**), expressed in °C) and salinity (on the right, (**b1–b3**), expressed in PSU) at 2.5 m (1), 30.0 m (2) and 150 m (3), computed for August 2020.

In February 2020, *PF* is bigger than *AF* after *day*2 (Figures 6 and 7): this means that after the first day the advantage of doing the forecast is clear over persisting the initial condition. It is interesting to note that for salinity this advantage grows more rapidly, i.e., for salinity the forecast overcomes the persistence even before the first day. Over the water column, temperature and salinity analysis and forecast exhibit a similar variability: higher error reaching a maximum around 0.6° Celsius and 0.5 PSU at *day*10 is shown for temperature (Figure 6(a1,a2)) and salinity (Figure 6(b1,b2)) in the subsurface up to 30 m; at 150 m, both variables exhibit a lower error (never higher than 0.1° Celsius and 0.3 PSU over the reference 10-day period).

In August 2020, errors at 30 m are higher than the ones computed in the previous analyzed month: starting from *day*2, in particular for temperature (Figure 7(a1,a2)), *PF* shows higher values than *AF*—almost 50% higher at 30 m, with maximum error of about 2.0° Celsius (Figure 7(a2)). This means that during the summer the forecast adds to persistence more than during winter: the dynamical model is clearly responsible for the proper changes in water column stratification even for few days. At 30 m and in August (Figure 7(a1)) the difference between *PF* and *AF* is the largest: we assume that this is an indication of the complex dynamics involved in the mixing and the CIL dynamics. Regarding salinity (Figure 7(b1–b3)), *PF* and *AF* errors are quite similar comparing to winter and of the order of 0.3–0.4 PSU.

4. Black Sea Diagnostics and Circulation Consistency

4.1. Temporal Evolution of the CIL and Stratification

The BSFS is able to represent the formation and evolution of the cold intermediate layer (CIL), which is a typical water mass property of the Black Sea basin, and well documented in the literature ([16] for a very comprehensive literature review and updated evaluation). It is identified by the 8.0 °C isotherm, arising at the surface during winter, penetrating the subsurface (typically at a depth range of 50–100 m), intruding the warmer zone, and persisting over time. The time versus depth diagrams at the observation location are shown for temperature and for salinity in Figure 8, between 2015–2019. Figure 8a shows the cold event of 2017, which is also captured by observations as demonstrated by the quite low difference between the model and observations (Figure 8b). It also shows the reduction in the CIL width because of the warming period from 2018 and its perforation and partial disappearance in 2019 [16]. The major differences of about 2 °C occur in the upper layer and are shown in Figure 8b. Salinity stratification is well represented in Figure 9a with small differences with respect to observations, as shown in Figure 9b. It is characterized by a 2-layer structure with low salt in the CIL zone down to 100 m in depth and saltier values of about 22 PSU or more in the intermediate to deepest levels.

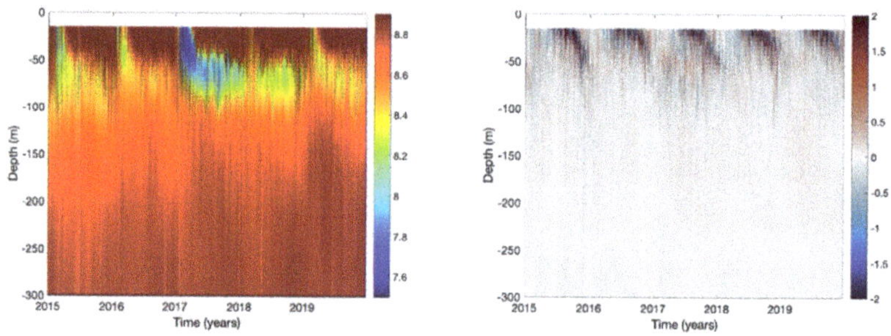

(a) (b)

Figure 8. Time versus depth diagram for BSFS temperature in the period 2015–2019 (**a**) at the observation location and (**b**) for the computed model minus observation value.

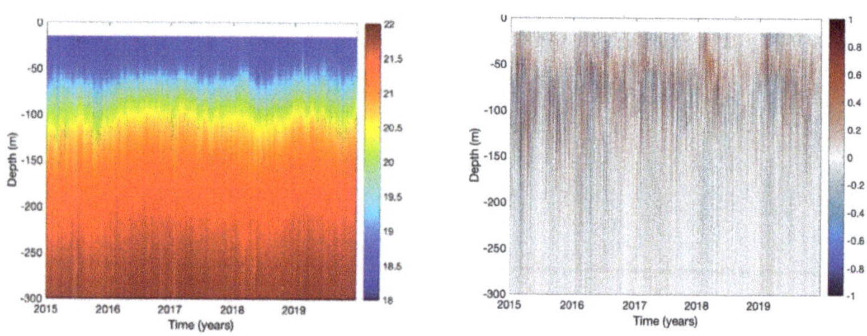

(a) (b)

Figure 9. Time versus depth diagram for BSFS salinity (PSU) in the period 2015–2019 (**a**) at the observation location and (**b**) for the computed model minus observation value.

4.2. Circulation

The surface mean circulation is shown in Figure 10 as monthly means for 2019. Considering the poor availability of observations, a robust validation exercise is extremely difficult, thus the main features emerging from the annual monthly means are described in detail with respect to those extensively described in [28–31].

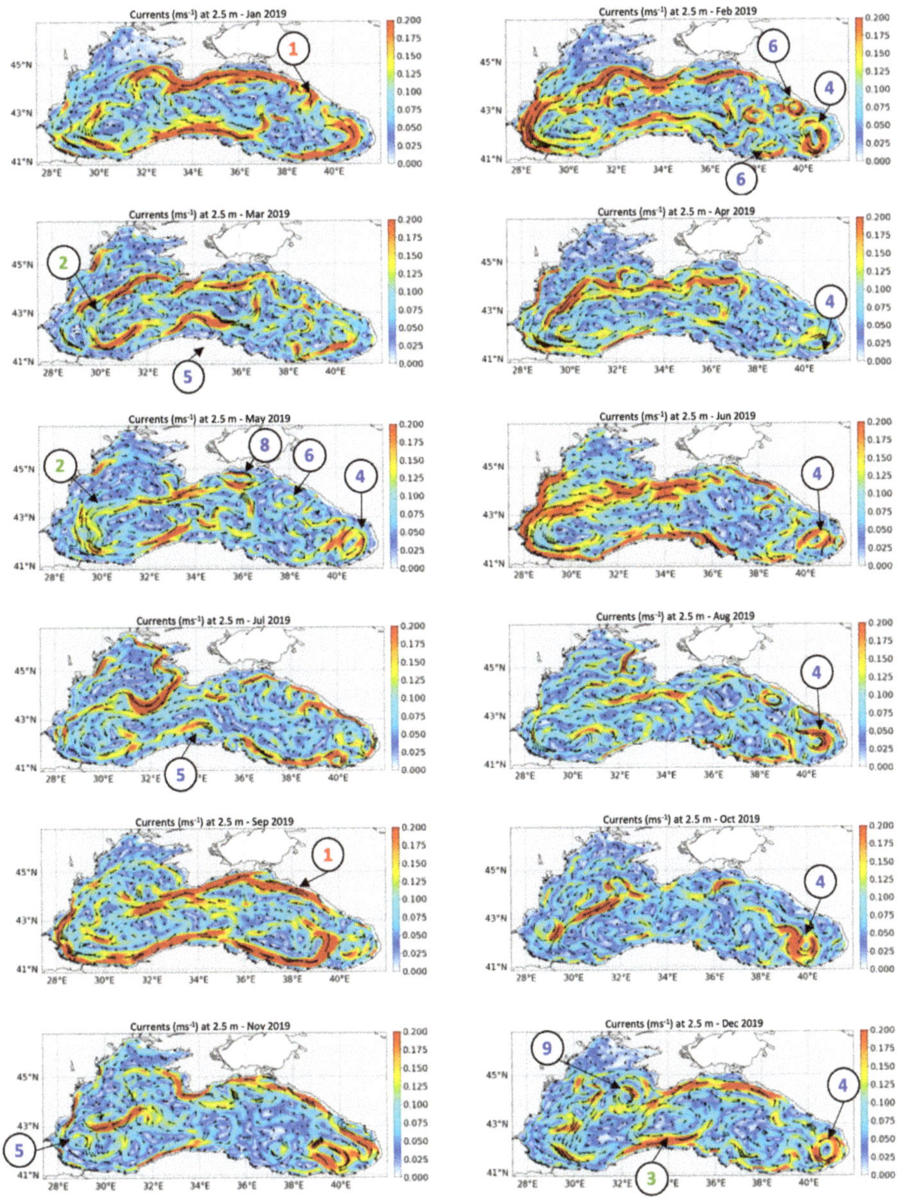

Figure 10. Surface currents in 2019: monthly means from Jan to Dec 2019. The reference month and depth are written on the top of each panel. Main structures: (1) Rim Current, (2) Western Gyre, (3) Central Gyre, (4) Batumi anticyclonic eddy, (5) anticyclonic coastal eddy, (6) Caucasus anticyclonic eddy, (7) cyclonic coastal eddy, (8) Kerch anticyclonic eddy, (9) Sevastopol anticyclonic eddy.

The Black Sea surface dynamics is characterized by a main cyclonic gyre, the Rim current, encompassing the basin and a variety of mesoscale eddies along the coast, some of them quasi-stationary. BSFS captures most of the particular dynamical structures in the basin, such as the Rim current, which persists over 2019, and small-scale structures such as coastal anticyclonic eddies, which appear along the Russian–Georgian coastline (one quite intense in February and progressively weaker in March) and coastal cyclonic eddies, which are much more intermittent over the year and weak). The Batumi Gyre, located near the Georgian coast, forms in April and lasts until June, when the circulation in that region is fragmented in more unstable and weak eddies until its regeneration in August–September. Progressively towards the southwestern region, small coastal eddies become part of the Rim current. On the Turkish coastline, cyclonic eddies like the one in the area of Trabzon (January–February; November–December) coexist with anticyclonic eddies, as in the Synop area (June–October). In the area between Istanbul and Burgas—the Bosporus region—an anticyclonic eddy appears from October to December. Close to the Danube, coastal currents are quite well captured by the model, then going further towards the continental slope (30° E–32° E and 44° N–45° N) the Sevastopol anticyclonic gyre starts in May (very weak), becoming stronger in November and December.

5. Conclusions

Within the framework of the CMEMS and the BS-MFC, the Black Sea physical analysis and forecasting system provides essential variables for understanding the physical processes and dynamics of the Black Sea basin. The BSFS has been operational since the end of 2016 and has been developed and maintained at CMCC in collaboration with the USOF (University of Sofia, Bulgaria, scientific partner in the BS-MFC consortium).

We have presented the BSFS ocean model based on NEMO v3.4 able to assimilate near real time, in-situ and satellite observational products using the OceanVar scheme. The Black Sea hydrodynamic model has about 3 km horizontal resolution and uses 31 levels with partial steps. It implements a closed boundary condition at the Bosporus Strait and is forced by ECMWF analysis and forecasting atmospheric fields.

The BSFS implements two production cycles, one daily (which includes 3 days analysis) and one weekly (based on 14 days analysis), the latter to assimilate a higher number of collected observations to provide the best quality initial conditions for the forecasting cycle: the processing system is completed every day by 1-day simulation and 10-day forecasts.

We have also described an operational dashboard for product quality monitoring which assesses the skill scores of the analysis products for 2018–2019 and gives an accuracy of around 1 °C in the sea surface temperature. Errors increase at the sub-surface (10–30 m layer): in particular, the thermocline experiences a maximum error of 2 °C during the summer period, while salinity reaches an error of about 0.5 PSU. Considering the sea level, the assimilation of the along-track satellite SLA guarantees an average error of about 2.3 cm. Such skills put the BSFS on the same track of quality and robustness as state-of-the-art regional configurations in the CMEMS framework. A regional website is operationally maintained at CMCC to provide daily bulletins and metrics for monitoring the lifecycle and performances of the system.

The future forecasting system for the Black Sea will include at least four new main components which will significantly improve the quality of the BSFS analysis fields and forecasting skill scores by means of improved hydrodynamical core model: (1) increased vertical resolution for a better representation of open ocean dynamics and mixing processes, combined with a revised bathymetry and coastline and data assimilation upgrades; (2) the Bosporus Strait will work as an open boundary, in order to improve the connection with the Mediterranean Sea. This will be achieved thanks to a novel implementation of the Marmara Sea model, based on an unstructured grid method that provides open boundary conditions to the Black Sea through the Bosporus and the Mediterranean Sea through the Dardanelles; (3) improved representation of the land forcing: in particular, the representation of the Danube River using historical discharge datasets provided by the NIHWM (National

Institute of Hydrology and Water Management, Romania—scientific partner in the BS-MFC consortium); (4) an online wave-current model to improve the small scales dynamics at the surface.

Author Contributions: S.A.C. coordinated the work in collaboration with G.C. and N.P. The ocean circulation model has been designed and improved by S.A.C., E.P., N.P., G.C., M.I.; the data assimilation scheme has been designed by E.J. with the collaboration of L.L.; R.L. contributed on upstream data access; Funding acquisition, A.P.; S.C. (Sergio Creti') and L.S. contributed on operational chain design and operational maintenance of the system. Validation has been developed in collaboration with E.J., D.A. and S.C. (Salvatore Causio). All authors have read and agreed to the published version of the manuscript.

Funding: This research was funded by the Copernicus Marine Environment Monitoring Service for the Black Sea Monitoring and Forecasting Center (Contract No. 72-CMEMS-MFS-BS).

Institutional Review Board Statement: Not applicable.

Informed Consent Statement: Not applicable.

Data Availability Statement: This work has been used CMEMS data from the BS-MFC, in particular Black Sea physical analysis and forecast (https://marine.copernicus.eu/ (accessed on 1 April 2021).

Conflicts of Interest: The authors declare no conflict of interest.

Appendix A

Surface fluxes are iteratively computed by using bulk formulae parameterization, as proposed by [11,32], for the Mediterranean Sea and revised for the Black Sea basin for the purposes of the operational forecasting system. They are used to handle operational ECMWF analysis and forecast products at resolution of 3–6 h.

The total heat flux Q_T is computed according to:

$$Q_T = Q_s + Q_b + Q_h + Q_e \tag{A1}$$

where Q_s is the shortwave radiation flux, Q_b is the net longwave radiation flux, Q_h is the sensible heat flux, Q_e is the latent heat flux. Such quantities depend upon the air temperature at 2 m (T_A), the sea surface temperature computed by the model itself (T_0), the total cloud cover (C), the relative humidity computed from the dew point temperature at 2 m (rh) and the 10 m wind velocity amplitude ($|\mathbf{V_w}|$).

The Q_s is computed by means of the [33]:

$$Q_s = \begin{cases} Q_T(1 - 0.62C + 0.0019\beta)(1 - \alpha)\ if\ C \geq 0.3 \\ Q_T(1 - \alpha)\ if\ C < 0.3 \end{cases} \tag{A2}$$

where β is the solar noon altitude in degrees and α is the ocean surface albedo. The albedo is computed from [34].

The Q_b is computed by means of the Brunt-Berliand formula as in [35]:

$$Q_b = \epsilon\sigma T_0^4(0.39 - 0.05\sqrt{e_A})(1 - 0.8C) + 4\epsilon\sigma T_0^3(T_0 - T_A) \tag{A3}$$

where ϵ is the ocean emissivity, σ is the Stefan-Boltzman constant, e_A is the atmospheric vapor pressure [36] given as function of the mixing ration of the air (w_A) and mean sea level pressure (p):

$$e_A = \left(\frac{w_A}{w_A + 0.622}\right)p \tag{A4}$$

In the Mediterranean Sea model, it is computed by [37].

The Q_h is computed as follows:

$$Q_h = -\rho_A C_P C_h|\mathbf{V_w}|(T_0 - T_A) \tag{A5}$$

where ρ_A is the density of the moist air, C_P is the specific heat capacity, C_h is the turbulent exchange coefficient for humidity set as a constant and equal to $1.3 \cdot 10^{-3}$.

The Q_e is computed as follows:

$$Q_e = -\rho_A L_e C_e \left|\overline{\mathbf{V}_w}\right| (q_0 - q_A) \tag{A6}$$

where L_e is the latent heat of vaporization, C_e is the turbulent exchange coefficient for temperature set as a constant and equal to $1.5 \cdot 10^{-3}$, q_0 is the specific humidity saturated at T_0 while q_A is the specific humidity of air.

The two constants, C_h and C_e, are computed according to empirical formulation as suggested by [38] and extensively described in [11].

The momentum flux is given by the wind stress components:

$$\left(\tau^{\lambda}, \tau^{\varphi}\right) = \rho_A C_D \left|\overline{\mathbf{V}_w}\right| \left(V_x, V_y\right) \tag{A7}$$

where V_x, V_y are the wind components, while C_D is the drag coefficient given as a function of wind speed and temperature difference $T_A - T_0$ according to [39].

The bulk formulation for the Black Sea are implemented in NEMO and to be used by the model it requires the following list of atmospheric fields in specific units:

- Zonal and Meridional components of the 10 m winds, expressed as ms^{-1}.
- Total Cloud Cover, expressed as %.
- 2 m Air Temperature, expressed as K degrees.
- 2 m Dew Point Temperature, expressed as K degrees.
- Total Precipitation, expressed as $kg/m^2/s$.
- Mean Sea Level Pressure, expressed as hPa.

References

1. Pinardi, N.; Cavaleri, L.; Coppini, G.; De Mey, P.; Fratianni, C.; Huthnance, J.; Lermusiaux, P.F.J.; Navarra, A.; Preller, R.; Tibaldi, S. From weather to ocean predictions: An historical viewpoint. *J. Mar. Res.* **2017**, *75*, 103–159. [CrossRef]
2. Brassington, G.B. Forecast Terrors, goodness and verification in Ocean forecasting. *J. Mar. Res.* **2017**, *75*, 403–433. [CrossRef]
3. Le Traon, P.Y.; Reppucci, A.; Alvarez Fanjul, E.; Aouf, L.; Behrens, A.; Belmonte, M.; Bentamy, A.; Bertino, L.; Brando, V.E.; Kreiner, M.B.; et al. From Observation to Information and Users: The Copernicus Marine Service Perspective. *Front. Mar. Sci.* **2019**, *6*, 234. [CrossRef]
4. Pinardi, N.; Coppini, G. Operational oceanography in the Mediterranean Sea: The second stage of development. *Ocean Sci.* **2010**, *6*, 263–267. [CrossRef]
5. Clementi, E.; Oddo, P.; Drudi, M.; Pinardi, N.; Korres, G.; Grandi, A. Coupling hydrodynamic and wave models: First step and sensitivity experiments in the Mediterranean Sea. *Ocean. Dyn.* **2017**, *67*, 1293–1312. [CrossRef]
6. Palazov, A.; Ciliberti, S.A.; Peneva, E.; Gregoire, M.; Staneva, J.; Lemieux-Dudon, B.; Masina, S.; Pinardi, N.; Vandenbulcke, L.; Behrens, A.; et al. Black Sea Observing System. *Front. Mar. Sci.* **2019**, *6*, 1–8. [CrossRef]
7. Madec, G.; The NEMO Team. *NEMO Ocean Engine, Note du Pôle de Modélisation*; Institute Pierre-Simon Laplace (IPSL): Paris, France, 2008; ISSN 1288–1619.
8. Stanev, E.V.; Rachev, N.H. Numerical study on the planetary Rossby waves in the Black Sea. *J. Mar. Syst.* **1999**, *21*, 283–306. [CrossRef]
9. Historical GEBCO Data Sets. Available online: https://www.gebco.net/data_and_products/historical_data_sets/#gebco_one (accessed on 1 April 2021).
10. Wessel, P.; Smith, W.H.G. A global, self-consistent, hierarchical, high-resolution shoreline database. *J. Geophys. Res.* **1996**, *101*, 8741–8743. [CrossRef]
11. Pettenuzzo, D.; Large, W.G.; Pinardi, N. On the corrections of ERA-40 surface flux products consistent with the Mediterranean heat and water budgets and the connection between basin surface total heat flux and NAO. *J. Geophys. Res.* **2010**, *115*, C06022. [CrossRef]
12. Roullet, G.; Madec, G. Salt conservation, free surface, and varying levels: A new formulation for ocean general circulation models. *J. Geophys. Res.* **2000**, *105*, 23927–23942. [CrossRef]
13. Jarosz, E.; Teague, W.J.; Book, J.W.; Beşiktepe, Ş. On flow variability in the Bosphorus Strait. *J. Geophys. Res.* **2011**, *116*, C08038. [CrossRef]
14. Özsoy, E.; Altıok, H.A. Review of Hydrography of the Turkish Straits System. In *The Sea of Marmara-Marine Biodiversity, Fisheries, Conservation and Governance*; Özsoy, E., Çağatay, N.M., Balkis, N., Balkis, N., Öztürk, B., Eds.; Turkish Marine Research Foundation (TÜDAV): Istanbul, Turkey, 2016; Publication #42.

15. Özsoy, E.; Altıok, H.A. Review of Water Fluxes across the Turkish Straits System. In *The Sea of Marmara-Marine Biodiversity, Fisheries, Conservation and Governance*; Özsoy, E., Çağatay, N.M., Balkis, N., Balkis, N., Öztürk, B., Eds.; Turkish Marine Research Foundation (TÜDAV): Istanbul, Turkey, 2016; Publication #42.
16. Stanev, E.V.; Peneva, E.; Chtirkova, B. Climate change and regional ocean water mass disappearance: Case of the Black Sea. *J. Geophys. Res.* **2019**, *124*, 4803–4819. [CrossRef]
17. Ünlüata, Ü.; Oğuz, T.; Latif, M.A.; Özsoy, E. On the physical oceanography of the Turkish Straits. In *The Physical Oceanography of Sea Straits*; NATO/ASI Series; Kluwer: Dordrecht, The Netherlands, 1990; pp. 25–60.
18. Kara, A.B.; Wallcraft, A.J.; Hurlburt, H.E.; Stanev, E.V. Airsea fluxes and river discharges in the Black Sea with a focus on the Danube and Bosphorus. *J. Mar. Syst.* **2008**, *74*, 74–95. [CrossRef]
19. Pinardi, N.; Bonaduce, A.; Navarra, A.; Dobricic, S.; Oddo, P. The mean sea level equation and its application to the Mediterranean Sea. *J. Clim.* **2014**, *27*, 442–447. [CrossRef]
20. Stanev, E.V.; Beckers, J.M. Barotropic and baroclinic oscillations in strongly stratified ocean basins. Numerical study for the Black Sea. *J. Mar. Syst.* **1999**, *19*, 65–112. [CrossRef]
21. Ludwig, W.; Dumont, E.; Meybeck, M.; Heussner, S. River discharges of water and nutrients to the Mediterranean Sea: Major drivers for ecosystem changes during past and future decades? *Prog. Oceanogr.* **2009**, *80*, 199–217. [CrossRef]
22. Dobricic, S.; Pinardi, N. An oceanographic three-dimensional variational data assimilation scheme. *Ocean Model.* **2008**, *22*, 89–105. [CrossRef]
23. Storto, A.; Dobricic, S.; Masina, S.; Di Pietro, P. Assimilating along-track altimetric observations through local hydrostatic adjustment in a global ocean variational assimilation system. *Mon. Weather Rev.* **2010**, *139*, 738–754. [CrossRef]
24. Waters, J.; Lea, D.J.; Martin, M.J.; Mirouze, I.; While, J. Implementing a variational data assimilation system in an operational 1/4 degree global ocean model. *Q. J. R. Meteorol.* **2014**, *141*, 333–349. [CrossRef]
25. Clementi, E.; Pistoia, J.; Escudier, R.; Delrosso, D.; Drudi, M.; Grandi, A.; Lecci, R.; Cretí, S.; Ciliberti, S.; Coppini, G.; et al. *Mediterranean Sea Analysis and Forecast (CMEMS MED-Currents, EAS5 System) [Data Set]*; Copernicus Monitoring Environment Marine Service (CMEMS); Centro Euro-Mediterraneo sui Cambiamenti Climatici (CMCC): Lecce, Italy, 2019. Available online: https://doi.org/10.25423/CMCC/MEDSEA_ANALYSIS_FORECAST_PHY_006_013_EAS5 (accessed on 9 December 2021).
26. Murphy, A.H. What is a good forecast? An essay on the nature of goodness in weather forecasting. *Weather Forecast.* **1993**, *8*, 281–293. [CrossRef]
27. Tonani, M.; Pinardi, N.; Fratianni, C.; Pistoia, J.; Dobricic, S.; Pensieri, S.; de Alfonso, M.; Nittis, K. Mediterranean Forecasting System: Forecast and analysis assessment through skill scores. *Ocean Sci.* **2009**, *5*, 649–660. [CrossRef]
28. Ciliberti, S.A.; Peneva, E.L.; Jansen, E.; Martins, D.; Cretí, S.; Stefanizzi, L.; Lecci, R.; Palermo, F.; Daryabor, F.; Lima, L.; et al. *Black Sea Analysis and Forecast (CMEMS BS-Currents, EAS3 System) (Version 1) [Data Set]*; Copernicus Monitoring Environment Marine Service (CMEMS); Centro Euro-Mediterraneo sui Cambiamenti Climatici (CMCC): Lecce, Italy, 2020. Available online: https://doi.org/10.25423/CMCC/BLKSEA_ANALYSIS_FORECAST_PHYS_007_001_EAS3 (accessed on 9 December 2021).
29. Özsoy, E.; Ünlüata, Ü. Oceanography of the Black Sea: A review of some recent results. *Earth-Sci. Rev.* **1997**, *42*, 231–272. [CrossRef]
30. Staneva, J.V.; Stanev, E.V. Oceanic response to atmospheric forcing derived from different climatic data sets. Intercomparison study for the Black Sea. *Oceanol. Acta* **1998**, *21*, 393–417. [CrossRef]
31. Gunduz, M.; Özsoy, E.; Hordoir, R. A model of Black Sea circulation with strait exchange (2008–2018). *Geosci. Model Dev.* **2020**, *13*, 121–138. [CrossRef]
32. Castellari, S.; Pinardi, N.; Leaman, K. A model study of air-sea interactions in the Mediterranean Sea. *J. Mar. Syst.* **1998**, *18*, 89–114. [CrossRef]
33. Reed, R.K. On estimating insolation over the ocean. *J. Phys. Oceanogr.* **1977**, *7*, 482–485. [CrossRef]
34. Payne, R.E. Albedo of the sea surface. *J. Atmos. Sci.* **1972**, *29*, 959–970. [CrossRef]
35. Rosati, A.; Miyakoda, K. A general circulation model for upper ocean simulation. *J. Phys. Oceanogr.* **1988**, *18*, 1601–1626. [CrossRef]
36. Lowe, P.R. An approximating polynomial for the computation of saturation vapor pressure. *J. Appl. Meteorol.* **1977**, *16*, 100–103. [CrossRef]
37. Bignami, F.; Marullo, S.; Santoleri, R.; Schiano, M.E. Long wave radiation budget on the Mediterranean Sea. *J. Geophys. Res.* **1995**, *100*, 2501–2514. [CrossRef]
38. Kondo, J. Air-sea bulk transfer coefficients in diabatic condition. *Bound. Layer Meteorol.* **1975**, *9*, 91–112. [CrossRef]
39. Hellerman, S.; Rosenstein, M. Normal monthly wind stress over the world ocean with error estimates. *J. Phys. Oceanogr.* **1983**, *23*, 1009–1039. [CrossRef]

Journal of
Marine Science and Engineering

Article

Monitoring and Forecasting the Ocean State and Biogeochemical Processes in the Black Sea: Recent Developments in the Copernicus Marine Service

Stefania A. Ciliberti [1,*], Marilaure Grégoire [2,*], Joanna Staneva [3,*], Atanas Palazov [4,*], Giovanni Coppini [1], Rita Lecci [1], Elisaveta Peneva [5], Marius Matreata [6], Veselka Marinova [4], Simona Masina [7], Nadia Pinardi [1,8], Eric Jansen [1], Leonardo Lima [7], Ali Aydoğdu [7], Sergio Creti' [1], Laura Stefanizzi [1], Diana Azevedo [1], Salvatore Causio [1], Luc Vandenbulcke [2], Arthur Capet [2], Catherine Meulders [2], Evgeny Ivanov [2], Arno Behrens [3], Marcel Ricker [3], Gerhard Gayer [3], Francesco Palermo [1], Mehmet Ilicak [9], Murat Gunduz [10], Nadezhda Valcheva [4] and Paola Agostini [1]

1 Ocean Predictions and Applications Division, Fondazione Centro Euro-Mediterraneo Sui Cambiamenti Climatici, 73100 Lecce, Italy; giovanni.coppini@cmcc.it (G.C.); rita.lecci@cmcc.it (R.L.); nadia.pinardi@unibo.it (N.P.); eric.jansen@cmcc.it (E.J.); sergio.creti@cmcc.it (S.C.); laura.stefanizzi@cmcc.it (L.S.); diana.azevedo@cmcc.it (D.A.); salvatore.causio@cmcc.it (S.C.); francesco.palermo@cmcc.it (F.P.); paola.agostini@cmcc.it (P.A.)
2 Faculty of Science, Department of Astrophysics, Geophysics and Oceanography (AGO), MAST (Modeling for Aquatic Systems), University of Liege, 4000 Liege, Belgium; luc.vandenbulcke@uliege.be (L.V.); acapet@uliege.be (A.C.); catherine.meulders@uliege.be (C.M.); evgeny.ivanov@uliege.be (E.I.)
3 Helmholtz-Zentrum Hereon, 21502 Geesthacht, Germany; arno.behrens@hereon.de (A.B.); marcel.ricker@hereon.de (M.R.); gerhard.gayer@hzg.de (G.G.)
4 Institute of Oceanology–Bulgarian Academy of Science, 9000 Varna, Bulgaria; marinova@io-bas.bg (V.M.); valcheva@io-bas.bg (N.V.)
5 Faculty of Physics, Department of Meteorology and Geophysics, Sofia University "St. Kliment Ohridski", 1504 Sofia, Bulgaria; elfa@phys.uni-sofia.bg
6 National Institute of Hydrology and Water Management, 013686 Bucarest, Romania; marius.matreata@hidro.ro
7 Ocean Modelling and Data Assimilation Division, Fondazione Centro Euro-Mediterraneo Sui Cambiamenti Climatici, 40127 Bologna, Italy; simona.masina@cmcc.it (S.M.); leonardo.lima@cmcc.it (L.L.); ali.aydogdu@cmcc.it (A.A.)
8 Department of Physics and Astronomy "Augusto Righi", University of Bologna Alma Mater Studiorum, 40126 Bologna, Italy
9 Eurasia Institute of Earth Sciences, Istanbul Technical University, Istanbul 34467, Turkey; milicak@itu.edu.tr
10 Institute of Marine Sciences and Technology, Dokuz Eylul University, Izmir 35220, Turkey; murat.gunduz@deu.edu.tr
* Correspondence: stefania.ciliberti@cmcc.it (S.A.C.); mgregoire@uliege.be (M.G.); joanna.staneva@hereon.de (J.S.); palazov@io-bas.bg (A.P.)

Citation: Ciliberti, S.A.; Grégoire, M.; Staneva, J.; Palazov, A.; Coppini, G.; Lecci, R.; Peneva, E.; Matreata, M.; Marinova, V.; Masina, S.; et al. Monitoring and Forecasting the Ocean State and Biogeochemical Processes in the Black Sea: Recent Developments in the Copernicus Marine Service. *J. Mar. Sci. Eng.* **2021**, *9*, 1146. https://doi.org/10.3390/jmse9101146

Academic Editor: Christos Tsabaris

Received: 15 September 2021
Accepted: 12 October 2021
Published: 18 October 2021

Abstract: The Black Sea Monitoring and Forecasting Center (BS-MFC) is the European reference service for the provision of ocean analyses, forecasts, and reanalyses in the Black Sea basin. It is part of the Copernicus Marine Environment and Monitoring Service (CMEMS) and ensures a high level of efficiency in terms of operations, science, and technology for predictions and the monitoring of physical and biogeochemical processes in the Black Sea. The operational BS-MFC framework is based on state-of-the-art numerical models for hydrodynamics, biogeochemistry, and waves; analysis, forecast, and reanalysis are provided on a spatial grid with about 3 km of horizontal resolution that covers the whole Black Sea basin (the Azov Sea is not included). The scientific assessment of BS-MFC products is performed by implementing a product quality dashboard that provides pre-qualification and operational model skills according to GODAE/OceanPredict standards. Novel interfaces based on high-resolution models are part of the scientific development plan to ensure a strong connection with the nearest seas from a modelling point of view, in particular with the Mediterranean Sea. To improve forecasting skills, dedicated online coupled systems are being developed, which involve physics, biogeochemistry, and waves together with the atmosphere and, in the future, with ensemble forecasting methodologies and river-ocean interfaces.

Keywords: Black Sea; operational oceanography; physical oceanography; biogeochemistry; waves; numerical modelling; data assimilation; forecasting; reanalysis; validation

1. Introduction

Understanding, predicting, and reconstructing the ocean state is globally one of the most challenging objectives for operational oceanography, especially to support user needs as well as down-stream services and applications for what is termed the Blue Growth and the Blue Economy for a Blue Society. In the European panorama, the Black Sea, together with the Azov Sea, the Marmara Sea, and the Mediterranean Sea, is part of a peculiar complex system of marginal seas that is functional to ocean monitoring and climate. Due to particular physical and biogeochemical processes, the Black Sea deserves a more dedicated focus so as to be able to set up a reliable modelling system and observing network for the next generation of operational ocean forecasting systems.

This baseline is one of the main priorities pursued within the Copernicus Marine Environment and Monitoring Service (CMEMS, [?]). The Black Sea Monitoring and Forecasting Center (BS-MFC) is the CMEMS operational service dedicated to the regional needs of the Black Sea, which has been operational since the end of 2016 and serves about 60 "active users" (i.e., users that regularly download CMEMS data) from academia (~30%), public sectors and organizations (~20%), and the business and private sectors (~50%). The main objective of the BS-MFC is to serve the specific needs of the Black Sea countries in decision-making by developing new operational services that are able to protect the marine ecosystem, contrasting pollution and environmental emergencies, supporting maritime safety and routing, and promoting a science-driven cooperation for user-driven applications from the regional to the coastal scale. Access to high-quality operational data related to forecasting and reconstruction of past ocean states in the Black Sea is a pre-requisite for achieving these targets. The BS-MFC is a joint venture between the Institute of Oceanology-Bulgarian Academy of Sciences (IO-BAS, Bulgaria) in the role of coordinator, and the following institutes and universities: the Euro-Mediterranean Center on Climate Change Foundation (CMCC, Italy), the University of Liege (ULiege, Belgium), the Helmholtz-Zentrum Hereon (HEREON, Germany), the Sofia University "St. Kliment Ohridski" (USOF, Bulgaria), and the National Institute of Hydrology and Water Management (NIHWM, Romania).

BS-MFC delivers products for analysis, simulation, forecast, and reanalysis through the CMEMS and maintains a level of efficiency and robustness in operations, ensuring new advances in science for the development of specific products. The BS-MFC catalogue offers products for the Blue Ocean (physics and waves) and the Green Ocean (biogeochemistry) at the regional scale. These include: (a) near-real-time (NRT) datasets—analysis and forecast fields at different frequencies for supporting daily services and down-stream applications; and (b) multi-year (MY) datasets—reanalysis and long-term simulations—at different frequencies to provide an estimate of climate variability and trends.

The present paper provides a complete overview of the current BS modelling systems together with a description of the main operational products delivered through CMEMS interfaces. A detailed description of the BS-MFC components (PHY, BIO, and WAV) is also provided, together with a description of the models and data assimilation capabilities used to improve the accuracy of multi-year and forecasting products (Section **??**). Product quality and monitoring capacity are described in Section **??** by showing the operational dashboard used to evaluate the accuracy of the BS-MFC products. The last part of this paper summarizes the main scientific challenges and ongoing developments for the next generation of BS-MFC operational systems (Section **??**) together with the conclusions (Section **??**).

2. The Black Sea MFC High-Level Architecture and Components

The BS-MFC systems provide NRT and MY products over the Black Sea domain (Figure **??**), excluding the Azov Sea and the Bosporus Strait, at a resolution of $1/27°$ in the zonal direction and $1/36°$ in the meridional direction.

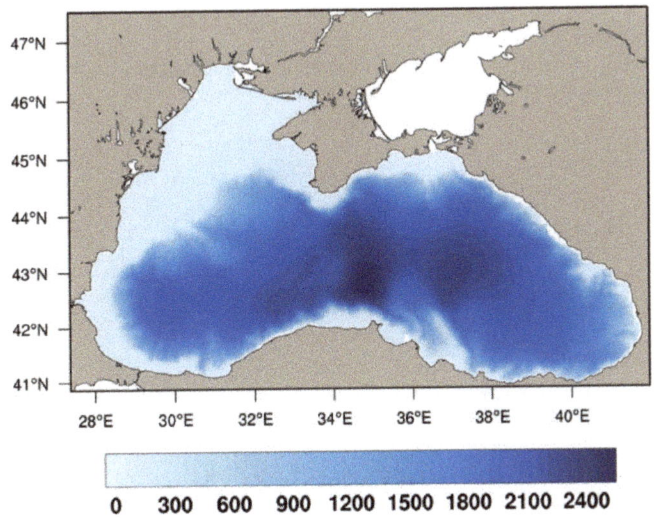

Figure 1. Black Sea spatial domain and bathymetry (in meters).

The BS-MFC uses an operational framework based on a high-level architecture, represented in Figure **??**. Three production units (PU) are responsible for the system and operational services, including evolutions of the physics (BS-PHY PU), biogeochemistry (BS-BIO PU), and wave (BS-WAV PU) components (run by CMCC, ULiege, and HEREON, respectively). Each PU is connected to its own archiving unit (AU) for the long-term storage of the BS-MFC products, and to a backup unit (BU) in the case of nominal operational failures and recoveries in order to guarantee the continuity of operational services. Additionally, each PU implements dedicated interfaces: (a) with the BS-MFC technical group, for the technical implementation of the product catalogue (BS TEC), (b) with the dissemination unit (DU) through the delivery buffer zone, for the operational delivery of the products (CMEMS CIS), and (c) with the BS-MFC service desk (BS LSD), for user support through the CMEMS service desk. Service evolution activities, aimed at R&D activities, are implemented at the PU level, with the collaboration of USOF and NIHWM (BS-EVO). Observations used by BS-MFC systems for assimilation and validation purposes are summarized in Table **??**.

Table 1. Upstream data dependency for the BS-MFC: for each type of observation, provider and product ID are given as well as the BS component—PHY, BIO, WAV—and product category—NRT and MY—which up-take it for assimilation (A) and/or validation (V).

Type of Observation	Provider	Product ID (If Available)	BS-PHY	BS-BIO	BS-WAV
Temperature and salinity profiles (in situ)	CMEMS SeaDataNet	INSITU_BS_NRT_OBSERVATIONS_013_034 INSITU_GLO_TS_REP_OBSERVATIONS_013_001_B NA, see [?]	A, V (NRT) A, V (MY) A (MY)	V (NRT) V (MY)	V (NRT) A, V (MY)
Sea surface temperature (satellite)	CMEMS	SST_BS_SST_L4_NRT_OBSERVATIONS_010_006 SST_BS_SST_L3S_NRT_OBSERVATIONS_010_013 SST_BS_SST_L4_REP_OBSERVATIONS_010_022	A (NRT) V (NRT) A, V (MY)		
Sea level anomaly (satellite)	CMEMS	SEALEV-EL_EUR_PHY_L3_NRT_OBSERVATIONS_008_059 SEALEV-EL_BS_PHY_L4_REP_OBSERVATIONS_008_042	A, V (MY) A, V (MY)		
Chlorophyll (satellite)	CMEMS	OCEANCOL-OUR_BS_CHL_L3_NRT_OBSERVATIONS_009_045 OCEANCOL-OUR_BS_OPTICS_L3_NRT_OBSERVATIONS_009_04 OCEANCOL-OUR_BS_CHL_L3_REP_OBSERVATIONS_009_079		A (NRT) V (NRT) V (MY)	
Significant wave height (satellite)	CMEMS	WAVE_GLO_WAV_L3_SWH_NRT_OBSERVATIONS_014_001			V (NRT), A, V (MY)

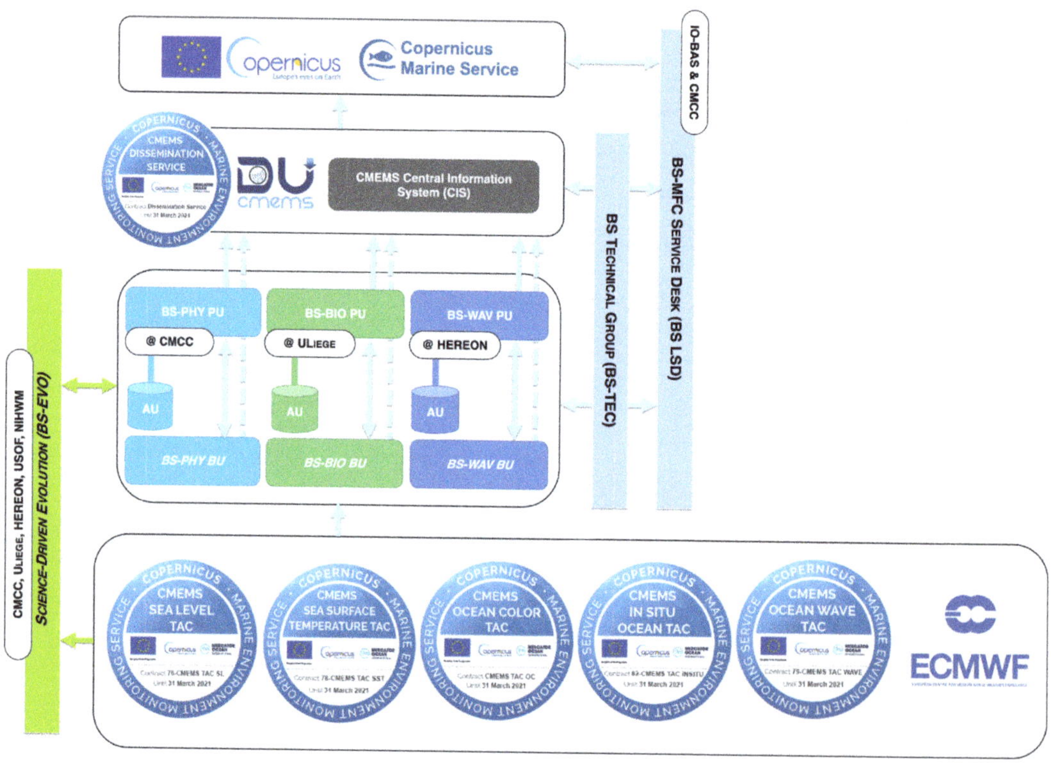

Figure 2. BS-MFC high-level architecture operationally implemented within the CMEMS framework since April 2018.

2.1. Physics

The Black Sea physical analysis and forecasting system (BS-PHY NRT) as well as the reanalysis system (BS-PHY MY) are free-surface versions of the NEMO ocean general circulation model (Nucleus for European Modelling of the Ocean, [?]), coupled online with a 3D-variational data assimilation scheme, the OceanVar [? ?]. The model's governing equations are discretized over a regular grid with about 3 km of horizontal resolution, using 31 z-levels with partial steps over the spatial domain, as shown in Figure ??.

The BS-PHY NRT [?] core model is based on NEMO v3.4. The bottom topography was reconstructed from the GEBCO 1 min resolution dataset [?]. It is forced by water, heat, and momentum fluxes, interactively computed by bulk formulae, implemented for the Mediterranean Forecasting System [?] and modified for the Black Sea to account for the Brunt–Berliand formula for the net longwave radiation, as in [?]. It uses monthly climatological precipitation from the GPCP dataset [? ?] and the 3–6 h and 0.125° operational analysis and forecast atmospheric fields provided by the European Centre for Medium-Range Weather Forecasts (ECMWF) through the Italian *Aeronautica Militare*.

The model includes 72 rivers, distributed as source points along the Black Sea coastline; major rivers such as the Danube, the Dnieper, and the Dniester are implemented over multi-grid points due to their wide delta and inflow contribution. The river discharge values are provided as monthly climatologies by [?] within the framework of the EU SESAME project, while a zero salinity value is accounted for at the river mouths. Initial conditions for the pre-operational run come from [?] as the climatology for January interpolated on the BS-PHY spatial grid. The model configuration is closed at the Bosporus Strait. Vertical mixing is parameterized according to the turbulent kinetic energy closure scheme.

The assimilation time window is one day. Background error covariances are decomposed into vertical covariances and horizontal correlations through 15-mode multivariate empirical orthogonal functions (EOFs). Observations, assimilated in the BS-PHY NRT, include (Table ??): (i) in situ temperature and salinity profiles (ARGO floats) from the CMEMS INS TAC. If profiles are disseminated at a high vertical resolution, a vertical thinning is applied to the profile before ingestion in OceanVar; (ii) along-track sea level anomaly L3 data, currently from AltiKa, Cryosat-2, and Jason-2/3, Sentinel3A/3B, distributed by the CMEMS SL TAC; (iii) gridded sea surface temperature SST L4 observations provided by the CMEMS SST TAC. Assimilation of SST satellite data is performed at the first model level. For satellite observations (SLA, SST), a horizontal thinning is also applied to retain approximately one observation only every 6 km.

The BS-PHY NRT processing system consists of two different cycles, run every day. One cycle consists of a 3-day analysis (e.g., a simulation performed with ECMWF analysis atmospheric forcing and assimilation correction), a 1-day simulation (e.g., a run performed with ECMWF analysis atmospheric forcing without assimilation correction), and a 10-day forecast (e.g., a run performed with ECMWF forecast atmospheric fields). The other cycle takes place once a week: the system performs a 14-day analysis in order to ingest a larger number of observations through the data assimilation and produce the best initial condition for the forecasting run. The system produces hourly and daily means of 3D temperature, salinity, and currents as well as 2D sea surface height, mixed layer depth, and bottom temperature, with the nominal start of the forecast at 00:00Z (i.e., averaged daily fields are centered at 12:00Z of each run day).

The BS-PHY MY [?] core model is based on NEMO v3.6 over the same spatial grid as the corresponding NRT. The bottom topography has been improved in relation to the NRT system and is based on a GEBCO 30″ resolution dataset [?], with the blending of a high-resolution bathymetric dataset around the Bosporus exit [?] to better represent the connection between the Black Sea and the Marmara Sea, and consequently, the Mediterranean Sea. BS-PHY MY system is forced by ECMWF ERA5 atmospheric reanalysis and GPCP monthly climatological precipitation [? ?].

Similarly to the NRT, the BS-PHY MY system imposes a closed boundary condition at the Bosporus Strait. To account for the Mediterranean water inflow into the Black Sea, the model solution is relaxed to temperature and salinity vertical profiles extracted from the high-resolution time series presented in [?]. The model is coupled online with OceanVar to assimilate (i) in situ temperature and salinity profiles (ARGO floats) pro-vided by the CMEMS INS TAC and the historical SeaDataNet dataset [?], (ii) along-track sea level anomaly L3 data distributed by CMEMS SL TAC. The SST is also relaxed to a gridded sea surface temperature SST L4 product produced by the CMEMS SST TAC. The time series covers the period of January 1993–December 2019 and provides monthly and daily means for the 3D temperature, salinity, current and 2D sea-surface height and mixed-layer depth.

2.2. Biogeochemistry

The Black Sea biogeochemical forecasting (BS-BIO NRT, [?]) and reanalysis (BS-BIO MY, [?]) systems are based on the online-coupled Biogeochemical Model for Hypoxic and Benthic Influenced areas (BAMHBI, [? ? ?]) and NEMO v3.6 ([?], version aligned with BS-PHY NRT). The coupled model is run over the same grid as used by BS-PHY NRT, with 31 vertical levels using z-layer vertical coordinates. BAMHBI describes the food web from bacteria to gelatinous carnivores through 24 state variables, including three groups of phytoplankton: diatoms, small phototrophic flagellates, and dinoflagellates; two zooplankton groups: micro- and mesozooplankton; two groups of gelatinous zooplankton: omnivorous and carnivorous forms; and an explicit representation of the bacterial loop: bacteria, labile and semi-labile dissolved organic matter, and particulate organic matter. The model simulates oxygen, nitrogen, silicate, and carbon cycling, and explicitly represents processes in the anoxic layer.

Biogeochemical processes in anaerobic conditions are represented using an approach similar to that used in the modelling of diagenetic processes in sediments, lumping together all the reduced substances in one state variable. Processes in the upper oxygenated layer are thus fully coupled with anaerobic processes in the deep waters, which enable long-term simulations to be performed. This full coupling between aerobic and anaerobic processes is key to performing the long-term reanalysis. Processes typical of anaerobic environments such as denitrification, anaerobic ammonium oxidation (ANAMMOX), and reduced decomposition efficiency are explicitly represented [? ?].

BAMHBI involves a module describing the carbonate dynamics based on the approach proposed by [?]. The model solves for DIC and the Excess Negative charge from which the Total Alkalinity is computed (considering the contribution of sulfide), as well as pH, the speciation of DIC ($[HCO_3]^-$, $[CO_3]^{2-}$, $[CO_2]$), and CO_2 air–sea flux. The model also includes a representation of diagenetic processes [?] using a computationally efficient representation, as proposed by [?].

The incorporation of a benthic module allows for a better representation of the impact of the sedimentary compartment in important biogeochemical processes such as sediment oxygen consumption (which is responsible for the generation of hypoxic conditions in summer), the active degradation of organic matter that determines the vigor of the shelf ecosystem (~30% of the primary production in shelf waters is degraded in the sediment), and the intense consumption of nitrate by benthic denitrification, which filters a substantial part (~50%) of the nitrogen carried by the north-western shelf rivers (the Danube being the most important one) and modulates primary production in the deep basin. In addition to a representation of diagenesis, the biogeochemical model represents the transport of sediments by waves. This is an important feature that is necessary to sustain the primary production of the deep basin.

Every day, the BS-BIO NRT system runs one day of analysis and 10 days of forecasts. Once a week, BS-BIO NRT performs a 10-day analysis during which the model assimilates daily L3 satellite chlorophyll observations (Table ??) via an Ocean Assimilation Kit (OAK) [?], developed as part of the SANGOMA project. The assimilation increments are three-dimensional, i.e., the analysis also modifies the model variables below the surface layer. The BS-BIO NRT system uses ECMWF analysis and forecast atmospheric fields to compute air–sea fluxes, while the BS-BIO MY system uses ECMWF ERA5 reanalysis. Atmospheric fields are used to force NEMO to compute the air–sea exchanges of O_2 and CO_2.

The atmospheric deposition of inorganic nitrogen [?] is also considered. Such a process has a similar order of magnitude as the river inputs and is needed in order to sustain the primary production in the deep basin. For the rivers, due to the absence of operational data, the BS-BIO NRT system uses climatological averages of river flows, inorganic nutrients, and organic material inputs computed from the long-term series of data provided by [?]. In its current configuration, BS-BIO systems (both NRT and MY) involve river inputs from six main rivers: the Danube (split into three branches), Dniepr, Dniestr, Rioni, Sakarya, and Kizilirmak. Water discharges are provided as monthly climatology to capture seasonal signal. The Bosporus Strait is considered an open boundary as in [?] and [?]. The velocity and salinity are determined in such a way that total sea water and salt are conserved in the Black Sea domain.

BS-BIO NRT and MY systems generate the following six datasets:

- NUTR: phosphorus, nitrate;
- PFTC: chlorophyll and phytoplankton biomass;
- BIOL: dissolved oxygen (O_2) concentrations and net primary production;
- CARB: pH, dissolved inorganic carbon (DIC), total alkalinity (TA);
- CO_2F: surface partial CO_2 pressure, surface CO_2 flux;
- OPT: photosynthetically active radiation (PAR) and attenuation coefficient (Kd) (only BS-BIO NRT).

2.3. Waves

The BS-WAV forecasting (BS-WAV NRT, [?]) and reanalysis systems (BS-WAV MY, [?]) are based on the WAM Cycle 6 Black Sea model, which replaced the former Cycle 4.6.2 (operational since April 2017) within CMEMS in December 2020. The wave model describes the ocean surface gravity waves (periods of 1.5–25 s). The regional wave model for the semi-enclosed Black Sea runs in shallow water mode with the same spatial resolution as BS-PHY. WAM calculates the two-dimensional energy density spectrum at each of the 44,699 active model grid points in the frequency and directional space. The solution of the energy balance equation is provided for 24 directional bands at 15°, starting at 7.5° and measured clockwise with respect to true north, and 30 frequencies logarithmically spaced from 0.042 Hz to 0.66 Hz at intervals of $\Delta f/f = 0.1$. Therefore, the prognostic part of the wave model covers approximately 25–1.5 s. In order to include the higher frequency waves into wave-growth/dissipation processes and the output wave characteristics, a parametric tail is fitted for frequencies above the spectral maximum [?]. A detailed description is given in [? ? ? ? ? ? ?]. The WAM Cycle 6 used for the Black Sea wave hindcast is an update of the former WAM Cycle 4. The basic physics and numerics are kept in the new release. The source function integration scheme by [?] and the model updates by [?] are incorporated. The wave model performance is discussed in [? ? ?].

The driving forces for the wave model are the 10 m wind fields provided by the ECMWF analysis and forecast fields, as for BS-PHY. The initial conditions of the BS-WAV NRT are constrained over successive cycles by including a 24 h hindcast run of the model prior to each forecast. The hindcast applies analyzed wind fields to the wave model so that the model is forced by the best available descriptions of the atmosphere and ocean. This prevents any drifts in the initial conditions of the wave model since the key response in the wave model is to the wind, and the analyzed forcing fields reduce the impact of any systematic drifts in the atmospheric model. The parameterization of the wave growth in the wind input source term is adapted to the driving wind fields.

The WAM model estimates the sea-state-dependent momentum and energy fluxes, and the Stokes Coriolis forcing diagnostics needed in order for it to be coupled to the ocean model [? ?]. The wave-induced processes have a significant impact on the skills of drifter estimations, e.g., [? ?]. WAM cycle 6.0 considers the new extreme wave diagnostics (maximal wave height and wave crest, [? ?]) that are included in the new BS-WAV products. In the new BS-WAV NRT wave-breaking parameterization is considered as well as the time-dependent depth and current fields from the BS-PHY NRT. A novel feature of BS-WAV MY data is that radar altimeter data are assimilated. Besides a significant wave height, the assimilation includes wind speed data. The data measured are assimilated into the wave model fields using an optimal interpolation (OI) scheme [?]. Given the lack of available systematic in situ observations in the Black Sea, the satellite data add value to the wave simulations.

3. Evaluating the Quality of the BS Products in the Operational Framework and for Monitoring

Evaluating the quality of BS-MFC products is key to providing users with a reliable service. Quality assurance of the product is coordinated within the CMEMS through a dedicated working group, while BS-MFC implements state-of-the-art metrics and supports the centralized product quality dashboard (https://pqd.mercator-ocean.fr/ accessed on 11 October 2021). BS-MFC product quality is based on GODAE/OceanPredict and MERSEA/MyOcean standards for the evaluation of product accuracy [?].

The following sections provide some examples of pre-qualification and operational capacities for the BS-MFC components.

3.1. Validation of BS-PHY Systems

Estimated accuracy numbers (EANs) for BS-PHY NRT were computed using the daily means analysis fields and compared with available observations (Table ??). Root mean square difference (RMSD) and bias were estimated over the period of 2019–2020. ???? show EANs computed in the period of 2019–2020 for temperature and salinity, respectively, using ARGO in situ profiles. The evaluation was performed in specific layers. Table ?? shows the averaged RMSD for temperature, which ranges from a maximum of 2.1 °C at 20–30 m to below 0.55 °C for depths greater than 75 m. The temperature bias is around zero over the water column. Regarding salinity (Table ??), the error is approximatively 0.26 PSU in the 5–50 m layer, and then increases to about 0.4 PSU in the halocline. Below 200 m, it is less than 0.1 PSU. Salinity bias is generally slightly negative on the subsurface, but never greater than 0.2 PSU. Regarding sea level, the system has an overall error of about 2.9 cm, while for sea surface temperature, the error is about 0.54 °C [?]. In order to highlight how BS-PHY NRT reproduces the main features of the circulation at the basin scale, ???? show surface currents as the mean in 2019 and 2020, respectively.

Both figures show the persistency of the Rim current and submesoscale eddies during the selected period—major eddies in the Western basin, the central cyclonic gyre particularly evident in 2020 and less pronounced in 2019, small coastal eddies, and bifurcation of the Rim along the Crimean peninsula. The full overview of the skills is also operationally available through the CMEMS Product Quality Dashboard https://pqd.mercator-ocean.fr/ (accessed on 11 October 2021). Additionally, BS-PHY uses a regional validation website (https://bsfs.cmcc.it/, accessed on 11 October 2021) to evaluate and monitor the quality of the near-real-time physical products.

Table 2. EANs for temperature (°C) in terms of bias and RMSD for 2019–2020, including number of observations used in the comparison.

	2019			2020		
Layer (m)	**Bias**	**RMSD**	**N. Observations**	**Bias**	**RMSD**	**N. Observations**
5–10	−0.13	1.11	1844	−0.05	0.78	2541
10–20	−0.04	1.87	3462	−0.23	1.57	4217
20–30	0.05	1.62	3484	−0.02	2.10	3925
30–50	0.04	0.91	6909	0.11	1.37	7367
50–75	0.01	0.3	8717	−0.02	0.72	8955
75–100	−0.02	0.17	8329	−0.3	0.27	7647
100–200	0.04	0.09	20,965	0.00	0.10	15,698
200–500	−0.02	0.04	39,114	−0.03	0.05	29,148
500–1000	−0.01	0.02	33,457	−0.01	0.02	18,786

Table 3. EANs for salinity (PSU) in terms of bias and RMSD for 2019–2020, including number of observations used in the comparison.

	2019			2020		
Layer (m)	**Bias**	**RMSD**	**N. Observations**	**Bias**	**RMSD**	**N. Observations**
5–10	−0.08	0.30	1844	−0.06	0.32	2541
10–20	−0.05	0.25	3462	−0.07	0.25	4217
20–30	0.00	0.23	3484	−0.04	0.22	3925
30–50	0.09	0.29	6909	0.05	0.25	7367
50–75	0.08	0.38	8717	0.14	0.38	8955
75–100	0.02	0.37	8329	0.04	0.40	7647
100–200	−0.02	0.22	20,965	0.01	0.20	15,698
200–500	0.00	0.08	39,114	0.00	0.08	29,148
500–1000	0.00	0.02	33,457	−0.01	0.03	18,786

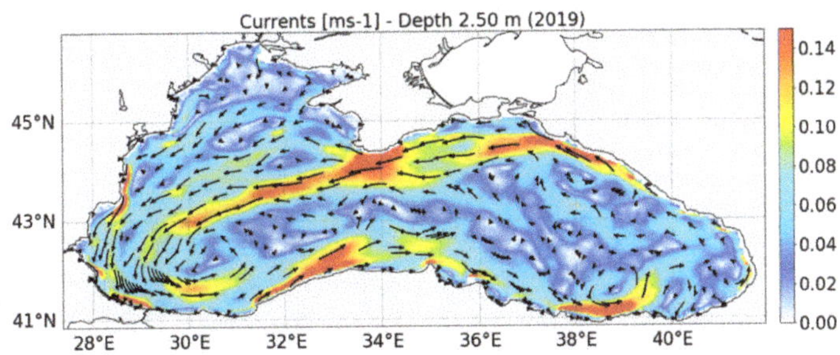

Figure 3. Surface currents in 2019.

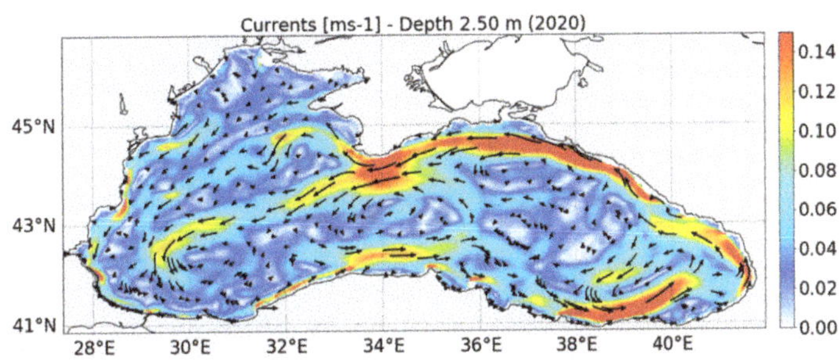

Figure 4. Surface currents in 2020.

In the BS-PHY MY, the highest temperature RMSDs are at the seasonal thermocline depths of around 20 m as the values exceed 1.5 °C (Figure ??a). The temperature bias is satisfactory, with a maximum absolute value of about 0.07 °C at the surface, and around zero from 60 down to 500 m, then it increases up to 0.12 °C below 500 m. Salinity has the largest RMSD on the surface, with a maximum of 0.7 PSU (Figure ??b). Similar to the temperature, salinity RMSDs are relatively high at the halocline depths. The Hovmöller diagram of temperature RMSD reveals a clear seasonal pattern: the values are low (high) in winter (summer) so that the highest errors are in the seasonal thermocline (Figure ??a). The Hovmöller diagram of salinity RMSD shows that there is a seasonal signal in the error on the surface: RMSD exceeds 1.5 PSU near the surface before 2008. At depths of 50–150 m, the error decreases up to around 0.8 PSU; however, it is almost zero below 150 m (Figure ??b).

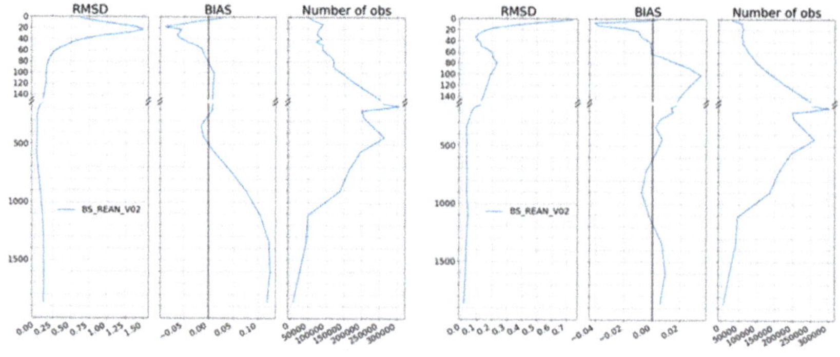

(a) (b)

Figure 5. Vertical profiles of the root mean square deviation (left panel), bias (middle panel), and number of observations (right panel) for (**a**) temperature (°C) and (**b**) salinity (PSU) by comparing the reanalysis results with in situ profilers in the Black Sea domain from 1 January 1993 to 31 December 2019.

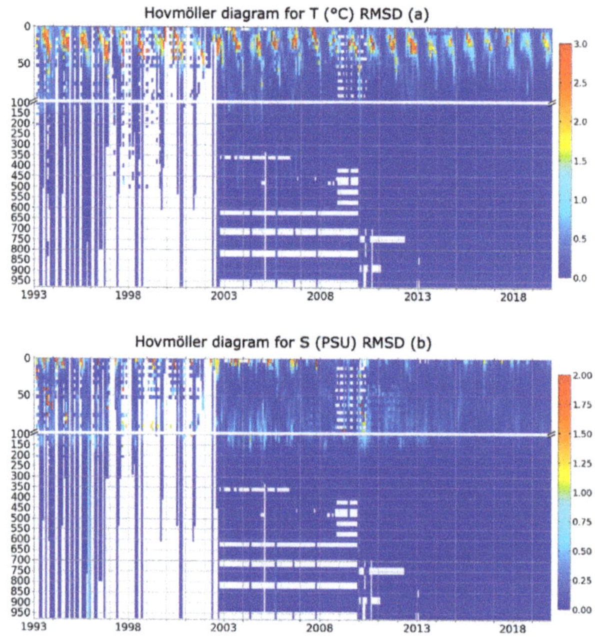

Figure 6. Hovmöller diagrams of root mean square difference for temperature in °C (**a**) and salinity in PSU (**b**), by comparing the BS-PHY MY results with in situ profilers in the Black Sea domain from 1 January 1993 to 31 December 2019.

The Hovmöller diagrams reveal the lack of in situ observations between 1997 and 2003 in the Black Sea, which means that for many years the reanalysis system was only constrained by altimeter observations. In comparison with along-track sea level anomaly observations, BS-PHY MY product had a mean sea level anomaly RMSD of 2.24 cm from 1993 to 2019. Finally, the geostrophic currents estimated from the time-averaged sea level (1993–2019) reveal the Rim current, which surrounds the entire Black Sea and forms a large-scale cyclonic gyre (Figure ??). Major details are provided in [?].

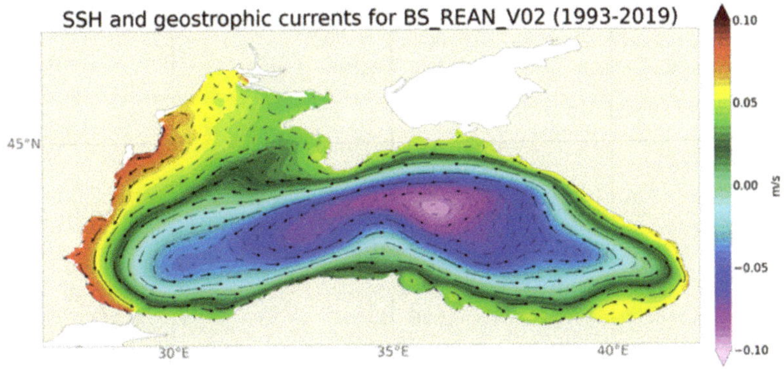

Figure 7. Mean sea-surface height and corresponding mean sea-surface circulation from the BS-PHY MY product evaluated for the period of 1993–2019.

3.2. Validation of BS-BIO Systems

The quality of the BS-BIO NRT and BS-BIO MY systems was assessed using all the observations available for the Black Sea in existing databases (e.g., CMEMS INS and OC TACs, World Ocean database [?], EMODnet at https://portal.emodnet-physics.eu/ (accessed on 11 October 2021), R/V KNORR at https://www.whoi.edu/multimedia/r-v-knorr/, accessed on 11 October 2021) using EANs in the form of BIAS_log10 between the model (M) and the observation (O), as follows (Table **??**):

$$\mathrm{BIAS}_{\log 10} = \log 10(\mathrm{M}) - \log 10(\mathrm{O}), \quad (1)$$

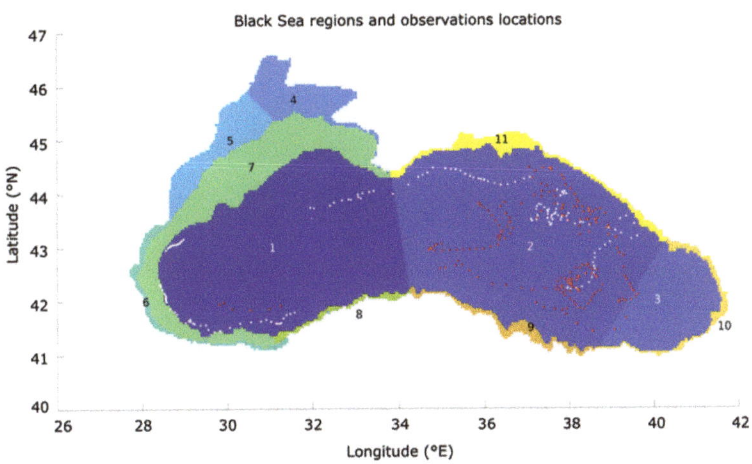

Figure 8. Sub-regions schematization from [?] used for computing the error statistics for chlorophyll. Regions 1–3 are the "open sea", regions 4–5 are the North-Western Shelf, and region 7 is located at the shelf break and is considered a transition region between eutrophic and oligotrophic conditions.

Table 4. $BIAS_{log10}$ statistics for chlorophyll obtained for BS-BIO NRT, considering the observation minus model prediction pairs for the various regions (1–11) shown in Figure ?? Data are from the L3 CMEMS satellite chlorophyll observations (Table ??).

Region	1	2	3	4	5	6
EAN	0.012	−0.012	−0.011	0.19	0.15	0.014
	7	**8**	**9**	**10**	**11**	
EAN	0.04	0.009	0.055	0.07	−0.07	

In the BS-BIO NRT, these error statistics can only be computed for oxygen, chlorophyll, Kd, and PAR, which are derived from satellites and BG-ARGO. Much of the validation has entailed the assessment of the model's capacity to simulate oxygen. Oxygen integrates the balance of physical and biogeochemical processes such as the formation of the cold intermediate layer, the long-term stability of the main pycnocline, dissolution, photosynthesis, and pelagic and benthic respiration. BS-BIO NRT simulates the main open sea oxycline but overestimates its vertical extension compared to BGC-ARGO data profiles that have a deeper injection of rich oxygen filaments (Figure ??).

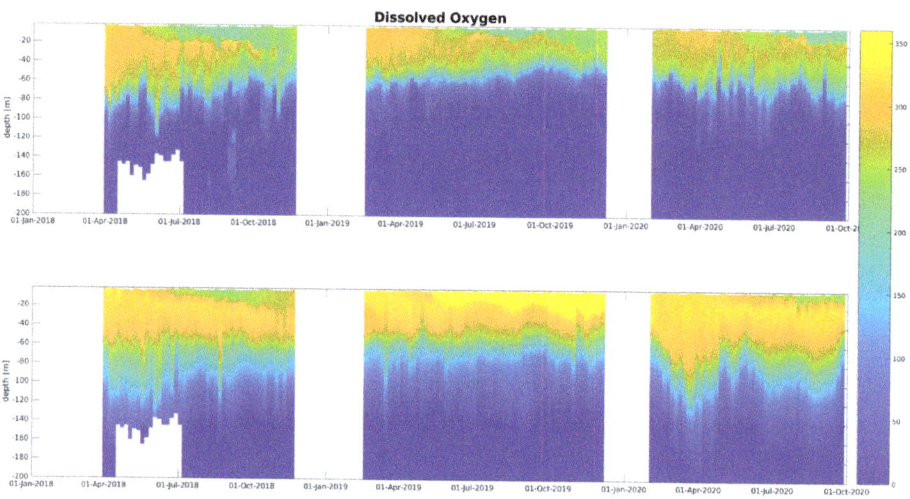

Figure 9. BGC-ARGO (**top**) and modelled (**bottom**) vertical profiles of oxygen from 2018 to 2020.

In April 2019, the dissolution of oxygen was overestimated, which is probably due to an underestimation of the SST. As shown in Figure ??, on the north-western shelf, the model (in orange) seems to capture the main characteristics of the observed seasonal cycle (in blue), with an oxygenation of the water column in winter due to ventilation and surface photosynthesis, and a reduction later in the year in response to decreased dissolution and the intensification of respiration. The model tends to underestimate the oxygen concentration on the surface in winter time (Jan–Mar) (Figure ?? left), and does the reverse in the bottom layer in summer–autumn time (Jun–Dec), except in winter when the water column is not stratified (Figure ?? right). The model underestimates the oxygenation of the water column in winter, while later in the year, the underestimation of the surface values and the overestimation of the bottom oxygen could indicate an underestimation of the level of primary production. The lower difference between surface and bottom values in the model as compared to the observations suggests excessive mixing in the model. From July to September, the model simulates hypoxia (O_2 < 63 µmol kg^{-1}) in the northernmost part of the shelf; however, the number of hypoxic records is underestimated.

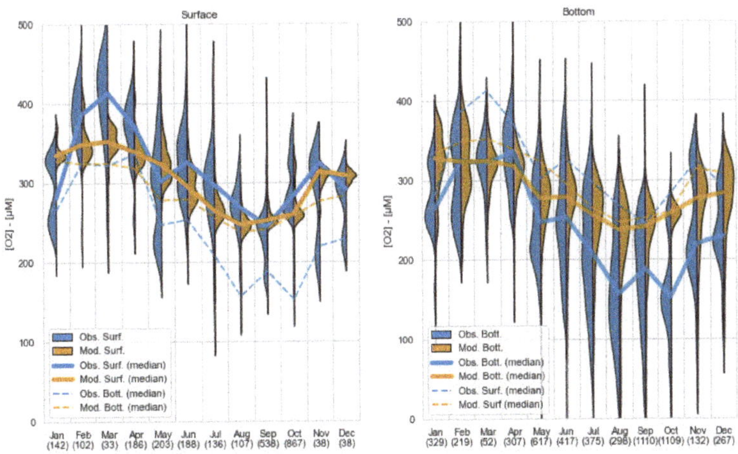

Figure 10. Seasonal cycle between 1992 and 2019 of the modelled (orange) and observed (blue) oxygen concentrations over the north-western shelf (at depths of 15–80 m) on the surface (**left**) and at the bottom (**right**). The number of observations for each month is reported along the *X*-axis.

A comparison of the chlorophyll-a (Chla) simulated by BS-BIO NRT and retrieved from the satellite for characteristic optical regions in the Black Sea, as in Figure **??** [?], is shown in Figure **??** (subplots from a to f are related to some crucial regions as shown in Figure **??**). Subplots in Figure **??** show that the model reproduces the winter–early spring bloom typical of the deep Black Sea (Figure **??**a,c,e) and the higher level of production on the north-western shelf with the presence of several peaks (Figure **??**b,d,f). Table **??** shows the log10 BIAS statistics for chlorophyll as the estimated accuracy number (EAN). Considering regions as in Figure **??**, EAN is around zero, except in the North-West coastal region (region 4) and at the Danube Delta (region 5), strongly impacted by water discharge (e.g., from the Danube, the Dniepr and the Dniester). Dynamics of the bloom are strongly conditioned by the amount of nutrients discharged on the shelf and the model has the tendency to slightly underestimate the process, as shown by the positive EAN. This is explained using climatological averaged values for the nutrients discharged by the rivers instead of NRT values. The quality of the boundary conditions at the interface with the rivers can hamper model performance.

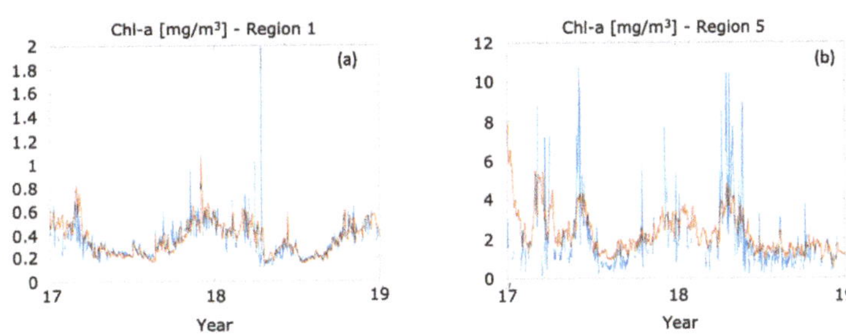

Figure 11. *Cont.*

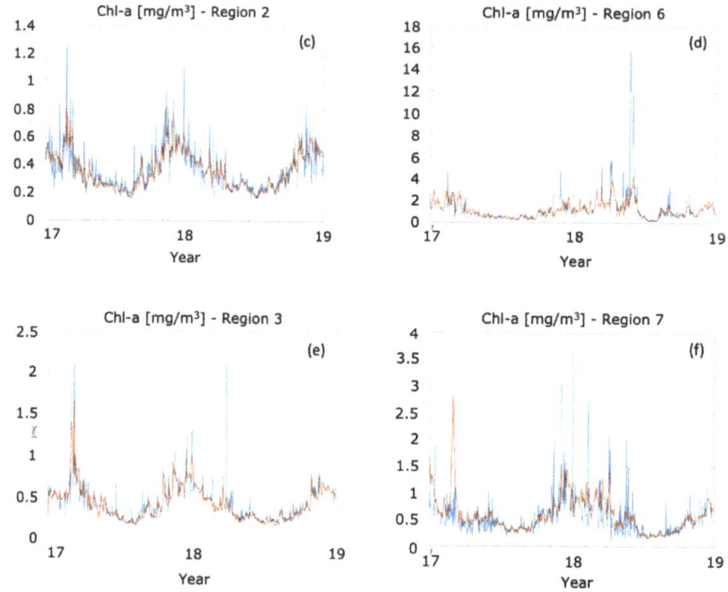

Figure 11. Time series of satellite (in blue) and model (in orange) chlorophyll spatial averages computed for optical characteristic regions as highlighted by [?].

3.3. *Validation of BS-WAV Systems*

This section describes the quality of the BS-WAV NRT and MY products. Validation is performed by comparing model results with satellite observations, as provided by Sentinel-3a, Sentinel-3b, Cryosat-2, Jason-3, and SARAL/Altika for the period between 01 July 2018 and 30 June 2020, and available in situ observations. The assessment of the corresponding wave hindcast is the best way to understand the validity of the WAM domain model underpinning these products, since the wave analysis-forecast system provided to CMEMS considers surface currents and water-level deviations from BS-PHY NRT. The data are incorporated into the WAM model grid. The growth of errors in the wave forecasts is dominated by growth errors in the forcing fields, which are the 10 m wind fields from the ECMWF IFS (Integrated Forecasting System).

In the new MY product (released to users in December 2020), wave breaking and the assimilation of measured satellite data were taken into account. The required radar altimeter data are available on the public server of AVISO at ftp-access.aviso.altimetry.fr (accessed on 11 October 2021) and CMEMS WAVE TAC, and includes significant wave height and also wind speed. The measured data are assimilated into the wave model fields using an optimal interpolation (OI) scheme. Satellites cross the Black Sea once or twice a day for less than two minutes, so very few measured values are available for assimilation into the wave and wind fields. The currents and water level deviations were not considered for the MYP.

Figure **??** depicts the scatter plots for a comparison between the modelled significant wave height (SWH) (provided by BS-WAV MY) and the one taken from satellites (Jason-1, Jason-2, Jason-3, and combinations thereof). The computed and measured mean values of SWH are generally all located between 0.9 and 1.0 m. The bias is nearly zero for all the merged satellite values, although the values for Jason-2 and Jason-3 show a small underestimation of the wave heights in WAM. The calculated biases of WAM for the different satellites confirm the good agreement between measurements and model results; in fact, they are in the range between −8 cm (Jason-3) and 3 cm (Jason-1). The RMSD varies from 17 cm (Jason-1) to 26 cm (Jason-3). Another deviation is in the difference in the simulated and modelled variability, given as the standard deviations in this article. The differences range from about 5 cm (Jason-2) to 6 cm (Jason-3). Note that the measurement errors and noise that our initial quality control has not filtered out can also impact metrics, thus potentially degrading the model's skill level. The wind fields also influence the variability and skills of the model as they force the wind-wave model WAM.

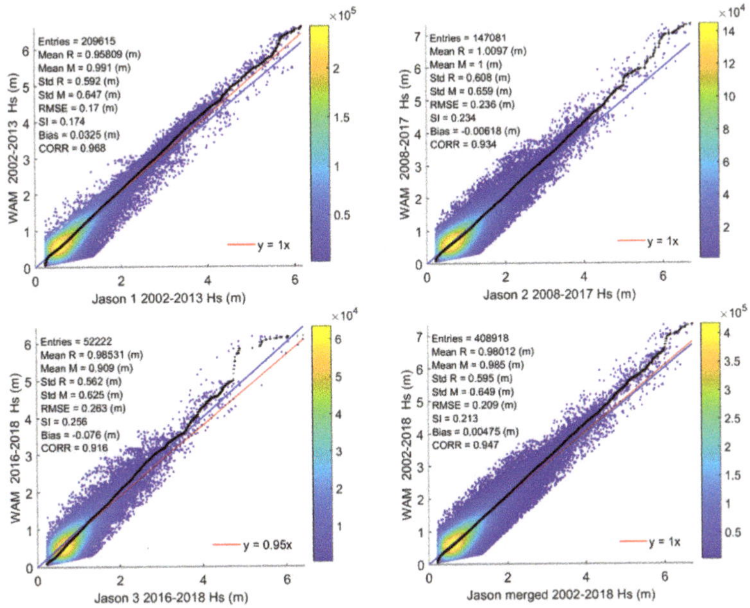

Figure 12. Scatter plots showing satellite measurements versus modelled significant wave heights (BS-WAV MY) for 2002–2013 (Jason-1), 2008–2017 (Jason-2), 2016–2018 (Jason-3), and 2002–2018 (all satellites merged). Additionally, the following are presented: the estimated bivariate probability density (colored area), the linear slope-fit regression of modelled and observed wave heights (red line), specific quantiles taken from the empirical cumulative density function (black line), and the diagonal (blue line). Summary statistics and skill scores are also included. R: reference (satellite) data, M: model data.

Figure **??** shows two examples of a comparison between BS-WAV NRT significant wave height (SWH) and Jason-3 satellite measurements. The figure includes the distribution of SWH combined with a track of the Jason-3 satellite (upper panel), and the corresponding time series of measured and modelled SWH along the satellite track (lower panel).

On the left is the ascending path on 30 December 2019 16:53:13–16:54:56 UTC, which touches the area of the maximum wave height of about 6 m (Figure **??** left). The second example describes the conditions for the path on 18 December 2018 21:49:15–21:50:58 UTC, which crosses the area of maximum wave height of around 4 m (Figure **??** left). Both comparisons show a good agreement between satellite measurements and model results, with a small overestimation by the model for the first example near the maximum SWH. Regarding the quarterly validation procedure for the BS-WAV system, the comparisons between modelled and measured satellite data were analyzed for the two quarters of 2018 and all four quarters of 2019 (Table **??**). Representative results are shown in Figure **??** for the last quarter of 2018 for the four different satellites SARAL/Altika, Sentinel-3a, Jason-3, and Cryosat-2. The statistics in Table **??** show an underestimation of the measured data by the wave model simulations, particularly revealing a negative bias in the period of 2019–2020.

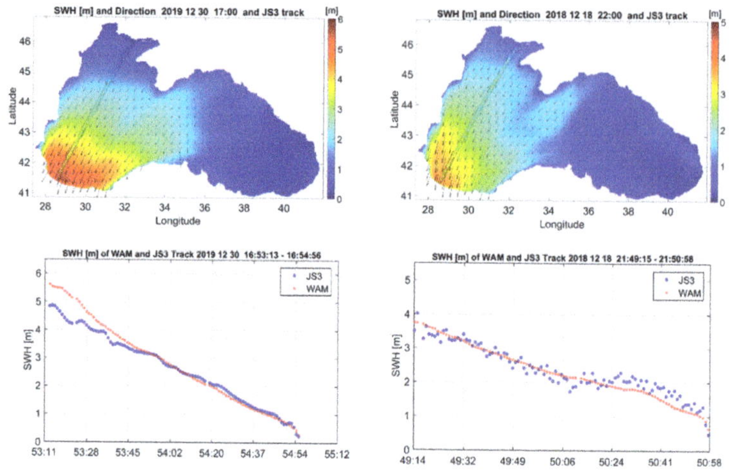

Figure 13. Left: distribution of BS-WAV NRT SWH on 30 December 2019 (17:00 UTC) and a Jason-3 satellite track also on 30 December 2019 (16:53:13–16:54:56 UTC). **Right**: distribution of SWH on 18 December 2018 (22:00 UTC) and a Jason-3 satellite track on 18 December 2018 (21:49:15–21:50:58 UTC).

Table 5. EANs for the current BS-WAV NRT system (all values in centimeters).

Satellite	Q1/2019		Q2/2019		Q3/2019		Q4/2019		Q1/2020		Q2/2020	
	Bias	RMSD	Bias	RMSD	Bias	RMSD	Bias	RMSD	Bias	RMSD	Bias	RMSD
Saral/Altika	−6.7	16.7	−5.9	16.8	−8.7	16.8	−6.1	21.8	−1.2	24.9	−4.6	16.6
CryoSat-2	−3.2	16.3	−1.1	15.1	−4.0	17.2	2.3	23.7	1.7	25.8	0.5	18.0
Jason-3	−6.4	17.2	−6.0	17.8	−7.2	18.8	−6.2	24.8	−3.9	19.6	−2.8	19.8
Sentinel-3A	−3.4	16.4	−5.4	17.5	−5.9	17.1	−4.7	21.8	−2.2	22.4	−9.6	22.3
Sentinel-3B	−4.1	16.5	−8.0	18.9	−9.5	19.2	−7.4	23.3	−5.4	25.5	−9.6	25.3

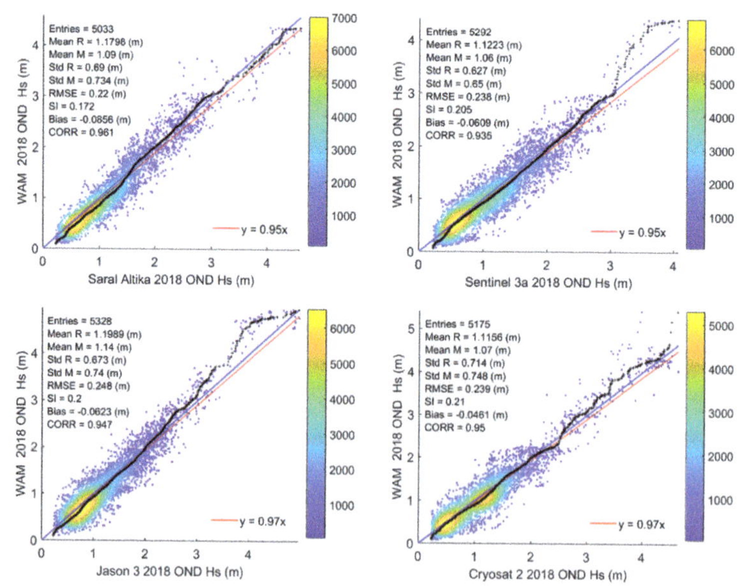

Figure 14. Scatter plots as in Figure ??, but for Q4 2018 and four different satellites.

Figure ?? includes a time series of statistical parameters of the BS-WAV MY product for the complete validation period between 2002 and 2019. On this annual scale, the metrics show that the skill level in all of the statistical parameters varies for the individual satellites. While the values for the bias are always around zero for Jason-1 and Jason-2, the bias for Jason-3 is slightly higher, as shown by the green curve in Figure ?? for the Annual Bias from 2016, and proves the underestimation of 7–10 cm of the measurements by the wave model.

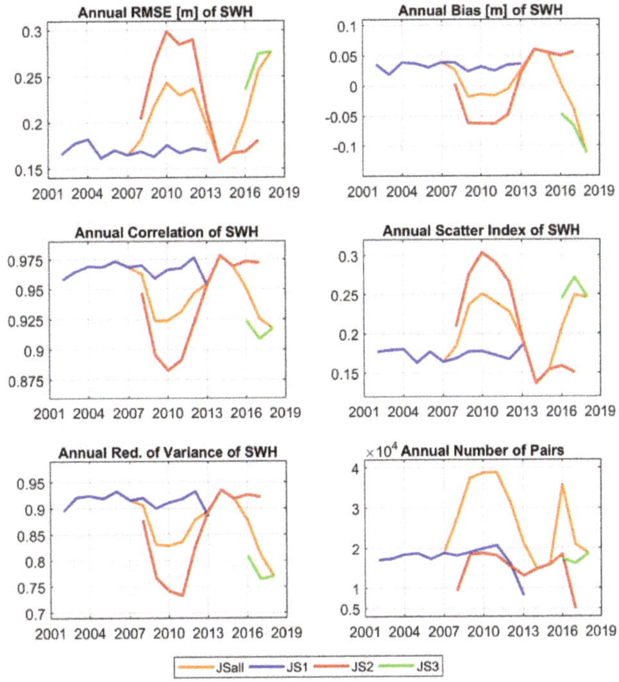

Figure 15. Yearly values of significant wave height metrics (using the BS-WAV MY system) for the Black Sea derived from individual and combined satellite measurements.

The very good behavior of all statistical parameters for Jason-1 is notable, as there is a very small RMSD value of about 17 cm, a scatter index of about 17%, a correlation of more than 0.96, and a reduction in variance with values above 0.9. That is of course due to the assimilation of Jason-1 data into the wave fields during the time period that they were available.

The parameters for Jason-2 went up for the RMSD (30 cm) around 2010, and down again after 2013 with a low of 17 cm. For the correlation, this is the other way around. The values went down from 0.94 to 0.88 in 2010, and then up again to around 0.97 after 2013. This was expected since after 2013, Jason-2 data were assimilated into the wave fields. The Jason-3 satellite provided data for a short time only. It started in 2016 and showed the most pronounced deviation between model results and the measurements. The RMSD was about 25 cm, the scatter index 25%, and the correlation 0.92. This also generated less satisfactory values for the combined satellite parameters between 2016 and 2019, which were fairly good for the previous years.

Overall, the skill scores depend to some extent on the number of collocated measurements. Clearly, with a decreasing number of observations, the values for the statistical parameters such as RMSD, bias, and scatter index go up, while those for the reduction of variance and the correlation go down. For 11 days, from 4 February 2012 to 15 February 2012, ADCP data were available at the location Pasha Dere (located at 28.03° E, 43.08° N) in the western part of the Black Sea near the Bulgarian coast. This period covers a storm event that occurred between 7 and 9 February 2012. A comparison of the significant wave height and peak period with WAM is shown in Figure **??**. The measured and modelled time series of the significant wave height and peak period show a very good agreement. The skill scores support the findings. The bias for the significant wave height is only about −2 cm, while the RMSD is 37 cm. The corresponding values for the peak period, which depend on the resolution of the model frequency, are also very good, with a bias of 0.1 s and an RMSD of about 1 s. The correlation is very high for both integrated parameters: about 0.97 for the significant wave height and 0.91 for the peak period. The peak of the storm was also simulated quite accurately and perfectly illustrates the capacity of the wave model to represent the arrival of a storm as captured by the observations.

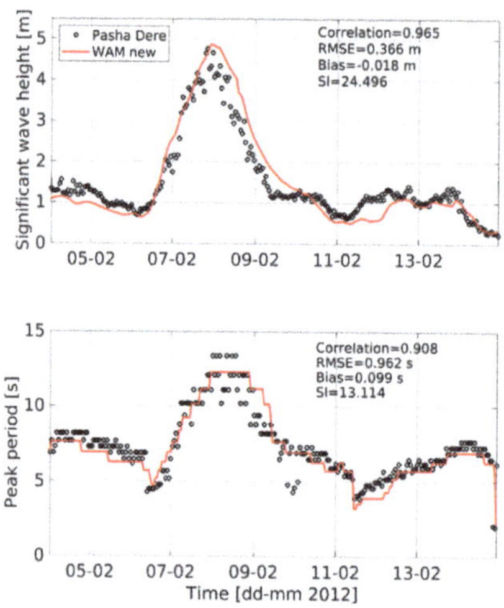

Figure 16. Significant wave height and peak period from WAM (using the BS-WAV MY system) and from the ADCP station Pasha Dere from 4 to 15 February 2012.

4. Summary and Conclusions

The BS-MFC products support users and down-stream services in terms of both sustainable industry and society, by means of accurate datasets, for a number of ocean variables such as temperature, salinity, currents for the physical core, nutrients, phytoplankton, chlorophyll, oxygen for the biogeochemical core, significant wave height and period, Stokes drift, and spectra for the wave core.

Since 2016, the BS-MFC has provided high-quality analyses, 10-day forecasts (as daily and hourly means, including hourly instantaneous for wave fields), and reanalyses (as monthly and daily means from 1993, including hourly predictions for wave fields available from 1979) of the ocean variables to describe the ocean state—both for physics and waves—and biogeochemical processes in the Black Sea region. This is part of the CMEMS product catalogue, available through https://marine.copernicus.eu/ (accessed on 11 October 2021). These systems benefit from constant improvements in the modelling components (e.g., structure, formulation, parameterization) and data assimilation capabilities. BS-MFC has developed reliable interfaces with upstream data providers—ECMWF for atmospheric forcing; CMEMS TACs for access to in situ and satellite observations, including SeaDataNet; EMODnet for historical in situ ones—and the CMEMS Dissemination Unit for the operational delivery of NRT and MY products.

The availability of in situ (i.e., ARGO, mooring stations, radar stations) and satellite observations (i.e., sea surface temperature, sea level anomaly, ocean color, significant wave height, etc.) is fundamental for the evolution of the BS-MFC systems, as reported in [?].

They are used for the performance of validation exercises and data assimilation. Observing capabilities in the Black Sea have significantly improved in recent years, thanks to national and international initiatives such as CMEMS by enforcing Black Sea thematic assembly centers. It is crucial that this effort is maintained in various parts of the sea; for example, in the north-western shelf, which receives vast river inflows, adequate monitoring of physical and biogeochemical variables is needed. BS-MFC models can be used to support the design phase of the observational network and to optimize it by using observing system (simulation) experiments.

In the short term, the BS-MFC implementation phase is going to further improve the overall product offer. The BS-PHY team is building new operational systems based on open boundary conditions at the Marmara Sea to improve the representation of the Bosporus Strait dynamics, including the use of historical data for the Danube River, and increased vertical resolution. BS-BIO plans to improve the data assimilation framework, moving towards ensemble approaches to better represent the uncertainty in the prediction. The BS-WAV team proposes new core models based on WAM Cycle 6.0; the NRT is now one-way offline coupled with hourly means of ocean surface currents and sea levels provided by BS-PHY NRT system. MY now assimilates SWH from SL TAC and, similarly to PHY and BIO, is forced by ECMWF ERA5, with which it will also be coupled in the future.

BS-MFC evolutions are carried out by science-driven activities based on the use of state-of-the-art modelling platforms—NEMO, BAMHBI, and WAM, and data assimilation schemes—devoted to solving physical and biogeochemical processes in an accurate way. For the next generation of BS-MFC systems, research and development efforts will focus on ensemble forecasting, refining the coupling strategy of Ocean-Atmosphere-Waves-Biogeochemistry, developing data assimilation capabilities according to newly-available high-resolution data, developing bio-optical models, and enforcing the optimal interface with the Mediterranean Sea using a high-resolution model for the Marmara Sea, which provides high-quality open boundary conditions for both basins.

Author Contributions: S.A.C., M.G. (Marilaure Grégoire) and J.S. designed the work and coordinated the preparation of the contributions in collaboration with A.P., G.C., and the following teams: E.J., E.P., L.L., A.A., S.C. (Sergio Creti'), L.S., D.A., S.C. (Salvatore Causio) for the BS-PHY; L.V., A.C., C.M., E.I. for the BS-BIO; A.B., M.R., G.G. for BS-WAV. Service management contribution was provided by R.L., V.M., F.P., N.V. Product quality activities were coordinated by E.P for the overall BS-MFC. M.I., M.G. (Murat Gunduz), M.M., S.M., G.C., N.P. contributed to the scientific discussions for the evolutions of the systems. N.V. and P.A. contributed to the overall project management. All authors have read and agreed to the published version of the manuscript.

Funding: This research was funded by the Copernicus Marine Environment Monitoring Service for the Black Sea Monitoring and Forecasting Center (Contract No. 72-CMEMS-MFS-BS).

Institutional Review Board Statement: Not applicable.

Informed Consent Statement: Not applicable.

Data Availability Statement: This work has been used CMEMS data from the BS-MFC, in particular Black Sea physical, biogeochemical and wave products (analysis, forecast, reanalysis) (https://marine.copernicus.eu/ (accessed on 11 October 2021)).

Conflicts of Interest: The authors declare no conflict of interest.

References

1. Le Traon, P.Y.; Reppucci, A.; Alvarez Fanjul, E.; Aouf, L.; Behrens, A.; Belmonte, M.; Bentamy, A.; Bertino, L.; Brando, V.E.; Kreiner, M.B.; et al. From Observation to Information and Users: The Copernicus Marine Service Perspective. *Front. Mar. Sci.* **2019**, *6*, 234. [CrossRef]
2. Pecci, L.; Fichaut, M.; Schaap, D. SeaDataNet, an enhanced ocean data infrastructure giving services to scientists and society. In Proceedings of the IOP Conference Series: Earth and Environmental Science 2020, 11th International Symposium on Digital Earth (ISDE 11), Florence, Italy, 24–27 September 2019; Volume 509. [CrossRef]
3. Madec, G.; The NEMO Team. *NEMO Ocean Engine*; Note du Pole de Modelisation; Institute Pierre-Simon Laplace (IPSL): Paris, France, 2008; ISSN 1288-1619.

4. Dobricic, S.; Pinardi, N. An oceanographic three-dimensional variational data assimilation scheme. *Ocean. Model.* **2008**, *22*, 89–105. [CrossRef]
5. Storto, A.; Dobricic, S.; Masina, S.; Di Pietro, P. Assimilating along-track altimetric observations through local hydrostatic adjustment in a global ocean variational assimilation system. *Mon. Weather Rev.* **2010**, *139*, 738–754. [CrossRef]
6. Ciliberti, S.A.; Peneva, E.L.; Jansen, E.; Martins, D.; Creti', S.; Stefanizzi, L.; Lecci, R.; Palermo, F.; Daryabor, F.; Lima, L.; et al. Black Sea Analysis and Forecast (CMEMS BS-Currents, EAS3 System) (Version 1) [Data Set]. Copernicus Monitoring Environment Marine Service (CMEMS). 2020. Available online: https://www.cmcc.it/doi/black-sea-physics-analysis-and-forecast-cmems-bs-currents-eas3-system (accessed on 15 August 2021).
7. Historical GEBCO Data Sets. Available online: https://www.gebco.net/data_and_products/historical_data_sets/#gebco_one (accessed on 11 October 2021).
8. Pettenuzzo, D.; Large, W.G.; Pinardi, N. On the corrections of ERA-40 surface flux products consistent with the Mediterranean heat and water budgets and the connection between basin surface total heat flux and NAO. *J. Geophys. Res. Ocean.* **2010**, *115*, C06022. [CrossRef]
9. Rosati, A.; Miyakoda, K. A general circulation model for upper ocean simulation. *J. Phys. Ocean.* **1988**, *18*, 1601–1626. [CrossRef]
10. Adler, R.F.; Huffman, G.J.; Chang, A.; Ferraro, R.; Xie, P.; Janowiak, J.; Rudolf, B.; Schneider, U.; Curtis, S.; Bolvin, D.; et al. The Version 2 Global Precipitation Climatology Project (GPCP) Monthly Precipitation Analysis (1979-Present). *J. Hydrometeor.* **2003**, *4*, 1147–1167. [CrossRef]
11. Huffman, G.J.; Adler, R.F.; Bolvin, D.T.; Gu, G. Improving the Global Precipitation Record: GPCP Version 2.1. *Geophys. Res. Lett.* **2009**, *36*, L17808. [CrossRef]
12. Ludwig, W.; Dumont, E.; Meybeck, M.; Heussner, S. River discharges of water and nutrients to the Mediterranean Sea: Major drivers for ecosystem changes during past and future decades? *Prog. Oceanogr.* **2009**, *80*, 199–217. [CrossRef]
13. Simonov, A.I.; Altman, E.N. Hydrometeorology and Hydrochemistry of the Seas. Vol IV: The Black Sea. Issue 1: Hydrometeorological conditions. *Hydrometeoizdat* **1991**. (In Russian)
14. Lima, L.; Aydoğdu, A.; Escudier, R.; Masina, S.; Ciliberti, S.A.; Azevedo, D.; Peneva, E.L.; Causio, S.; Cipollone, A.; Clementi, E.; et al. Black Sea Physical Reanalysis (CMEMS BS-Currents) (Version 1) [Data Set]. Copernicus Monitoring Environment Marine Service (CMEMS). 2020. Available online: https://resources.marine.copernicus.eu/product-detail/BLKSEA_MULTIYEAR_PHY_007_004/INFORMATION (accessed on 15 August 2021).
15. Weatherall, P.; Marks, K.M.; Jakobsson, M.; Schmitt, T.; Tani, S.; Arndt, J.E.; Rovere, M.; Chayes, D.; Ferrini, V.; Wigley, R. A new digital bathymetric model of the world's ocean. *Earth Space Sci.* **2015**, *2*, 331–345. [CrossRef]
16. Gürses, Ö. Dynamics of the Turkish Straits System. A Numerical Study with a Finite Element Ocean Model Based on an Unstructured Grid Approach. Ph.D. Thesis, Middle East Technical University, Mersin, Turkey, 2016.
17. Aydoğdu, A.; Pinardi, N.; Özsoy, E.; Danabasoglu, G.; Gürses, Ö.; Karspeck, A. Circulation of the Turkish Straits System under interannual atmospheric forcing. *Ocean Sci.* **2008**, *14*, 999–1019. [CrossRef]
18. Grégoire, M.; Vandenbulcke, L.; Capet, A. Black Sea Biogeochemical Analysis and Forecast (CMEMS Near-Real Time BLACKSEA Biogeochemistry) (Version 1) [Data Set]. Copernicus Monitoring Environment Marine Service (CMEMS). 2020. Available online: https://resources.marine.copernicus.eu/product-detail/BLKSEA_ANALYSIS_FORECAST_BIO_007_010/INFORMATION (accessed on 15 August 2021).
19. Grégoire, M.; Vandenbulcke, L.; Capet, A. Black Sea Biogeochemical Reanalysis (CMEMS BS-Biogeochemistry) (Version 1) [Data Set]. Copernicus Monitoring Environment Marine Service (CMEMS). 2020. Available online: https://resources.marine.copernicus.eu/product-detail/BLKSEA_REANALYSIS_BIO_007_005/INFORMATION (accessed on 15 August 2021).
20. Grégoire, M.; Raick, C.; Soetaert, K. Numerical modeling of the deep Black Sea ecosystem functioning during the late 80's (eutrophication phase). *Prog. Oceanogr.* **2008**, *76*, 286–333. [CrossRef]
21. Grégoire, M.; Soetaert, K. Carbon, nitrogen, oxygen and sulfide budgets in the Black Sea: A biogeochemical model of the whole water column coupling the oxic and anoxic parts. *Ecol. Model.* **2010**, *221*, 2287–2301. [CrossRef]
22. Capet, A.; Meysman, F.J.R.; Akoumianaki, I.; Soetaert, K.; Grégoire, M. Integrating sediment biogeochemistry into 3D oceanic models: A study of benthic-pelagic coupling in the Black Sea. *Ocean Model.* **2016**, *101*, 83–100. [CrossRef]
23. Soetaert, K.; Hofmann, A.; Middelburg, J.; Meysman, F.; Greenwood, J. The effect of biogeochemical processes on pH. *Mar. Chem.* **2007**, *105*, 30–51. [CrossRef]
24. Soetaert, K.; Middelburg, J.; Herman, P.; Buis, K. On the coupling of benthic pelagic biogeochemical models. *Earth-Sci. Rev.* **2000**, *51*, 173–201. [CrossRef]
25. Vandenbulcke, L.; Barth, A. A stochastic operational forecasting system of the Black Sea: Technique and validation. *Ocean Model.* **2015**, *93*, 7–21. [CrossRef]
26. Kanakidou, M.; Duce, R.A.; Prospero, J.M.; Baker, A.R.; Benitez-Nelson, C.; Dentener, F.J.; Hunter, K.A.; Liss, P.S.; Mahowald, N.; Okin, G.S.; et al. Atmospheric fluxes of organic N and P to the global ocean. *Glob. Biogeochem. Cycles* **2016**, *26*, GB3026. [CrossRef]
27. Stanev, E.V.; Staneva, J.V.; Roussenov, V.M. On the Black Sea water mass formation. Model sensitivity study to atmospheric forcing and parameterization of physical processes. *J. Mar. Syst.* **1997**, *13*, 245–272. [CrossRef]
28. Stanev, E.V.; Beckers, J.M. Barotropic and baroclinic oscillations in strongly stratified ocean basins. Numerical study for the Black Sea. *J. Mar. Syst.* **1999**, *19*, 65–112. [CrossRef]

29. Staneva, J.; Behrens, A.; Ricker, M.; Gayer, G. Black Sea Waves Analysis and Forecast (CMEMS BS-Waves) (Version 2) [Data Set]. Copernicus Monitoring Environment Marine Service (CMEMS). 2020. Available online: https://resources.marine.copernicus.eu/product-detail/BLKSEA_ANALYSISFORECAST_WAV_007_003/INFORMATION (accessed on 15 August 2021).
30. Staneva, J.; Behrens, A.; Ricker, M.; Gayer, G. Black Sea Waves Reanalysis (CMEMS BS-Waves) (Version 2) [Data Set]. Copernicus Monitoring Environment Marine Service (CMEMS). 2020. Available online: https://www.cmcc.it/doi/black-sea-waves-reanalysis-cmems-blk-waves (accessed on 15 August 2021).
31. The WAMDI Group. The WAM Model-A Third Generation Ocean Wave Prediction Model. *J. Phys. Oceanogr.* **1988**, *18*, 1775–1810. [CrossRef]
32. Staneva, J.; Ricker, M.; Carrasco Alvarez, R.; Breivik, Ø.; Schrum, C. Effects of Wave-Induced Processes in a Coupled Wave-Ocean Model on Particle Transport Simulations. *Water* **2021**, *13*, 415. [CrossRef]
33. Komen, G.J.; Cavaleri, L.; Donelan, M.; Hasselmann, K.; Hasselmann, S.; Janssen, P.A.E.M. *Dynamics and Modelling of Ocean Waves*; Cambridge University Press: Cambridge, UK, 1994; 532p.
34. Gunther, H.; Hasselmann, S.; Janssen, P.A.E.M. *The WAM Model Cycle 4.0. User Manual. Technical Report No. 4*; Deutsches Klimarechenzentrum: Hamburg, Germany, 1992; 102p.
35. Janssen, P.A.E.M. Progress in ocean wave forecasting. *J. Comput. Phys.* **2008**, *227*, 3572–3594. [CrossRef]
36. Staneva, J.; Behrens, A.; Wahle, K. Wave modelling for the German Bight coastal-ocean predicting system. *J. Phys. Conf. Ser.* **2015**, *633*, 233–254. [CrossRef]
37. Staneva, J.; Behrens, A.; Gayer, G. Predictability of large wave heights in the western Black Sea during the 2018 winter storms. *J. Oper. Oceanogr.* **2020**, *13*. [CrossRef]
38. Hersbach, H.; Janssen, P.A.E.M. Improvement of the Short-Fetch Behavior in the Wave Ocean Model (WAM). *J. Atmos. Ocean. Technol.* **1999**, *16*, 884–892. [CrossRef]
39. Bidlot, J.-R.; Janssen, P.; Abdalla, S. *A Revised Formulation of Ocean Wave Dissipation and Its Model Impact*; ECMWF Tech. Memo. 509; ECMWF: Reading, UK, 2007; 27p.
40. Behrens, A. Development of an ensemble prediction system for ocean surface waves in a coastal area. *Ocean Dyn.* **2015**, *63*, 469–486. [CrossRef]
41. Staneva, J.; Alari, V.; Breivik, Ø.; Bidlot, J.-R.; Mogensen, K. Effects of wave-induced forcing on a circulation model of the North Sea. *Ocean Dyn.* **2017**, *67*, 81–191. [CrossRef]
42. Staneva, J.; Grayek, S.; Behrens, A.; Günther, H. GCOAST: Skill assessments of coupling wave and circulation models (NEMO-WAM). In Proceedings of the Journal of Physics: Conference Series, 9th International Conference on Mathematical Modeling in Physical Sciences (IC-MSQUARE) 2020, Tinos Island, Greece, 7–10 September 2020; Volume 1730, p. 012071. [CrossRef]
43. Lewis, H.W.; Castillo Sanchez, J.M.; Siddorn, J.; King, R.R.; Tonani, M.; Saulter, A.; Sykes, P.; Pequignet, A.-C.; Weedon, G.P.; Palmer, T.; et al. Can wave coupling improve operational regional ocean forecasts for the north-west European Shelf? *Ocean Sci.* **2019**, *15*, 669–690. [CrossRef]
44. Benetazzo, A.; Barbariol, F.; Pezzutto, P.; Staneva, J.; Behrens, A.; Davison, S.; Bergamasco, F.; Sclavo, M.; Cavaleri, L. Towards a unified framework for extreme sea waves from spectral models: Rationale and applications. *Ocean Eng.* **2021**, *219*, 108263. [CrossRef]
45. Bruciaferri, D.; Tonani, M.; Lewis, H.W.; Siddorn, J.R.; Saulter, A.; Castillo Sanchez, J.M.; Valiente, N.G.; Conley, D.; Sykes, P.; Ascione, I.; et al. The impact of ocean-wave coupling on the upper ocean circulation during storm events. *J. Geophys. Res. Ocean.* **2021**, *126*, e2021JC017343. [CrossRef]
46. Hernandez, F.; Blockley, E.; Brassington, G.B.; Davidson, F.; Divakaran, P.; Drevillon, M.; Ishizaki, S.; Garcia-Sotillo, M.; Hogan, P.J.; Lagemma, P.; et al. Recent progress in performance evaluations and near real-time assessment of operational ocean products. *J. Oper. Oceanogr.* **2015**, *8*, 221–238. [CrossRef]
47. Lima, L.; Ciliberti, S.A.; Aydoğdu, A.; Masina, S.; Escudier, R.; Cipollone, A.; Azevedo, D.; Causio, S.; Peneva, E.; Lecci, R.; et al. Climate signals in the Black Sea from a multidecadal eddy-resolving reanalysis. *Front. Mar. Sci.* **2021**. [CrossRef]
48. Boyer, T.P.; Baranova, O.K.; Coleman, C.; Garcia, H.E.; Grodsky, A.; Locarnini, R.A.; Mishonov, A.V.; Paver, C.R.; Reagan, J.R.; Seidov, D.; et al. World Ocean Database 2018; NOAA Atlas NESDIS 87. Available online: https://www.ncei.noaa.gov/sites/default/files/2020-04/wod_intro_0.pdf (accessed on 11 October 2021).
49. Kopelevich, O.V.; Sheberstov, S.V.; Sahling, I.V.; Vazyulya, S.V.; Burenkov, V.I. Bio-optical characteristics of the Russian Seas from satellite ocean color data of 1998–2012. In Proceedings of the VII International Conference "Current Problems in Optics of Natural Waters (ONW 2013)", St. Petersburg, Russia, 10–14 September 2013.
50. Palazov, A.; Ciliberti, S.; Peneva, E.; Grégoire, M.; Staneva, J.; Lemieux-Dudon, B.; Masina, S.; Pinardi, N.; Vandenbulcke, L.; Behrens, A.; et al. Black Sea Observing System. *Front. Mar. Sci.* **2019**, *6*, 315. [CrossRef]

Journal of
Marine Science and Engineering

MDPI

Article

Impacts of an Altimetric Wave Data Assimilation Scheme and Currents-Wave Coupling in an Operational Wave System: The New Copernicus Marine IBI Wave Forecast Service

Cristina Toledano [1,*], Malek Ghantous [2,3], Pablo Lorente [1,4], Alice Dalphinet [2], Lotfi Aouf [2] and Marcos G. Sotillo [4,5]

1 Nologin Consulting SL, NOW Systems, 50018 Zaragoza, Spain; pablo.lorente@nologin.es
2 Meteo France, 31057 Tolouse, France; mghantous@groupcls.com (M.G.); alice.dalphinet@meteo.fr (A.D.); lotfi.aouf@meteo.fr (L.A.)
3 Collect Localization Satellites, 31520 Ramonville-Saint-Agne, France
4 Puertos del Estado, Área de Medio Físico, 28042 Madrid, Spain; marcos@puertos.es
5 Mercator Ocean International, 31400 Tolouse, France
* Correspondence: cristina.toledano@nologin.es

Abstract: The Copernicus Marine IBI-MFC (Iberia–Biscay–Ireland Monitoring and Forecasting Centre) has delivered operational wave forecasts since 2017. The operational application is based on a MFWAM model (Meteo-France WAve Model) set-up, running at a 1/20° grid (5-km). The research presented here was conducted to improve the accuracy of the IBI-MFC wave model products, by means of (i) including a new wave data assimilation scheme and (ii) developing a new coupled ocean-wave modelling framework. Evaluation of these set-up upgrades, in terms of improvements in IBI wave model system capabilities, is here presented. All the model sensitivity test runs, performed for the year 2018, are assessed over the whole IBI domain, using the available in-situ (from 49 mooring buoys) and independent satellite wave observation. The results show that the most relevant improvement is due to the data assimilation, while the impact of surface ocean currents, although less significant, also improves the wave model qualification over the IBI area. The demonstrated benefit, related to the herein proposed upgrades, supported the IBI-MFC decision to evolve its operational wave system, using (since the March 2020 Copernicus Marine Release) the resulting wave model set-up, with data assimilation and currents-wave coupling for operational purposes.

Keywords: forecasting; wave modeling; data assimilation; current–wave coupling; current forcing; model validation; wave altimetric products; Copernicus Marine IBI

Citation: Toledano, C.; Ghantous, M.; Lorente, P.; Dalphinet, A.; Aouf, L.; Sotillo, M.G. Impacts of an Altimetric Wave Data Assimilation Scheme and Currents-Wave Coupling in an Operational Wave System: The New Copernicus Marine IBI Wave Forecast Service. *J. Mar. Sci. Eng.* **2022**, *10*, 457. https://doi.org/10.3390/jmse10040457

Academic Editor: Liliana Rusu

Received: 11 February 2022
Accepted: 22 March 2022
Published: 24 March 2022

Publisher's Note: MDPI stays neutral with regard to jurisdictional claims in published maps and institutional affiliations.

1. Introduction

Waves constitute the interface between ocean and atmosphere and have an important role in terms of exchanges through this interface [1]. Their representation is necessary to accurately compute the different air–sea fluxes of heat and momentum [2].

There is a widespread worldwide offer of accurate and reliable wave forecast services. A variety of operational wave forecast services, ranging from global to local coastal scales, are run by different operational oceanographic centres (some of them national weather offices); the wave forecast products benefit different end-users, supporting day-to-day operations at sea and contributing to warning systems that minimize potential risks for marine safety (among others). The authors in [3], in their review of European Operational oceanographic capacities, indicated how several wave models (i.e., WAM (Wave Model), SWaN (Simulation Waves Nearshore), WaveWatch-iii, WWM-II, etc.) [4–6] are used in the forecast services delivered by the operational oceanographic centres. Some of these operational services use operational assimilation schemes to account for near real-time observational wave information, especially from satellite altimeters [7].

Upper ocean dynamics are strongly affected by sea-state dependent processes, inducing the impact of waves on the ocean's small- and large-scale circulation.

On the one hand, waves affect the ocean surface layer through different processes [8]: Waves induce surface current via the Stokes drift, adding a term on the Coriolis effect in the momentum equation (the so-called Stokes–Coriolis force). Part of the atmospheric wind stress contributes to the wind-wave growth, thus, subtracting a quantity of energy to the ocean currents. Furthermore, during wave breaking, turbulent kinetic energy is produced and affects the upper ocean surface layer, enhancing the turbulent mixing. Recent studies have attempted to determine the impacts of wave effects on the representation of the ocean surface layer at different spatio-temporal scales. Among others: wave-induced mix-layer depths representation [9], relevant impacts on the atmospheric surface temperature, pressure, and precipitation [10,11], modifications in wind stress by the rise of roughness length and friction velocity [12]. This is especially true during storm events, when wave–current interactions might represent a leading order process of the upper ocean. In this context, ref. [13] strongly recommends using an ocean-waves-atmosphere coupled system to improve the representation of tropical cyclones' intensity, structure and motion. Indeed, ref. [14] studied the effect of sea waves on the typhoon Imodu (15–19 July 2003). Moreover, ref. [15] demonstrated how a coupled system simulates more accurate surface dynamics than uncoupled models, with larger improvement on the shelf, showing that (especially during extreme events) ocean-wave coupling improves the accuracy of the surface dynamics, with larger improvements in the simulation of ocean currents over the shelf due to the synergy between strong tidal currents and more mature decaying waves.

On the other hand, the presence of ocean currents affects the waves, changing their amplitude, frequency and direction. This is generally due to the energy bunching, accounted in the wave energy balance when the velocity of the wave energy propagates across the current, the energy transfer between waves and currents, the frequency shifting (including Doppler shifting) and current-induced refraction [16]. Ref. [17] accounts for significant wave height changes in the Baltic Sea due to the impact of ocean currents (up to 20% in specific severe storm conditions, mostly in shallower waters and when waves and surface currents propagate in opposite directions [18,19]). The Copernicus Marine Service [20,21], one of the streamlined six thematic streams of the Copernicus Services (Atmosphere, Marine, Land, Climate Change, Security and Emergency) [22,23] and internationally recognized as one of the most advanced service capabilities in terms of ocean monitoring and forecasting, provides regular systematic reference information on the physical, biogeochemical and sea-ice state for the European regional seas and the global ocean. This service recently included, in its product portfolio, essential ocean variables related to the sea state, and near-real-time wave forecasts, and multi-year wave reanalysis products were progressively incorporated (along the 2015–2018 development phase) in the Copernicus Marine Service offer. The Copernicus Marine Service high-level strategy includes a roadmap with associated Research and Development (R&D) priorities [24], which identifies some developments per thematic area that are key for the future service evolution. Among others, (i) upgrade of data assimilation schemes (to improve the analysis and reanalysis capabilities) and (ii) enhancement of the representations of coupling effects between ocean-wave-sea-ice-atmosphere-land components (to improve forecast model solutions) are seen as prioritized research lines for any Copernicus Marine Monitoring and Forecasting Centres (MFC).

Specifically, for the European Atlantic Façade, the Copernicus Marine IBI-MFC (Iberia–Biscay–Ireland Monitoring Forecasting Centre) delivers daily ocean model estimates and forecasts of different physical and biogeochemical parameters, including, since 2016, hourly wave forecasts and multi-year products [25].

The present work focuses on the research performed to develop the current operational version of the IBI-MFC wave model application. This research was mainly conducted to improve the accuracy of these IBI-MFC wave model products, by means of developing a new coupled ocean-wave modelling framework that also includes wave data assimilation. In that sense, this study has two specific objectives:

- To assess and quantify the potential added value, in terms of accuracy gain, that the assimilation of altimetric significant wave height satellite observation has on the IBI wave model solution.
- To analyse the impacts in the IBI wave model solution related to the use of surface current–wave coupling, evaluating the contribution of surface ocean currents in the wave energy balance.

To address these questions, different wave model sensitivity tests are performed. Several wave model simulations generated with the IBI-MFC wave model set-up, and only differing from the operational version (available in 2018) for the activation of the new data assimilation scheme and in the degree of the ocean current forcing applied, are run. The assessment of these model simulations is conducted using several local available in-situ and satellite wave observations.

The paper is organised as follows: Section 2 provides a description of the Copernicus IBI-MFC wave model system and the different model sensitivity tests performed, together with the model assessment proposed. Section 3 presents the main results, with an analysis of the proposed updates. Finally, in Section 4, the impacts of the current forcing interactions and the data assimilation scheme proposed are discussed, providing a look ahead to related benefits on the IBI wave operational forecast capabilities.

2. Methodology and Sensitivity Tests for Copernicus Marine IBI-MFC Wave System

2.1. The IBI Area and IBI-MFC Wave Model

The Copernicus Marine IBI-MFC (Iberia–Biscay–Ireland Monitoring Forecasting Centre) o ffers a comprehensive portfolio of regular and systematic regional information on the state of the ocean for the European Atlantic façade, supporting all kinds of marine applications. As part of this IBI-MFC service, a short term (10-days) high-resolution wave forecast is updated twice a day for the IBI area. Hourly instantaneous data for significant wave height, wave direction, wave period variables, together with wind sea and swell (primary and secondary partitioned wave spectra) parameters are delivered as part of this regional Copernicus Marine IBI wave product.

The MFWAM model configuration for IBI MFC is implemented on the IBI domain (26–56° N and −19–5° E; see geographical domain in Figure 1) with a horizontal resolution of 5 km approximately (1/20°).

The wave model used as base of this IBI-MFC operational system is the MFWAM [26]. This MFWAM model is based on the IFS-ECWAM 41R2 code [27], with changes regarding the dissipation by wave breaking and the swell damping source terms as developed by [28]. The current version of the model includes major improvements achieved within the FP7 European research MyWave Project [2,29]. The IBI-wave model performs a partitioning technique on wave spectra over all ocean grid points of the IBI domain. The partitioning technique is based on the watershed method developed for image processing [30]. This process effectively treats the wave spectrum as a topographic map from which individual peaks in wave energy can be identified to define the separate wave components. First, wave spectrum is split in wind sea and swell wave spectra. Then, partitioning is applied for the swell wave spectrum. The peaks on the spectrum are isolated and they are considered as partitions. Afterward, classification of swell partitions in primary and secondary swell is performed depending on the mean energy of each partition.

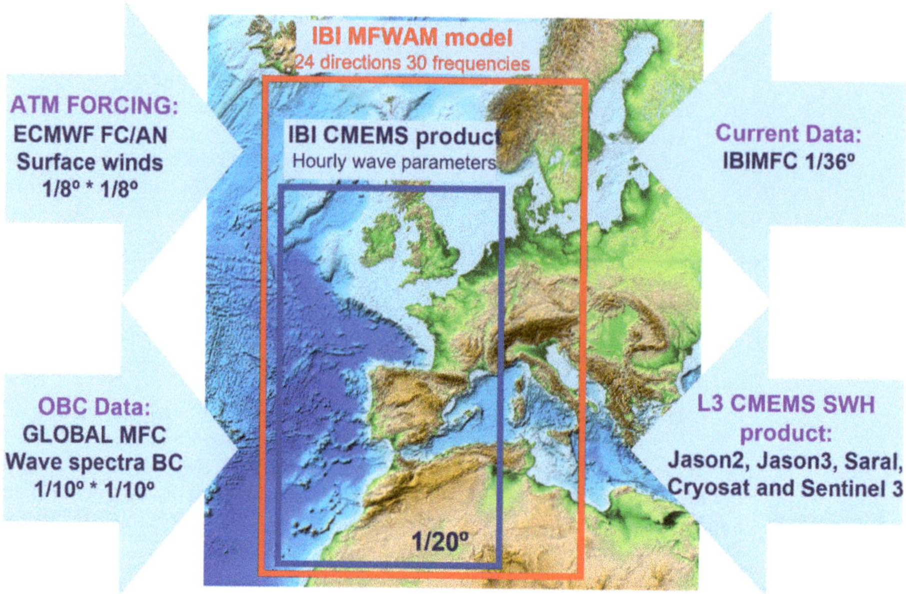

Figure 1. The CMEMS IBI-MFC wave Forecast/Analysis model Application. Model and product service spatial coverages, and details on the forcing, Open Boundary conditions and external data sources used for the ocean current forcing and the Data Assimilation applied.

The IBI-MFC wave model was upgraded (January 2018 Operational release) to improve the drag coefficient variation with the wind speed, resulting in positive impacts in the surface stress characterization. This improvement of the surface stress is certainly needed for the coupling with the IBI ocean currents here tested. To this end, a new setting on the wave dissipation term, the sheltering parameter, and the use of Phillips spectrum tail for the high-frequency part of the wave spectrum was also implemented. Moreover, the minimum water depth is taken as 5 m (instead of the 1 m value used in former IBI wave model versions). Associated to these upgrades, slight improvements in terms of significant wave height were obtained (reduced scatter index, around 1.9% when comparing model results with observations from altimeters [31]).

The bathymetry used is derived from the 1 arc minute ETOPO1 ocean bathymetry by National Geophysical Data (NOAA) [32]. The wave spectrum is discretized in 24 directions and 30 frequencies ranging from 0.035 Hz to 0.56 Hz. The Copernicus Marine IBI wave model is driven by short-range forecasted and analyzed IFS-ECMWF winds [33] at 1/8° hourly winds, which are used for the first 90 h, decreasing time frequency to 3-hourly until T + 144 h, and 6-hourly forecasts until T + 240 h. It uses as boundary conditions (wave spectra) Copernicus Marine GLOBAL wave data at 1/10° spatial resolution [34].

An IBI-MFC wave forecast product (10 days hourly data updated twice a day) is delivered to end-users for the IBI service domain (see spatial coverage in Figure 1), together with a 2-year historic timeseries, composed of IBI best estimates (analysis data for the Day-1 date).

2.2. Model Sensitivity Test: The Proposed IBI Wave Model Upgrades

The Copernicus Marine IBI-MFC identified, among others, the following two major shortcomings in its wave forecast service:

(i) The lack of IBI wave analyses to initialize IBI forecast cycles
(ii) The absence of any coupling of the IBI wave model data with ocean currents.

Bridging these 2 identified gaps in the Copernicus IBI wave service was considered as a major goal for the IBI-MFC service evolution planned for the last Copernicus-1 Service Phase (2018–2021) and 2 different research working lines were followed to achieve the objective. These specific research lines were fully aligned with the Copernicus Marine Evolution Strategy [20], and their scientific research priorities (implementation of data assimilation schemes and enhancement of the model coupling between different earth system components) were two of the major amelioration axes proposed for the Copernicus Marine products and services.

The present work aims to quantify the potential added value of new IBI wave analyses (to be generated by means of a new data assimilation scheme implemented to assimilate altimetric significant wave height satellite observations) with respect to the IBI best model wave estimates (note that traditionally, the IBI-MFC was delivered as historic best model estimate 2 years of model hindcast wave data, wave run for day D–1 forced with analyzed winds). Likewise, the impact of including the contribution of the surface ocean currents on the IBI wave model solution is evaluated.

The impacts on the IBI wave solution of both the new IBI wave DA system and the current–wave coupling are evaluated through specific IBI-like wave model scenarios. To this aim, different model sensitivity test runs, based on the IBI-MFC operational model configuration, have been designed and run. Table 1 shows an overview of the four proposed IBI wave model scenarios: the Control run (IBI-CO) was performed using the same wave model set-up that was used in the IBI operations in December 2019. Two more IBI wave sensitivity test runs were performed: one with the current forcing activated, but without Data Assimilation (the IBI-CU run) and another with data assimilation activated, but without any coupling contribution (the IBI-DA run). Finally, a last model scenario that includes the two novelties (data assimilation and ocean current forcing) was assessed. This model scenario (hereafter named as IBI-OP) is consistent with the new IBI-MFC operational CMEMS IBI wave model system (in operations since July 2020).

Table 1. IBI wave model scenarios: overview of model runs performed to test the 2 proposed novelties (the current forcing and the new Data Assimilation scheme). The Control run and all the test runs using the IBI wave model set-up used in operations in Copernicus release in December 2019. The last sensitivity test, switching both the DA and the currents coupling, to be proposed as base of the new Copernicus IBI wave forecast system (in operations since July 2020).

IBI-Wave Model Scenarios	Currents Coupling	Data Assimilation
IBI CONTROL RUN (IBI-CO)	-	-
IBI Run with currents forcing on (IBI-CU)	x	-
IBI Run with Data Assimilation on (IBI-DA)	-	x
New IBI Operational Wave system (IBI-OP)	x	x

The four different model scenarios proposed were run over the year 2018. The IBI wave model set-up novelties were assessed, and all the proposed sensitivity model runs were validated using in-situ and remote sensing observations (i.e., from coastal and deep-water mooring buoys, and altimetric SWH). A complete description of the different sensitivity model tests performed to improve the operational wave model application that comprise the IBI-MFC operational wave forecast service is provided below. Results from the different model test runs are provided in the following section.

2.2.1. The Altimetric Wave DA Scheme Proposed for IBI

The data assimilation scheme proposed to be applied in the IBI wave service and tested through the IBI-DA test run presented here, is based on the optimal interpolation scheme described by [35] and it is the same scheme used in the Copernicus Marine GLOBAL wave system. The variable to be assimilated is the significant wave height, Hs. Because Hs is not a state variable of the system, this introduces an extra complication in that the energy must be repartitioned from a frequency and direction integrated parameter (the Hs) to the full directional frequency spectrum. This involves making several assumptions and is by nature inexact, but in practice performs well [7]. What follows is a brief description of the method, which has been adjusted for the ST4 physics used in the IBI-wave model.

For a state vector x, optimal interpolation seeks a weighted combination of the background (or mode forecast), denoted by x^f, with observations, y^o, in order to produce an analysis x^a. The fundamental equation is:

$$x^a = x^f + \mathbf{K}\left(y^o - \mathbf{H}x^f\right), \tag{1}$$

where **H** is the observation matrix. **K** is a weighting matrix.

$$\mathbf{K} = \mathbf{PH}^{\mathrm{T}}\left(\mathbf{HPH}^{\mathrm{T}} + \mathbf{R}\right)^{-1}, \tag{2}$$

where the matrices **P** and **R** are respectively the model and observation error covariance matrices.

In MFWAM these matrices are expressed as correlation matrices:

$$\mathbf{K} = \mathbf{CH}^{\mathrm{T}}\left(\mathbf{HCH}^{\mathrm{T}} + \mathbf{I}\right)^{-1}, \tag{3}$$

The ratio of background and observation errors is kept constant over the IBI domain and set to 1 (i.e., model and observation errors are assumed to be equal everywhere). Although a different ratio may be theoretically justifiable, we have found that this value works best for this model in this domain; it is the same as that used in the global configuration. Observational errors are additionally assumed to be uncorrelated. With these assumptions, R is none other than the identity matrix, I. P becomes the correlation matrix C defined in terms of the correlation length λ_c:

$$c_{ij} = e^{-\left(\frac{d_{ij}}{\lambda_c}\right)^a}, \tag{4}$$

where d_{ij} is the great circle distance between points i and j and a is a tuning parameter.

With this simplified OI scheme the only tunable parameters are in the correlation function. For the IBI configuration used here the correlation length λ_c is set to 170 km, significantly less than the 300 km length used in the global configuration, and the tuning parameter a is set to 1. We performed some experiments with λ_c and a, in particular testing values determined from a correlation study of the global model divided by basin [36]. These did not result in an improvement in performance, however, so the original values were kept. The cutoff distance, beyond which observations are not included in the analysis, is set to 650 km. The analysis, Equation (1), produces a corrected estimate for the significant wave height. Since *Hs* is not a state vector of the wave model, but rather an integrated parameter, in order to correct the model itself MFWAM redistributes the energy in the wave spectrum using the method of [35], which is based on empirical wave growth laws. The analyzed spectrum is expressed as:

$$F^a = ABF^f(Bf, \theta), \tag{5}$$

where F denotes the wave spectrum, f the frequency and θ the direction, and the superscripts a and f refer to analysis and background respectively. A is the ratio of analysis to background energy, which can be expressed as $(H_s{}^a/H_s{}^f)^2$, where A determines the overall energy correction to the spectrum, while the effect of B is to rescale the frequencies.

Two different methods for computing B are used, depending on whether the spectrum is determined to be primarily a swell spectrum (the energy of the swell accounts for more than 1/4 of the total energy of the spectrum) or a wind–sea spectrum. If the spectrum is predominantly wind–sea, B is computed from the mean frequencies of the background and analysis as:

$$B = \overline{f}^f / \overline{f}^a, \tag{6}$$

The choice of mean frequency was for computation reasons; the peak frequency would be just as valid a choice (if not more so), as both are in any case approximations. If the spectrum is predominantly swell, the average steepness of the waves is assumed constant. Therefore:

$$B = \overline{f}^f / \overline{f}^a = \sqrt{H_s^a / H_s^f}, \quad (7)$$

The calculation of $\overline{f}^f$ and $H_s{}^f$ is taken care of by the model, but fa and $H_s{}^a$ have to be estimated from the *Hs* analysis. This is done by exploiting the empirically derived duration-limited growth laws. By defining the non-dimensional energy, mean frequency and duration, respectively:

$$\epsilon' = u_*^4 \epsilon / g^2, \quad (8)$$

$$f' = u_* f / g, \quad (9)$$

$$t' = tg / u_*, \quad (10)$$

where g is gravitational acceleration, and ϵ = *Hs2/4*. The growth relations are:

$$\epsilon'(t) = 1877\left(t'\left[t' + 5.440 * 10^5\right]\right)^{1.9}, \quad (11)$$

$$\epsilon\left(\overline{f}\right) = 5.054 * 10^{-4} \overline{f}^{-2.959}, \quad (12)$$

From the background friction velocity and *Hs* we can use Equation (9) and the growth law equations to estimate an updated ϵ' and f'. These in turn give us $\overline{f}^a$ and $H_s{}^a$, and with these we can calculate *B* and *A* and produce the updated spectrum (5).

The data are combined to generate super-observations. Within a set time window around the analysis time, the data are assigned to model grid points closest to them, and any data sharing a model point are averaged together and treated as one observation. This reduces the number of data to assimilate, lightening the computational load, and it helps to smooth out potential errors in the observations. Outlying data are rejected in this step.

This data assimilation method results in a final *Hs* corrected mostly in the wind sea part of the frequency spectrum [7].

2.2.2. Wave–Current Coupling Proposed for IBI

To incorporate surface ocean currents forcing in the IBI wave model system, surface current data from the CMEMS-IBI analysis and forecast ocean model system were used as inputs in the IBI wave model system for wave refraction.

The presence of current may change the amplitude, frequency, and direction of waves. This is generally due to the energy bunching that is readily accounted for the energy balance equation of waves using the velocity of the wave energy propagating across the current, the energy transfer between waves and currents, the frequency shifting (including Doppler shifting) and current-induced refraction [16].

MFWAM model equations include the depth and current refraction. The propagation velocity in the relative frequency space should be computed according to [37]:

$$c_\sigma = \frac{\partial \sigma}{\partial d}\left[\frac{\partial d}{\partial t} + \vec{u} \cdot \vec{\nabla} d\right] - c_g \vec{k} \cdot \frac{\partial \vec{u}}{\partial t}, \quad (13)$$

where σ is the relative frequency, c_σ the propagation velocity in the relative frequency space, t is time coordinate, s is the space coordinate in the direction of propagation, $\vec{\nabla}$ the gradient operator in the geographical space, d is the total depth, $\vec{u}$ is the current, c_g is the group velocity and $\vec{k}$ is the wave number vector. As in MFWAM model current and water depth are time-independent, the term $\partial d/\partial t$ in Equation (13) is not present.

The offline method of coupling with surface currents takes the files needed for the whole forecast period from the IBI-MFC ocean forecast system. The files used are not exactly the ones delivered through the Copernicus Marine catalogue, but rather the native IBI NEMO model outputs (at the 1/36° ORCA grid), which the IBI wave system interpolates from the IBI ocean model grid into the 1/20° regular grid used for the MFWAM model.

2.2.3. Assessment of Model Runs: Evaluation Criteria against In-Situ and Altimeter Data

To assess the performance of the numerical model applying both novelties and to identify the main sources of uncertainty linked to the Wave–Current coupling and the application of the data assimilation in the IBI wave model simulations, the four different model scenarios were performed over the year 2018. The significant wave height and mean period fields, resulting from the different IBI sensitivity runs, are validated by means of comparison with different in-situ and satellite remoted sensed observational data sources.

In-situ measurements of significant wave height, H_s, and mean wave period, Tm_{02}, were extracted from mooring buoys available in the IBI region, compiled in the product delivered by the Copernicus Marine IBI INSITU-TAC [38]. The mean wave period (Tm_{02}) is defined as follows:

$$Tm_{02} = 2\pi \left(\frac{\iint \omega^2 E(\omega,\theta) d\omega d\theta}{\iint E(\omega,\theta) d\omega d\theta} \right)^{\frac{-1}{2}}, \tag{14}$$

where $E(\omega,\theta)$ is the variance density and w the absolute radian frequency.

Measurements for Hs and Tm_{02} variables are available for the examined year in the area at 49 and 45 buoys, respectively (see list and locations in Figure 2).

To validate the model through model output and buoy data collocation, the time series were taken at the model grid point nearest to the buoy location. For regional validation purposes, the IBI domain is split into different sub-regions of interest, being validation metrics gathered for the whole IBI service domain and for each sub-region (see spatial domains in Figure 2). Likewise, different metrics are computed separately using coastal and deep-water mooring buoys.

The different model sensitivity test runs were validated with satellite observations of significant wave height, Hs. However, since data from Jason-2, Jason-3, Saral, Cryosat-2 and Sentinel3 altimeters are now assimilated into the model (information from this mission included in the L3 CMEMS altimeter data product used for assimilation), an independent data source is needed. Thus, the diagnostic after data assimilation is performed by comparing the model to the HY-2A satellite altimeter processed by the French Space Agency CNES. The validation procedure with altimetric observations begins with pre-processing the Hs data, rejecting Hs data lower than 0.5 m or higher than 12 m and eliminating big jumps in terms of Hs value from one observation to the next one (the biggest value of steps higher than 1 m in case of positive steps and −2 m, in case of negative steps, is rejected). To validate the model runs with altimetric observations we used the HY-2A satellite data for both DA and reference runs, ensuring that a data source independent from the assimilated data, HY-2A Hs is biased [39], so a calibration to reference mission such as Jason-3 has been implemented on crossover locations. This leads to a linear for HY-2A Hs, which is expressed as follows:

$$H_s = 0.9476 \times H_s^{biased} - 0.0230, \tag{15}$$

The modelled Hs is also post-processed with an upscaling to a 0.1 degree resolution (the nearest grid point for altimeters), in order to closer match the observations, and it is limited to values above 0.5 m. The final step to prepare for validation is then to find the points of modelled Hs that correspond to the observation points. The validation statistics are then computed using these two values for each point where a valid observation exists.

Apart from bias, root mean square difference (RMSD) and correlation (CCOR), validation against altimeters includes the statistical quantity scatter index ($SI2$) used for the wave model statistics defined as:

$$SI2 = \frac{\sqrt{\sum_{i=1}^{N}[(x_i - \overline{x}) - (y_i - \overline{y})]^2}}{\sum_{i=1}^{N} y_i}, \tag{16}$$

where y_i is the observation, x_i the model value corresponding to the *i*th observation, N is the total number of observations, and the overbars refer to the population mean. This definition of scatter index differs from others in that the observations are not squared before taking the mean, so is only valid for quantities such as *Hs* which are always positive.

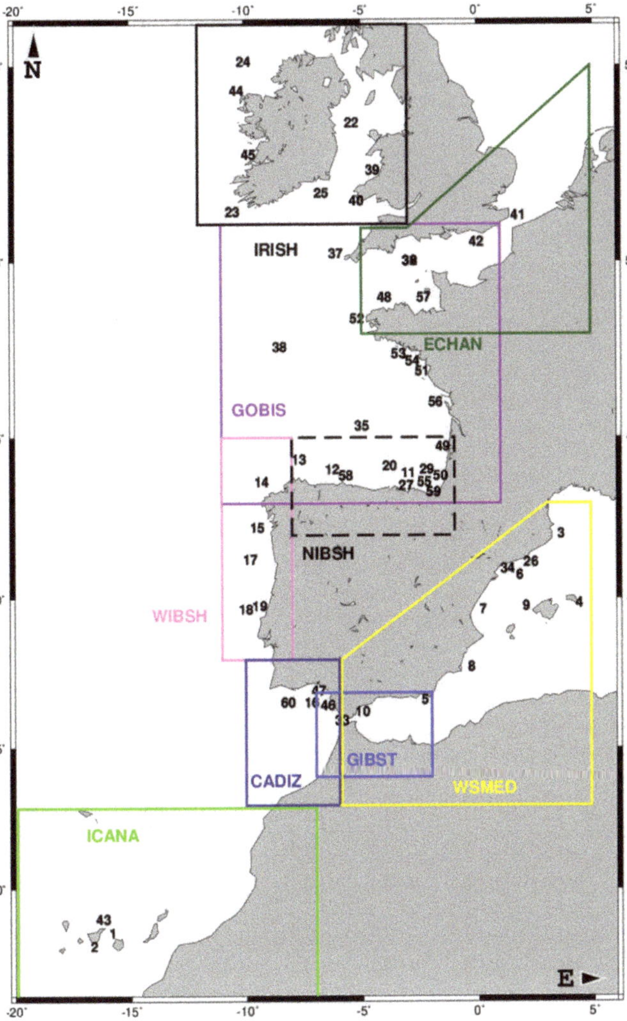

No.	ID / name	Location
1	1300130	DeepWater
2	1300131	DeepWater
3	6100196	DeepWater
4	6100197	DeepWater
5	6100198	DeepWater
6	6100280	DeepWater
7	6100281	DeepWater
8	6100417	DeepWater
9	6100430	DeepWater
10	6101404	Coastal
11	6200024	DeepWater
12	6200025	DeepWater
13	6200082	DeepWater
14	6200083	DeepWater
15	6200084	DeepWater
16	6200085	DeepWater
17	6200191	DeepWater
18	6200192	DeepWater
19	6200199	DeepWater
20	6201030	DeepWater
22	6200091	DeepWater
23	6200092	DeepWater
24	6200093	DeepWater
25	6200094	DeepWater
26	Barcelona coast buoy	Coastal
27	Bilbao coast buoy	Coastal
29	Donostia buoy	DeepWater
32	Melilla coast buoy	Coastal
33	Tarifa coast buoy	Coastal
34	Tarragona coast buoy	Coastal
35	6200001	DeepWater
36	6200103	DeepWater
37	6200107	DeepWater
38	6200163	DeepWater
41	6200304	DeepWater
42	6200305	DeepWater
43	Tenerife coast buoy	Coastal
44	AMETSBerthWaveBuoy	Coastal
45	WestwaveWaveBuoy	Coastal
46	Cadiz coast buoy	Coastal
47	[illegible] coast buoy	Coastal
49	6200064	Coastal
50	6200066	Coastal
51	6200067	Coastal
52	6200069	Coastal
53	6200074	Coastal
54	6200078	Coastal
55	6200079	Coastal
56	6200080	DeepWater
58	Gijon coast buoy	Coastal
59	Pasajes coast buoy	Coastal
60	6200200	DeepWater

Figure 2. Locations of all the mooring buoys used in the model validation performed for the year 2018. In the accompanying table, the seven-digit WMO identifier (or buoy name provided by CMEMS) is followed by the information on the location of the mooring buoy (coastal or deep water). The 8 different sub-regions of interest for validation purposes in which the IBI service domain is divided are shown in the map (see polygons of different colors). From the list of 60 buoys in the IBI region, only the 49 available for year 2018 have been used.

A lower scatter index is not always a reliable gauge for model performance [40]. The symmetrically normalized root square error (*HH*), also used for validation against altimeters, provides more accurate information on the accuracy of simulation.

The error indicator HH proposed by Hanna and Henold, ref. [41], is defined as the RMSE divided by the absolute value of the mean of the product of the observations and modelled values:

$$HH = \sqrt{\frac{\sum_{i=1}^{N}(x_i - y_i)^2}{\sum_{i=1}^{N} y_i x_i}}, \quad (17)$$

Then, in the next section, the impact for the two proposed novelties (current forcing and data assimilation) is assessed with tests performed in 2018, following the evaluation criteria against in-situ and altimeter data described above.

3. Results

3.1. Impacts in the IBI Wave Solution Related to the Use of Surface Current–Wave Coupling

For the assessment of the surface current–wave coupling, model outputs from the control run (IBI-CO; run with no currents coupling and no data assimilation) and the no data assimilation run with current–wave coupling (IBI-CU) were compared with in-situ observations, corresponding to the year 2018, focusing the model validation on the wave parameters of significant wave height, Hs, and mean wave period, Tm_{02}.

The error metrics from both model configuration test runs are summarized in Tables 2 and 3. Table 2 shows results of a comparison between the wave hindcast best estimate and in-situ observations, and Table 3 shows a similar comparison with altimeter data for the Hs parameter, with skill metrics differentiated between CMEMS L3 altimeters and HY-2A. The general trend of validation for Tables 2 and 3 is very similar due to the good correlation ($\geq$0.98) between buoys measurements and altimeters, as the comparison performed by [42] reveals.

The results shown in both tables do not show substantial improvement between the coupled and uncoupled model, but differences are definitively evident in terms of mean period. Coupling effects are stronger for coastal buoys locations, in some cases with worse metrics due to overestimation in the peaks of Tm_{02} time series. On the other hand, the systematic negative bias in Hs, caused by the tendency of the IBI wave system to underpredict, is reduced with the current surface implementation.

Metrics in Strait of Gibraltar (CADIZ an GIBST subregions) have different solutions for coastal and deep-water buoys. In general, results are slightly better for the coupled model IBI-CU. This is due to current coupling increasing values of wave height, but in some coastal locations, when the H_s is slight, an unrealistic ocean current overestimates the Tm_{02}.

Figure 3 depicts the pattern of agreement between the IBI wave system with wave–current coupling (IBI-CU) and no coupling (IBI-CO). As Table 2 shows, the western part of the IBI region (WIBSH, GOBIS and IRISH subregions) has analogous solutions: slightly better results for significant wave height and better metrics for the mean period, in the case of the coupled model (IBI-CU). At buoy 6200083, located on the western coast of Galicia (depth of 386 m), the comparison of model results serves as an explicit example of the general IBI performance. The effect of the current coupling is not very sensitive in the Hs, but bias for the Tm_{02} drops from 0.53 to 0.46 and the RMSD from 0.72 to 0.60 (note the time series agreement except for the mean period on 9 December 2018). Similar error reductions are found throughout the year.

Table 2. Error metrics for the test runs with and without current–wave coupling (IBI-CU (CU) and IBI-CO (CO) test runs, respectively) computed with hourly observations at mooring buoys. Time period: 2018 Variables: Significant wave height (SWH) and mean wave period (TM02). Metrics computed for the whole IBI service domain and for the 9 Validation regions (i.e., IRISH, ECHAN, GOBIS, NIBSH, WIBSH, GIBST, CADIZ, WSMED, ICANA) used by the Copernicus Marine IBI-MFC service. Metrics are gathered using all the available buoys in each region, and also using exclusively Coastal and Deep-water Buoys (CB and DB, respectively). Each error metric (Bias, Root-Mean-Square differences (RMSD) Correlation (CCOR)) provided for each model solution. N counts the size of the sample. Bold numbers highlight the best performing dataset. Mooring buoys in the English Channel area give the zero-crossing wave period (Tz) instead of Tm_{02}, so mean period measurements for the ECHAN region are not provided for this validation.

			SWH						TM02					
		N	BIAS (m)		RMSD (m)		CCOR		BIAS (s)		RMSD (s)		CCOR	
			CU	CO	CU	CO	CU	CO	CU	CO	CU	CO	CU	CO
IRISH	CB (2)	13,784	**−0.22**	−0.24	**0.45**	0.45	**0.97**	0.97	**0.23**	0.23	0.59	0.58	**0.93**	0.93
	DB (4)	20,758	**−0.02**	−0.05	**0.32**	0.33	**0.97**	0.96	**−0.27**	−0.33	**0.57**	0.61	**0.92**	0.92
	ALL (6)	34,542	**−0.09**	−0.11	**0.36**	0.37	**0.97**	0.97	**−0.10**	−0.15	**0.58**	0.60	**0.92**	0.92
ECHAN	CB (1)	17,085	**0.19**	0.19	**0.43**	0.43	**0.96**	0.96	-	-	-	-	-	-
	DB (3)	21,956	0.34	0.33	**0.47**	0.47	**0.91**	0.91	-	-	-	-	-	-
	ALL (4)	39,041	0.30	0.29	**0.46**	0.46	0.93	0.92	-	-	-	-	-	-
GOBIS	CB (8)	95,441	**−0.01**	−0.01	**0.37**	0.37	**0.96**	0.96	**0.34**	0.45	**1.03**	1.11	**0.89**	0.89
	DB (12)	111,202	**0.03**	0.04	0.36	0.35	**0.97**	0.96	−0.45	−0.33	**1.12**	1.20	**0.88**	0.87
	ALL (20)	206,643	**0.01**	0.02	0.37	0.36	**0.96**	0.96	−0.13	−0.02	**1.09**	1.16	**0.88**	0.88
NIBSH	CB (5)	58,313	−0.18	−0.15	0.34	0.33	0.96	0.96	**0.38**	0.61	**1.05**	1.20	**0.89**	0.89
	DB (5)	34,954	−0.13	−0.09	0.32	0.31	0.97	0.97	**0.13**	0.38	**0.70**	0.83	**0.93**	0.93
	ALL (10)	93,267	−0.16	−0.12	0.33	0.32	0.97	0.97	**0.26**	0.49	**0.88**	1.02	**0.91**	0.91
WIBSH	CB (1)	5294	−0.12	−0.11	**0.38**	0.38	**0.96**	0.96	−0.13	−0.03	**0.64**	0.66	**0.92**	0.92
	DB (4)	32,258	**−0.04**	−0.07	**0.31**	0.31	**0.98**	0.98	**0.08**	0.16	**0.62**	0.66	**0.95**	0.95
	ALL (5)	37,552	**−0.06**	−0.07	**0.32**	0.33	**0.98**	0.97	**0.04**	0.12	**0.62**	0.66	**0.94**	0.94
GIBST	CB (3)	26,173	0.10	**0.09**	0.31	**0.29**	0.85	**0.86**	0.48	**0.28**	1.15	**0.98**	0.62	**0.68**
	DB (1)	8126	**−0.03**	−0.04	**0.23**	0.23	**0.95**	0.95	**−0.07**	−0.09	**0.47**	0.47	0.84	**0.85**
	ALL (4)	34,299	0.07	0.06	0.29	0.28	0.88	0.89	0.34	0.18	0.98	0.85	0.68	0.72
CADIZ	CB (1)	8743	0.32	**0.27**	0.51	**0.47**	0.77	**0.79**	1.05	**0.65**	1.76	**1.45**	0.58	**0.61**
	DB (2)	14,159	**−0.04**	−0.10	**0.25**	0.27	**0.96**	0.96	0.15	−0.00	0.72	0.68	0.91	0.92
	ALL (3)	22,902	0.08	**0.02**	0.34	**0.33**	0.90	0.90	0.45	**0.21**	1.07	**0.93**	0.80	**0.81**
WSMED	CB (4)	56,832	**−0.06**	−0.06	**0.21**	0.21	**0.90**	0.90	0.04	**−0.03**	0.75	**0.70**	0.71	**0.74**
	DB (7)	33,184	**−0.13**	−0.14	**0.29**	0.29	**0.93**	0.93	**−0.21**	−0.24	0.59	0.58	0.83	0.84
	ALL (11)	90,016	**−0.11**	−0.11	**0.26**	0.26	**0.92**	0.92	**−0.12**	−0.16	0.65	**0.63**	0.78	**0.80**
ICANA	CB (1)	6767	**−0.25**	−0.26	**0.31**	0.32	**0.84**	0.84	**−0.17**	−0.27	**1.09**	1.12	**0.35**	0.33
	DB (2)	15,857	**−0.16**	−0.18	**0.24**	0.26	**0.92**	0.92	0.28	0.23	**0.86**	0.87	**0.66**	0.66
	ALL (3)	22,624	**−0.19**	−0.21	**0.26**	0.28	**0.89**	0.89	0.13	0.07	**0.94**	0.95	**0.56**	0.55
TOTAL	CB (18)	186,928	**−0.05**	−0.05	**0.34**	0.34	**0.93**	0.93	**0.24**	0.27	**0.93**	0.95	**0.81**	0.81
	DB (31)	228,392	**−0.03**	−0.03	**0.32**	0.32	**0.95**	0.95	−0.29	−0.27	**0.87**	0.90	**0.86**	0.86
	ALL (49)	422,451	**−0.03**	−0.04	**0.33**	0.33	**0.94**	0.94	−0.10	−0.07	**0.89**	0.92	**0.84**	0.84

Table 3. Error metrics for the test runs with and without current–wave coupling (IBI-CU and IBI-CO test runs, respectively) computed with satellite observations. Time period: 2018. Variable: Significant Wave Height (SWH). Metrics computed for the whole IBI service domain and for the 9 Validation regions (i.e., IRISH, ECHAN, GOBIS, NIBSH, WIBSH, GIBST, CADIZ, WSMED, ICANA) used by the Copernicus Marine IBI-MFC service. Metrics are computed using the available L3 CMEMS altimeter data (Janson-2, Janson-3, Saral, Cryosat-2 and Sentinel3) and HY-2A satellite data. Each error metric (Bias, Root-Mean-Square differences (RMSD) Correlation (CCOR) and Scatter Index (SI2)) provided for each model solution. N counts the size of available sample after the SWH data pre-process. Bold numbers highlight the best performing dataset.

		SWH									
		N		BIAS (m)		RMSD (m)		CCOR		SI2 (%)	
		IBI-CU	**IBI-CO**	**IBI-CU**	**IBI-CO**	**IBI-CU**	**IBI-CO**	**IBI-CU**	**IBI-CO**	**IBI-CU**	**IBI_CO**
IRISH	CMEMS L3	11,265	11,206	**−0.10**	−0.11	**0.34**	0.35	**0.98**	0.98	**12.27**	11.43
	HY-2A	2893	2890	**−0.05**	−0.07	**0.42**	0.43	**0.96**	0.96	**15.27**	15.41
ECHAN	CMEMS L3	8992	8996	**−0.14**	−0.14	**0.35**	0.36	**0.93**	0.93	**19.64**	19.91
	HY-2A	2172	2167	**−0.08**	−0.09	**0.38**	0.39	**0.92**	0.92	**22.01**	22.18
GOBIS	CMEMS L3	36,110	36,074	**−0.07**	−0.08	**0.30**	0.31	**0.98**	0.98	**11.30**	11.48
	HY-2A	9578	9574	**−0.00**	−0.01	**0.32**	0.33	**0.98**	0.98	**12.37**	12.50
NIBSH	CMEMS L3	5411	5411	−0.10	**−0.08**	0.29	0.28	**0.98**	0.98	**11.39**	11.47
	HY-2A	1482	1482	−0.02	**0.00**	**0.30**	0.30	**0.98**	0.98	**12.84**	12.91
WIBSH	CMEMS L3	9154	9151	**−0.07**	−0.09	**0.30**	0.32	**0.98**	0.97	**10.81**	11.15
	HY-2A	2409	2409	**−0.02**	−0.04	**0.31**	0.32	**0.98**	0.97	**11.52**	11.75
GIBST	CMEMS L3	3051	3029	**−0.10**	−0.12	**0.29**	0.29	**0.93**	0.93	**18.58**	18.65
	HY-2A	569	566	**−0.02**	−0.04	**0.36**	0.37	**0.90**	0.90	**25.57**	25.56
CADIZ	CMEMS L3	8288	8238	**−0.12**	−0.13	**0.27**	0.29	**0.97**	0.97	**11.95**	12.32
	HY-2A	2004	1999	**−0.04**	−0.05	**0.31**	0.31	**0.96**	0.96	**14.31**	14.52
WSMED	CMEMS L3	14,073	14,047	**−0.20**	−0.20	**0.33**	0.33	**0.94**	0.94	18.38	**18.31**
	HY-2A	3027	3024	**−0.13**	−0.13	**0.35**	0.35	0.93	0.93	21.67	**21.64**
ICANA	CMEMS L3	35,047	35,022	**−0.09**	−0.12	**0.24**	0.26	**0.97**	0.96	**10.49**	10.63
	HY-2A	8557	8554	**−0.02**	−0.05	**0.25**	0.24	**0.96**	0.95	**12.07**	12.30
TOTAL	CMEMS L3	221,521	221,321	**−0.09**	−0.11	**0.30**	0.31	**0.98**	0.98	**10.94**	11.05
	HY-2A	57,865	57,845	**−0.02**	−0.04	**0.32**	0.32	**0.98**	0.98	**12.15**	12.28

Predictably, current coupling has no influence in severe storms, where strong wind forcings control the wave model output, with good accuracy in both cases: IBI-CU and IBI-CO solutions. Figure 4 shows the time series at locations for the three biggest storms in western IBI area in 2018: Carmen, the 1 January in the Cantabrian Sea, Emma, the 28 February in the Gulf of Cadiz and Ali, the 19 September on the Irish coast. In these storms, current refraction has a limited impact on wave height patterns, with results (IBI-CU and IBI-CO) more similar than the usual state.

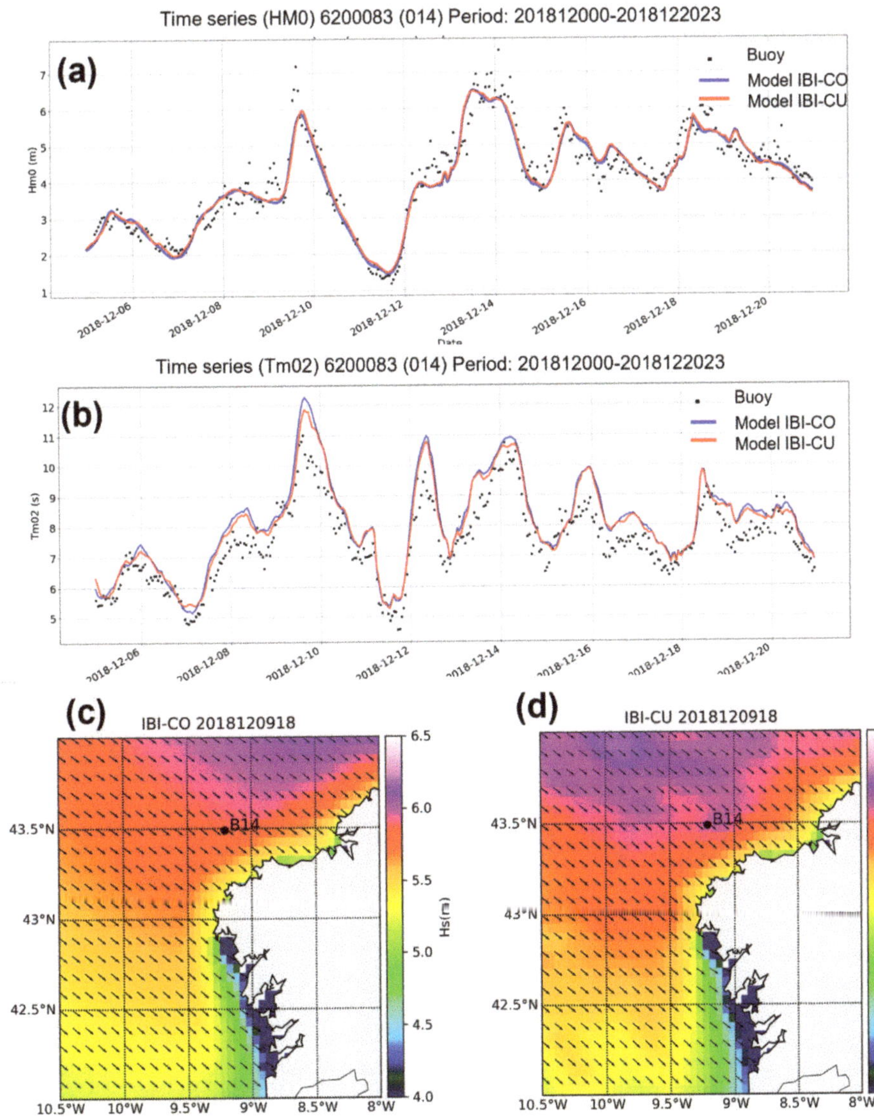

Figure 3. Typical time series of significant wave height (**a**) and mean period (**b**) at the buoy 6200083 for a period of 2 weeks (5–12 December 2018). The observed values are represented by the black dots. Two model results are shown, one including current coupling (red line, IBI-CU) and the other without currents (blue, IBI-CO). On the bottom, example of the modeled situation at 18:00 UTC 9 December 2018 for the western coast of Galicia for the coupled model, IBI-CU (**d**) and no coupled model, IBI-CO (**c**). Point B14 is the location of the buoy 6200083.

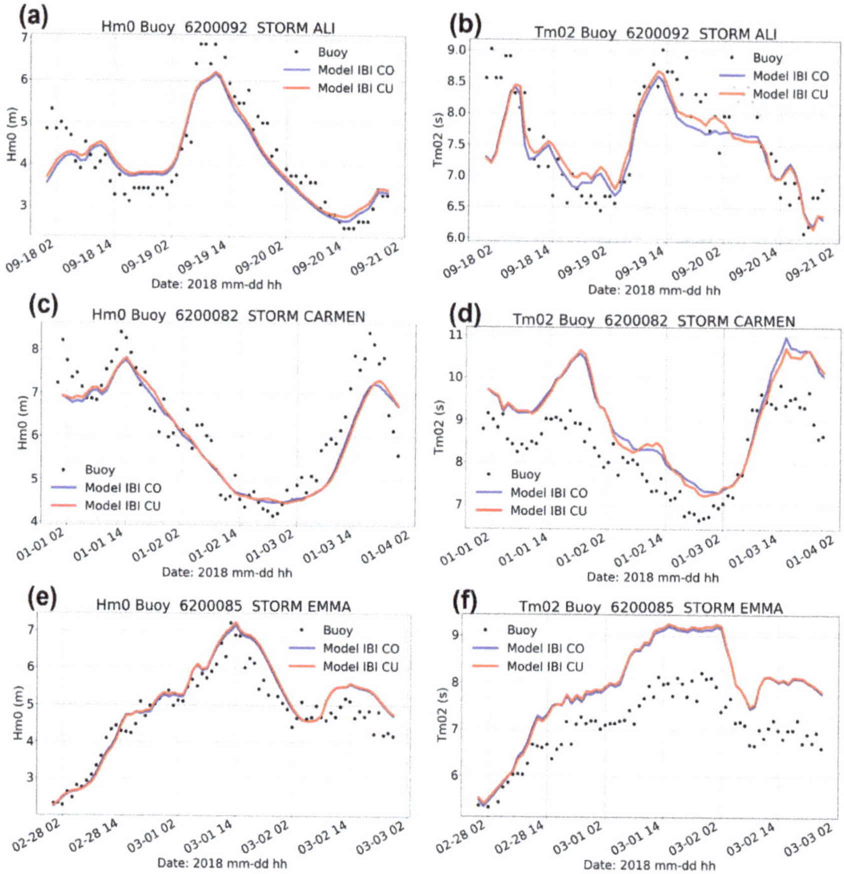

Figure 4. Comparison of coupled model IBI-CU, (red line) and no coupled, IBI-CO (blue line) against mooring buoys for different Storms: Ali ((**a**) Hs and (**b**) Tm_{02} at buoy 620092 on Irish coast), Carmen ((**c**) *Hs*, (**d**)Tm_{02} at the buoy 6200082, Gijon coastal buoy at Cantabrian Sea) and Emma ((**e**) *Hs*, (**f**) Tm_{02} at buoy 620085 in Gulf of Cadiz). The observed values are represented by the black dots.

3.2. Evaluation of Data Assimilation Performance: Validation of New IBI Wave Analysis

Data available for the year 2018 were assimilated to produce the hindcast. Analyses were performed every hour with a data window of 3 h (i.e., data 1.5 h either side of the analysis time were assimilated). A consequence of this arrangement is that the data are not all independent between analyses; however, in testing, we found that this performed better than insisting on the strict independence of the data. This is because with such a short analysis time step, a narrower window does not always contain enough data for an effective analysis. It also leads to a kind of smoothing of the data in time, which could potentially benefit the analysis because, if the data are too different between analyses, and lead to too big a disparity between observation and model in an ensuing analysis step, the data could be rejected. These are not general principles in data assimilation and depend on the assimilation algorithm used. Because optimal interpolation treats all data assimilated for a given analysis as being observed at the same time as the analysis, the correlation of data between analyses is not a concern, as it would be for a time-dependent assimilation algorithm, such as the Kalman filter.

For comparison, the 2018 model runs were also performed without data assimilation. We refer to the runs with data assimilation and no currents as the IBI-DA runs, and those without as the reference runs (IBI-CO).

Bulk statistics for the entire 2018 study period show that the model is closer to the observations with data assimilated than without. These results are summarized in Table 4. Interestingly, the bias, while shrinking in magnitude, changes from being negative to being positive with the assimilation of data. On the other hand, the scatter indices are reduced by DA in the whole IBI region, particularly in CADIZ subregion, where SI drops by 1.8%. In this area, however, the impact of DA on the model during storm Emma (26 February–7 March 2018) is more like the whole IBI region.

Table 4. Biases and scatter indices for control run (IBI-CO) and Data Assimilation simulation (IBI-DA) for the whole IBI area and the CADIZ subregion. In the last case, metrics computed not only for the year 2018, but specifically for the period of the Emma storm (26 February–7 March 2018).

	SWH					
	N		BIAS (m)		SI	
	IBI-CO	IBI-DA	IBI-CO	IBI-DA	IBI-CO	IBI-DA
TOTAL IBI (2018)	58,972	59,027	−0.05	0.02	12.2	11.2
CADIZ (2018)	7651	7656	−0.04	0.03	11.9	10.1
CADIZ (Storm Emma)	212	212	0.21	0.23	8.92	7.99

The scatter plots in Figure 5 gather all the observation–model data pairs. They were generated for the entire two-year period and allow us to examine the validation in more fine-grained detail. The IBI-DA run is visually more concentrated about the center of mass, the black line representing a one-to-one correspondence between the observations and model. The colored squares on that axis also appear hotter in color, indicating that more data pairs are concentrated on it. Finally, the linear regression of the IBI-DA run (the red line) has a gradient closer to one than that of the reference run, though again, for very low Hs, the regression line of the reference run is closer to the center of mass.

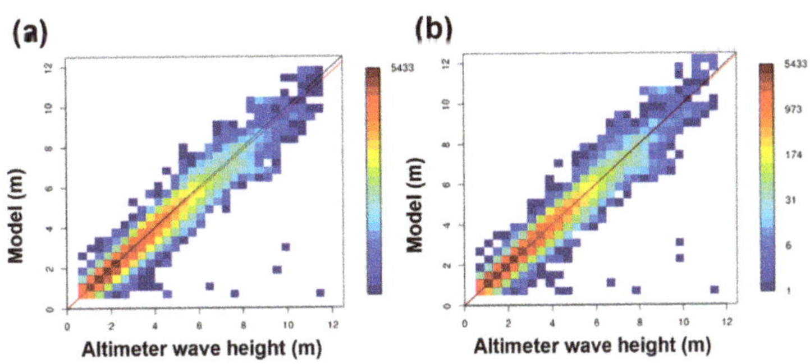

Figure 5. Density scatter plots comparing observed Hs against modelled Hs for all of 2018. The control run (IBI-CO, (**b**)) is on the left, the IBI-DA ((**a**)) run on the right. Each box represents the points found within the range of Hs it covers, and the color scale indicates the number of validated points within the box. The red line is a linear regression, with the black line representing an ideal 1-1 correspondence between model and data.

The quantile–quantile plots in Figure 6 allow us to compare the statistical distributions of the observed and modelled *Hs*, without and with data assimilation. These plots were produced using only the model points that correspond to observations. Assuming reliable data, a straight line of gradient equal to one would imply that the model produces an identical distribution of *Hs* to the observations. Comparing the plot for the IBI-DA run with that of the control run, we see that data assimilation helps bring the model's Hs distribution closer to that of the observations for most of the range of *Hs*. This is especially true for *Hs* between 6 and 9 m. For low *Hs*, up to around 2 m, the IBI-DA run's representativity is slightly worse, but the difference is small, and these wave heights are of much less interest to seafarers, so the inaccuracy here can be forgiven. At extremely high Hs, the distribution of the IBI-DA run is skewed high. There are not many data in this extreme regime to begin with, and the control run already suggests this by oscillating around the ideal unit gradient. Furthermore, the modelling of extremely high waves is even less reliable, though we should be cautious, given the paucity of data in the regime. One possible explanation for this over-correction could be because the reference model is under-estimating moderately high *Hs*, which are great in number, so when the analysis corrects these upward, it inadvertently increases the extremely high Hs as well. In other words, the moderately high *Hs*, because of their larger number, are weighting the analysis more than the extreme *Hs*. In the simple OI data assimilation scheme implemented here, where model errors are constant and covariances are defined solely based on the distance between points, there is no way for it to selectively apply the correction, in such a way as to avoid incorrectly increasing these extreme wave heights.

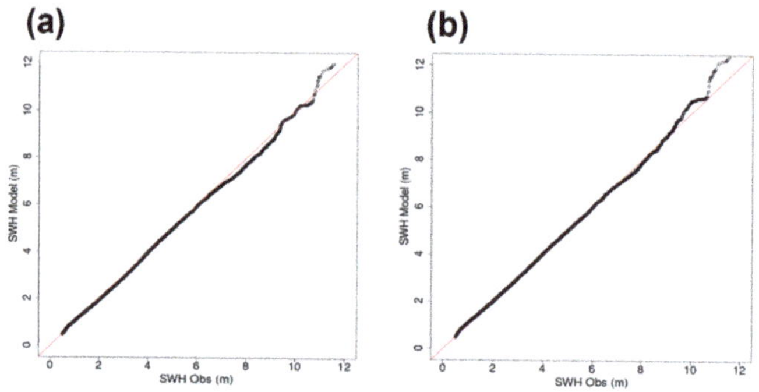

Figure 6. Quantile–quantile plots of observed Significant Wave Height (SWH), *Hs*, against modelled *Hs* for all of 2018. On the left is the IBI-CO run (**a**), on the right the IBI-DA run (**b**). The red line is the unit gradient line.

The bar plots for the monthly means in Figure 7 show, briefly, a consistent reduction in scatter index with data assimilated; it is most reduced in December, March and April, and least in January and September. The bar plots for bias reflect the shift from negative overall bias to positive overall bias, but in some instances, the absolute bias is greater with data assimilation—especially so in July, where the control run's bias is already positive. The fact that the data assimilation always results in the bias tending positive suggests that a bias remains in the data. The monthly diagnostics reveals a moderate seasonal signature in scatter index, with higher scatter indices in the summer months, conserved in the IBI-DA run. The seasonal signature for the bias is almost inverted in the DA run, with higher absolute biases in summer than in winter (apart from November and February). With all this said, the highest absolute bias for all months in the IBI-DA run is only just over 5 cm, which, to put it into perspective, is about the same as the bias for the control run for the whole year.

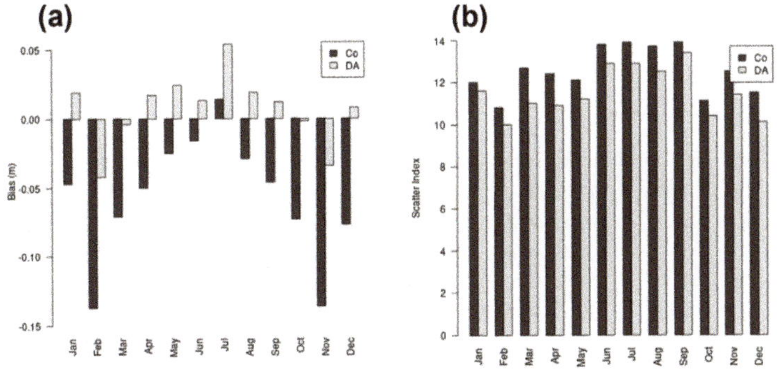

Figure 7. Model (IBI-DA and IBI-CO) Bias (**a**) and Scatter Index (**b**) compared with HY-2A altimeter Hs observations.

4. Discussion

In the previous section, the main results from the IBI wave model sensitivity tests performed (i.e., IBI-CU and IBI-DA) are shown. All the IBI wave simulations are validated with observational data sources for significant wave height and period, and the two test runs are compared with the control run (IBI-CO), which is performed using the same model set-up that has been used in the IBI operations since 2018.

The main aim of these sensitivity tests is to assess the impacts that potential upgrades in the wave model set-up have on the IBI wave model solution. The scientific qualification performed (based on comparisons of the IBI test model solutions with the control run, and the available observations) should provide enough information to decide if some of the proposed set-up novelties may be included in a new operational IBI-MFC wave system release. To make a decision on the adequateness of the proposed set-up upgrades to be launched as part of the next IBI operational wave system release, it would be necessary to verify if the new model solution improves or degrades, with respect to the control (and at what level).

Overall, skill computed along the whole test year do not show substantial improvement with wave–current coupling implementation, in terms of significant wave height accuracy. Model coupling performance (IBI-CU) is a little better for the mean wave period, but in any case, for both parameters, Hs and Tm_{02}, the improvement with wave–current coupling is small and does not exceed 0.2%. The scatter index *SI2* for the significant wave height (Table 3) and the 3% RMSD for the mean wave period computed with hourly observations at mooring buoys (Table 2). However, as it is well known, and extensively described in the literature [43–48], there are very specific situations when this current forcing improves the wave simulation (mostly in shallower coastal waters and related to the same case, where local waves propagate in the opposite direction to prevailing surface currents) differences on mean period have been observed, reducing the quality of the ocean-coupling model. There are some special cases, marked by differences, mainly in the mean period, representing a decrease in the wave model skill when coupling is activated, that should be discussed.

For instance, even though current forcing increases values of wave height so metrics are slightly better for the coupled model (IBI-CU), we see how, in some regions, such as the Gulf of Cadiz and Strait of Gibraltar (CADIZ and GIBST region), a significant decrease in quality in the coupled simulation (and particularly in the period) is detected.

This is seen in the two coastal buoys located close to the strait of Gibraltar (buoys 6101404 and Tarifa-coast-buoy), where the difficulty of obtaining a good ocean current performance produces worse results: they are strongly influenced by an unrealistic ocean current overestimate, the Tm_{02}, when *Hs* is low (less than 0.5 m). Figure 8 illustrates this pattern at buoy 6200085 (Cadiz deep buoy; see locations in the maps depicted in Figure 2) for a whole month, September 2018. This case is used because of the availability of surface current observations (the buoy location is covered by the Puertos del Estado HF Radar System) [49]. Skill metrics for mean period Tm_{02} weaken in the coupled model (correlation decreased from 0.83 for control simulation to 0.74 for IBI-CU in September). The time series depicted in Figure 8, help in analyzing this decrease in the validation scores. Figure 8b shows how differences in the wave period between the coupled and control run mainly occur on some specific days (events depicted in the figure by the dark blue squares). It is important to note that these days, when overestimated values for Tm_{02} in the IBI-CU solution are identified, show very low significant wave height values (Hs values being lower than 0.5 m; see Figure 8a). The IBI model surface currents can be validated with HF Radar observation for the day 3 September 2018. Comparison of daily averaged IBI model currents (Figure 8c) with the HF radar observed ones (Figure 8d) show how the unrealistic simulation of surface current at this location (P16 point indicates location of Buoy 6200085 used for the wave validation) spuriously introduces energy into the wave model, generating the increments of mean period in these cases, marked by low wave heights.

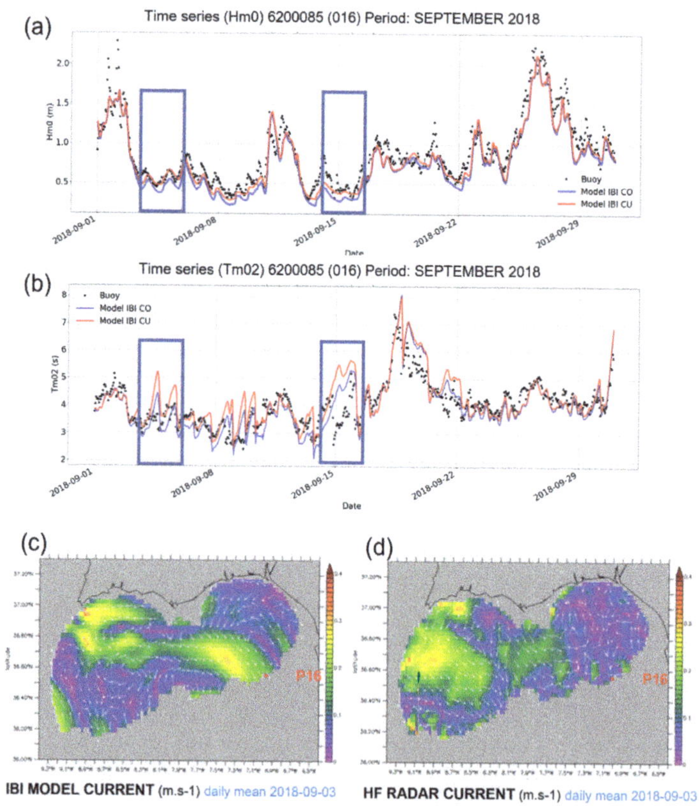

Figure 8. Monthly time series for Hs (**a**) and Tm_{02} (**b**) for the simulations of coupled model IBI-CU, (red line) and no coupled, IBI-CO (blue line) at buoy 6200085 in Gulf of Cadiz, observed data in black dots. The square in dark blue square encloses the overestimated values for Tm_{02} in IBI-CU solution. On the bottom, images of the validation tool NARVAL (North Atlantic Regional Validation; [50,51]): daily mean of current velocity and direction for the IBI model solution (**c**) and HF radar (**d**). Point P16 is the location of the buoy 6200085.

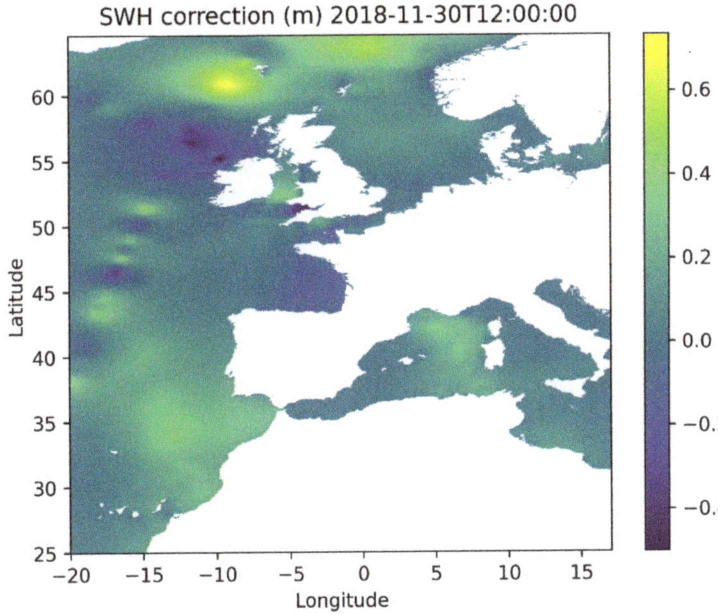

Figure 9. Difference of Significant Wave Height, Hs, (in meters) from IBI-DA and IBI-CO runs during a North-East Atlantic storm (on the 30 November 2018 at 12 UTC).

These results emphasize the importance, when coupling ocean current with waves, of having realistic high-quality model surface current fields. As shown in this example, the use of model currents inputs, locally affected by unrealistic model features, may spoil the wave model performance, especially in very low energetic situations outside of the main storm events. As such, ref. [52] shows a similar misfit induced by model currents in Southern Ocean.

The assimilation of the altimeter data showed a significant reduction in the bias and scatter index on significant wave height. On average, the scatter index of Hs is improved by roughly 8% in open ocean. We noticed that the assimilation is skilled to efficiently correct the wave model errors related to the uncertainties of the wind forcing in the North Atlantic, especially during storm events, as illustrated in Figure 9. This clearly brings better initial conditions for swell propagation to coastal areas, as revealed at Cadiz buoy. However, there is still room for improvement in the assimilation scheme, namely better estimates of the covariance model errors, by taking into account the variability of the sea state in the IBI domain. Moreover, the use of variable correlation length, depending on whether the sea state is wind sea or swell dominated, will induce a better spread of the assimilation correction on the model grid points.

Results from the new IBI wave systems (proposed to be the new IBI wave operational release, named here IBI-OP) can be seen in Table 5.

Table 5. Estimated Accuracy Numbers related to Significant Wave Height (year 2018) for the IBI wave model scenarios. On the right (column IBI-OP), the new IBI operational wave system. Observational source reference: HY-2A satellite-derived product.

	TOTAL IBI SWH			
	IBI-CO	**IBI-CU**	**IBI-DA**	**IBI-OP**
N	57,845	57,865	57,899	57,921
BIAS	−0.04	−0.02	0.02	0.03
CORR	0.98	0.98	0.98	0.98
RMSE	0.32	0.32	0.29	0.29
HH (%)	10.95	10.72	9.73	9.67
SI2 (%)	12.28	12.15	11.13	11.14

The statistical comparison for the entire year 2018 of the four scenarios, defined in Table 1, and the HY-2A altimeter shows a good evolution for each model upgrade (Table 5). However, the most relevant improvement is due to the data assimilation implementation (IBI-DA), with a scatter index *SI2* and error indicator *HH*, decreasing more than 1%. The impact of ocean current, although less significant, improves the model qualification, increasing the significant wave height and with better accuracy of mean period (Tables 2 and 3).

5. Conclusions

Two major conclusions emerge from these tests. First, the significant improvement on the wave model set-up that was used in the IBI operations in December 2019 (IBI-CO), activating current forcing (IBI-CU), data assimilation (IBI-DA), or both (IBI-OP), with the major upgrade in the performances, including data assimilation, with the scatter error indicator decreasing by more than 1% (Table 5). Second, the importance of having realistic high-quality forcings fields. In particular, the use of unrealistic currents inputs may spoil the wave model performance (Figure 8).

The research performed, combining information from different wave model solutions and several observational data sources, benefited the IBI-MFC operational wave forecast service, supporting the decision-making process, related to their latest major operational upgrade (occurred in the Copernicus Marine Operational Release March 2020). The extensive assessment performed with the IBI-DA and IBI-CU run, presented here, was useful to demonstrate, for each proposed IBI model set-up novelty, the significance of the associated improvements (in terms of IBI wave forecasting local added value). After verifying the impacts that these two proposed novelties have on the IBI wave solution (and always using, as a referential solution, the IBI wave model configuration in operations), a new IBI wave configuration (combining both the data assimilation and the current–wave coupling) were prepared to be transitioned into operations.

The significant improvement achieved by this IBI-OP configuration, with respect to the control one (IBI-CO, the IBI set-up currently in operations), in terms of wave solution over the IBI area, throughout the year 2018, supports the IBI-MFC decision to use this set-up for operational purposes, to be implemented in the IBI Wave, operational from March 2020. Furthermore, similar performance metrics were achieved by the systems in operations (see information from a 2-year validation in the Quality Information Document of the Copernicus Marine IBI wave product, Toledano et al. 2021 [31]). Finally, note that this research was useful not only to evolve the IBI-MFC operational wave forecast service, but also has been helpful to open new working lines prioritized in the Copernicus Marine service roadmap:

- Currently, the IBI-MFC is in the process of enhancing the coupling, in both directions, between waves and ocean dynamics. In the December 2020 release, the upgraded IBI wave system included improved computations of coupling parameters, such as the surface stress, and wave-breaking-induced turbulence in the ocean mixed layer. These parameters are used to drive a coupling run with the IBI MFC ocean forecast run, based on a NEMO model application. The sensitivity coupling tests of using the Stokes–Coriolis forcing, the sea-state-dependent stress momentum fluxes and including wave-breaking energy flux in the vertical mixing, are all validated and the impact on improving the forecast is addressed by [8,53,54].

- With respect to data assimilation, directional wave spectra data assimilation based on the optimal interpolation method, applied to mean wave parameters (total energy and wave number components) of each wave train composing the wave spectrum [55], will be implemented, in order to increase the impact of assimilation [7]. Likewise, assimilation of wave data from swath altimetry is expected to be tested in the future.

Author Contributions: Conceptualization, M.G.S., L.A., C.T. and M.G.; methodology, C.T. and M.G.; software, C.T., M.G., A.D. and P.L.; validation, C.T., M.G., P.L. and A.D.; formal analysis, C.T., M.G., L.A. and M.G.S.; investigation, C.T. and M.G.; data curation, C.T., M.G., P.L. and A.D.; writing—original draft preparation, C.T., M.G., M.G.S. and L.A.; supervision, L.A. and M.G.S.; project administration, M.G.S. and L.A. All authors have read and agreed to the published version of the manuscript.

Funding: This research was funded in the framework of the E.U. Copernicus Marine Service for the IBI region (CMEMS Contract No. 100-IBI-OPER-CMEMS and Copernicus-2 Marine service contract 21002L6-COP-MFC IBI-5600).

Institutional Review Board Statement: Not applicable.

Informed Consent Statement: Not applicable.

Data Availability Statement: The Copernicus Regional IBI-MFC Wave solution can be downloaded at Copernicus Marine Environmental Monitoring Service (CMEMS) (http://marine.copernicus.eu/, accessed on 18 January 2022).

Acknowledgments: This study has been conducted using E.U. Copernicus Marine Service Information. Specifically, from their in-situ TAC Product (INISITU_IBI_NRT_OBSERVATIONS_013_033) and CMEMS L3 wave altimetric product (WAVE_GLO_WAV_l3_SWH_NTRT_OBSERVATIONS_014_001). Likewise, satellite observations from the HY2A mission, provided by the CNES, have been used.

Conflicts of Interest: The authors declare no conflict of interest.

References

1. Cavaleri, L.; Fox-Kemper, B.; Hemer, M. Wind waves in the climate coupled system. *Bull. Am. Meteorol. Soc.* **2012**, *93*, 1651–1661. [CrossRef]
2. Janssen, P.A.E.M.; Aouf, L.; Behrens, O.; Breivik, A.; Korres, G.; Cavalieri, L.; Christiensen, K. *Final Report of Work-Package I of 7th Framework Program Mywave Project*; European Commission: Brussels, Belgium, 2014.
3. Capet, A.; Fernández, V.; She, J.; Dabrowski, T.; Umgiesser, G.; Staneva, J.; Mészáros, L.; Campuzano, F.; Ursella, L.; Nolan, G.; et al. Operational modeling capacity in European Seas—An EuroGOOS perspective and recommendations for improvement. *Front. Mar. Sci.* **2020**, *7*, 129. [CrossRef]
4. The WAMDI group. The Wam model–A third generation ocean wave prediction model. *J. Phys. Ocenography* **1988**, *18*, 1775–1810. [CrossRef]
5. Booij, N.; Holthuijsen, L.H.; Ris, R.C. The SWAN wave model for shallow water. In Proceedings of the 25th International Conference on Coastal Engineering, Orlando, FL, USA, 2–6 September 1996; pp. 668–676. [CrossRef]
6. Tolman, H.L. User manual and system documentation of WAVEWATCH III TM version 3.14. *Tech. Note MMAB Contrib.* **2009**, *276*, 220.
7. Aouf, L.; Lefèvre, J.-M. On the impact of the assimilation of SARAL/AltiKa wave data in the operational wave model MFWAM. *Mar. Geodesy* **2015**, *38*, 381–395. [CrossRef]
8. Breivik, Ø.; Mogensen, K.; Bidlot, J.; Balmaseda, M.A.; Janssen, P.A.E.M. Surface wave effects in the NEMO ocean model: Forced and coupled experiments. *J. Geophys. Res. Oceans* **2015**, *120*, 2973–2992. [CrossRef]
9. Fan, Y.; Griffies, S.M. Impacts of parameterized Langmuir turbulence and nonbreaking wave mixing in global cli-mate simulations. *J. Clim.* **2014**, *27*, 4752–4775. [CrossRef]
10. Babanin, A.V.; Ganopolski, A.; Phillips, W.R.C. Wave-induced upper-ocean mixing in a climate model of intermediate complexity. *Ocean. Model.* **2009**, *29*, 189–197. [CrossRef]
11. Janssen, P.A.E.M. Ocean wave effects on the daily cycle in SST. *J. Geophys. Res. Earth Surf.* **2012**, *117*, C11. [CrossRef]
12. Thévenot, O.; Bouin, M.-N.; Ducrocq, V.; Brossier, C.L.; Nuissier, O.; Pianezze, J.; Duffourg, F. Influence of the sea state on Mediterranean heavy precipitation: A case-study from HyMeX SOP1. *Q. J. R. Meteorol. Soc.* **2016**, *142*, 377–389. [CrossRef]

13. Ginis, I. Atmophere-Ocean coupling in tropical cyclone. In Proceedings of the ECMWF Workshop on Ocean-Atmosphere Interactions, Reading, UK, 10–12 November 2008.
14. Liu, C.; Qi, Y.; Liang, J. The effect of sea waves on the typhoon Imodu. In Proceedings of the High Resolution Modelling CAWCR Workshop, Melbourne, Australia, 25–28 November 2008.
15. Bruciaferri, D.; Tonani, M.; Lewis, H.W.; Siddorn, J.R.; Saulter, A.; Sanchez, J.M.C.; Valiente, N.G.; Conley, D.; Sykes, P.; Ascione, I.; et al. The impact of Ocean-Wave coupling on the upper Ocean circulation during storm events. *J. Geophys. Res. Oceans* **2021**, *126*, e2021017343. [CrossRef]
16. Holthuijsen, L.H. *Waves in Oceanic and Coastal Waters*; Cambridge University Press (CUP): Cambridge, UK, 2007.
17. Viitak, M.; Maljutenko, I.; Alari, V.; Suursaar, Ü.; Rikka, S.; Lagemaa, P. The impact of surface currents and sea level on the wave field evolution during St. Jude storm in the eastern Baltic Sea. *Oceanologia* **2016**, *58*, 176–186. [CrossRef]
18. Echevarria, E.R.; Hemer, M.A.; Holbrook, N.J. Global implications of surface current modulation of the wind-wave field. *Ocean. Model.* **2021**, *161*, 101792. [CrossRef]
19. Law-Chune, S.; Aouf, L.; Dalphinet, A.; Levier, B.; Drillet, Y.; Drevillon, M. WAVERYS: A CMEMS global wave reanalysis during the altimetry period. *Ocean. Dyn.* **2021**, *71*, 357–378. [CrossRef]
20. Le Traon, P.Y.; Reppucci, A.; Fanjul, E.A.; Aouf, L.; Behrens, A.; Belmonte, M.; Bentamy, A.; Bertino, L.; Brando, V.E.; Kreiner, M.B.; et al. From observation to information and users: The Copernicus marine service perspective. *Front. Mar. Sci.* **2019**, *6*, 234. [CrossRef]
21. Von Schuckmann, K.; Le Traon, P.-Y.; Smith, N.; Pascual, A.; Brasseur, P.; Fennel, K.; Djavidnia, S.; Aaboe, S.; Fanjul, E.A.; Autret, E.; et al. Copernicus marine service ocean state report. *J. Oper. Oceanogr.* **2018**, *11*, S1–S142. [CrossRef]
22. Sarvia, F.; De Petris, S.; Borgogno-Mondino, E. Mapping ecological focus areas within the EU CAP controls framework by Copernicus Sentinel-2 data. *Agronomy* **2022**, *12*, 406. [CrossRef]
23. Aschbacher, J. ESA's earth observation strategy and Copernicus. In *Satellite Earth Observations and Their Impact on Society and Policy*; Springer: Singapore, 2017; pp. 81–86.
24. CMEMS Scientific and Technical Advisory Committee (STAC). Copernicus Marine Environment Monitoring Service (CMEMS) Service Evolution Strategy: R&D Priorities. Version 4. 2018. Available online: https://marine.copernicus.eu/sites/default/files/CMEMS-Service_evolution_strategy_RD_priorities_V4.pdf (accessed on 13 January 2022).
25. Sotillo, M.G.; Aznar, R.; Gutknecht, E.; Levier, B.; Reffray, G.; Lorente, P.; Barrera, E.; Dabrowski, T.; Aouf, L. The CMEMS IBI-MFC along Copernicus-1): Evolution & achievements. Included in the special issue: The Copernicus Marine Service from 2015 to 2021: Six years of achievements by LeTraon et al. *Spec. Issue Mercator Océan J.* **2021**, *57*, 147–154. [CrossRef]
26. Lefèvre, J.M.; Aouf, L.; Bataille, C. Apport d'un nouveau modèle de vagues de 3ème génération à Météo-France. In Proceedings of the Actes de Conférence des Ateliers de Modélisation de l'Atmosphère; 2009; pp. 27–29.
27. ECMWF. PartVII: ECMWF wave model. *IFS Doc CY41R1* **2015**, 1–83.
28. Ardhuin, F.; Rogers, E.; Babanin, A.V.; Filipot, J.-F.; Magne, R.; Roland, A.; Van Der Westhuysen, A.; Queffeulou, P.; Lefevre, J.-M.; Aouf, L.; et al. Semiempirical dissipation source functions for Ocean Waves. Part I: Definition, calibration, and validation. *J. Phys. Oceanogr.* **2010**, *40*, 1917–1941. [CrossRef]
29. Cavaleri, L.; Abdalla, S.; Benetazzo, A.; Bertotti, L.; Bidlot, J.R.; Breivik, Ø.; Carniel, S.; Jensen, R.E.; Portilla-Yandun, J.; Rogers, W.E.; et al. Wave modelling in coastal and inner seas. *Prog. Oceanogr.* **2020**, *167*, 164–233. [CrossRef]
30. Gerling, T.W. Partitioning sequences and arrays of directional ocean wave spectra into component wave systems. *J. Atmos. Ocean. Technol.* **1992**, *9*, 444–458. [CrossRef]
31. Toledano, C.; Dalphinet, A.; Lorente, P.; Alfonso, M.; Ghantous, M.; Aouf, L.; Sotillo, M.G. Quality Information Document for Atlantic—Iberian Biscay Irish—Wave Analysis and Forecast Product. 2021, pp. 1–49. Available online: https://catalogue.marine.copernicus.eu/documents/QUID/CMEMS-IBI-QUID-005-005.pdf (accessed on 7 January 2021). [CrossRef]
32. NOAA National Geophysical Data Center. ETOPO1 1 Arc-Minute Global Relief Model. *NOAA Natl. Cent. Environ. Inf.* **2009**. [CrossRef]
33. Haiden, T.; Janousek, M.; Bidlot, J.; Buizza, R.; Ferranti, L.; Prates, F.; Vitart, F. *Evaluation of ECMWF Forecasts, including the 2018 Upgrade*; European Centre for Medium Range Weather Forecasts: Reading, UK, 2018.
34. Aouf, L. Quality Information Document for Global Ocean Wave Analysis and Forecast Product. 2021, pp. 1–21. Available online: https://catalogue.marine.copernicus.eu/documents/QUID/CMEMS-GLO-QUID-001-027.pdf (accessed on 11 January 2021). [CrossRef]
35. Lionello, P.; Günther, H.; Janssen, P.A.E.M. Assimilation of altimeter data in a global third-generation wave model. *J. Geophys. Res. Earth Surf.* **1992**, *97*, 14453–14474. [CrossRef]
36. Dalphinet, A.; MeteoFrance, Tolouse, France; Aouf, L.; MeteoFrance, Tolouse, France. Personal Communication, 2020.
37. Tolman, H.L. The influence of unsteady depths and currents of tides on wind-wave propagation in shelf seas. *J. Phys. Oceanogr.* **1990**, *20*, 1166–1174. [CrossRef]
38. Wehde, H.; Schuckmann, K.V.; Pouliquen, S.; Grouazel, A.; Bartolome, T.; Tintore, J.; De Alfonso, M.; Carval, T.; Racapé, V.; The INSTAC Team. Quality Information Document for Near Real Time In Situ Products. 2021, pp. 1–36. Available online: https://catalogue.marine.copernicus.eu/documents/QUID/CMEMS-INS-QUID-013-030-036.pdf (accessed on 7 January 2021). [CrossRef]

39. Wang, J.; Aouf, L.; Jia, Y.; Zhang, Y. Validation and calibration of significant wave height and wind speed retrievals from HY2B altimeter based on deep learning. *Remote Sens.* **2020**, *12*, 2858. [CrossRef]
40. Mentaschi, L.; Besio, G.; Cassola, F.; Mazzino, A. Problems in RMSE-based wave model validations. *Ocean. Model.* **2013**, *72*, 53–58. [CrossRef]
41. Hanna, S.R.; Heinold, D.W. *Development and Application of a Simple Method for Evaluating Air Quality Models*; American Petroleum Institute: Washington, DC, USA, 1985.
42. Liu, Q.; Babanin, A.V.; Guan, C.; Zieger, S.; Sun, J.; Jia, Y. Calibration and validation of HY-2 altimeter wave height. *J. Atmos. Ocean. Technol.* **2016**, *33*, 919–936. [CrossRef]
43. Moon, I.-J. Impact of a coupled ocean wave–tide–circulation system on coastal modeling. *Ocean. Model.* **2005**, *8*, 203–236. [CrossRef]
44. Jorda, G.; Bolaños, R.; Espino, M.; Sánchez-Arcilla, A. Assessment of the importance of the current-wave coupling in the shelf ocean forecasts. *Ocean. Sci.* **2007**, *3*, 345–362. [CrossRef]
45. Ardhuin, F.; Roland, A.; Dumas, F.; Bennis, A.-C.; Sentchev, A.; Forget, P.; Wolf, J.; Girard, F.; Osuna, P.; Benoit, M. Numerical wave modeling in conditions with strong currents: Dissipation, refraction, and relative wind. *J. Phys. Oceanogr.* **2012**, *42*, 2101–2120. [CrossRef]
46. Romero, L.; Lenain, L.; Melville, W.K. Observations of surface wave–current interaction. *J. Phys. Oceanogr.* **2017**, *47*, 615–632. [CrossRef]
47. Ardhuin, F.; Gille, S.T.; Menemenlis, D.; Rocha, C.B.; Rascle, N.; Chapron, B.; Gula, J.; Molemaker, J. Small-scale open ocean currents have large effects on wind wave heights. *J. Geophys. Res. Oceans* **2017**, *122*, 4500–4517. [CrossRef]
48. De León, S.P.; Soares, C.G. Extreme waves in the Agulhas current region inferred from SAR wave spectra and the SWAN model. *J. Mar. Sci. Eng.* **2021**, *9*, 153. [CrossRef]
49. Mulero-Martínez, R.; Gómez-Enri, J.; Mañanes, R.; Bruno, M. Assessment of near-shore currents from CryoSat-2 satellite in the Gulf of Cádiz using HF radar-derived current observations. *Remote Sens. Environ.* **2021**, *256*, 112310. [CrossRef]
50. Lorente, P.; Sotillo, M.G.; Amo-Baladrón, A.; Aznar, R.; LeVier, B.; Aouf, L.; Dabrowski, T.; De Pascual, Á.; Reffray, G.; Dalphinet, A.; et al. The NARVAL software toolbox in support of ocean models skill assessment at regional and coastal scales. In *Computational Science—ICCS 2019, Proceedings of the International Conference on Computational Science, Las Vegas, NV, USA 5–7 December 2019*; Springer Science and Business Media LLC: Cham, Switzerland, 2019; pp. 315–328.
51. Lorente, P.; Piedracoba, S.; Sotillo, M.G.; Aznar, R.; Amo-Baladrón, A.; Pascual, Á.; Soto-Navarro, J.; Álvarez-Fanjul, E. Ocean model skill assessment in the NW Mediterranean using multi-sensor data. *J. Oper. Oceanogr.* **2016**, *9*, 75–92. [CrossRef]
52. Derkani, M.H.; Alberello, A.; Nelli, F.; Bennetts, L.G.; Hessner, K.G.; MacHutchon, K.; Reichert, K.; Aouf, L.; Khan, S.; Toffoli, A. Wind, waves, and surface currents in the Southern Ocean: Observations from the Antarctic Circumnavigation Expedition. *Earth Syst. Sci. Data* **2021**, *13*, 1189–1209. [CrossRef]
53. Chune, S.L.; Aouf, L. Wave effects in global ocean modeling: Parametrizations vs. forcing from a wave model. *Ocean. Dyn.* **2018**, *68*, 1739–1758. [CrossRef]
54. Staneva, J.; Grayek, S.; Behrens, A.; Günther, H. GCOAST: Skill assessments of coupling wave and circulation models (NEMO-WAM). *J. Phys. Conf. Ser.* **2021**, *1730*, 012071. [CrossRef]
55. Voorrips, A.C.; Makin, V.K.; Hasselmann, S. Assimilation of wave spectra from pitch-and-roll buoys in a North Sea wave model. *J. Geophys. Res. Earth Surf.* **1997**, *102*, 5829–5849. [CrossRef]

Journal of
Marine Science and Engineering

MDPI

Article

A Regional Operational Model for the North East Atlantic: Model Configuration and Validation

Hazem Nagy [1,2,*], Kieran Lyons [1], Glenn Nolan [1], Marcel Cure [3] and Tomasz Dabrowski [1]

1 Marine Institute, Rinville, Oranmore, Co Galway, Galway, Ireland; Kieran.lyons@marine.ie (K.L.); Glenn.Nolan@Marine.ie (G.N.); tomasz.dabrowski@marine.ie (T.D.)
2 Oceanography Department, Alexandria University, Alexandria 21500, Egypt
3 The Numerics Warehouse Ltd. Tyrone, Kilcolgan, Co. Galway, Galway, Ireland; marcel.cure@numericswarehouse.com
* Correspondence: hazem.nagy@marine.ie or hazem.nagi@alexu.edu.eg; Tel.:+353-894-985-494

Received: 15 July 2020; Accepted: 27 August 2020; Published: 1 September 2020

Abstract: An operational model for an area of the northeast Atlantic that encompasses all of Ireland's territorial waters has been developed. The model is an implementation of the Regional Ocean Modelling System (ROMS) and uses operationally available atmospheric and boundary forcing, and a global tide solution for tidal forcing. River forcing is provided by climatological daily discharge rates for 29 rivers across Ireland, west Britain, and west France. It is run in an operational framework to produce 7-day hindcasts once a week, and daily 3-day forecasts which are published in a number of formats. We evaluated the model skill by comparing with measured data and calculating statistics such as mean error, root mean square error (RMSE), and correlation coefficient. The observations consist of satellite Sea Surface Temperature (SST), total surface velocity fields from satellite, water level time series from around the Irish coast, and temperature and salinity data from Array for Real-Time Geostrophic Oceanography (ARGO) and Conductivity Temperature Depth (CTD) profiles. The validation period is from 1 January 2016 until 31 December 2019. The correlation coefficient between the model and satellite SST is 0.97 and recorded in March and April 2018. The model error is about 5% of the total M_2 amplitude in the Celtic Sea recorded at Dunmore East tide gauge station. The maximum RMSE between the model and the CTD temperature profiles is 0.8 °C while it is 0.17 PSU for salinity. The model correctly defines the shelf water masses around Ireland. In 2019 the Irish Coastal Current (ICC) was very strong and well defined along most of the western Irish coast. The model results have well reproduced the ICC front for the whole simulation period.

Keywords: Northeast Atlantic; ROMS; ARGO; tide gauges; Irish coastal current

1. Introduction

Ireland is located in a hydrographically interesting part of the northeast Atlantic Ocean where the regimes of the deep ocean and the continental shelf interact. The North Atlantic Current (NAC) divides southwest of the Rockall Bank and produces several branches flowing roughly to the northeast or southeast [1]. The Rockall Trough is one pathway by which heat is transported into the Nordic Seas [2,3] (see Figure 1). The continental shelf edge has a poleward current associated with it, which is seasonally variable [4–6] and is the interface for the ocean-shelf exchange of nutrients and carbon [7]. On the Irish shelf, the Irish Shelf Front (ISF) separates fresher Irish coastal waters from the oceanic Eastern North Atlantic Water (ENAW) at ca. 11° W [8]. The Shannon River on the west coast of Ireland is the largest river in the region with a mean annual discharge of 186 $m^3 s^{-1}$ [9], but there are also significant freshwater discharges from rivers on the south coast of Ireland and west coast of Britain as well as from further afield such as the Loire in France. This induces vertical density stratification on the shelf while thermal stratification is an important process during the summer months. In contrast, mixing due to tide and wind creates areas with well-mixed water, such as the Irish Sea. Fronts occur between these stratified and well-mixed areas, such as the Celtic Sea front, which is located at the southern entrance to the Irish Sea [10]. A seasonal density-driven current, the Irish Coastal Current (ICC) has been described by [10,11]. The ICC is linking Land's End in southwest England to the northwest coast of Ireland and this may provide an important pathway for harmful phytoplankton, which has significant consequences for the aquaculture industry on the west coast of Ireland and onwards to the coast of west Scotland. The Scottish Coastal Current flows northwards from the Irish Sea along the west coast of Scotland [12].

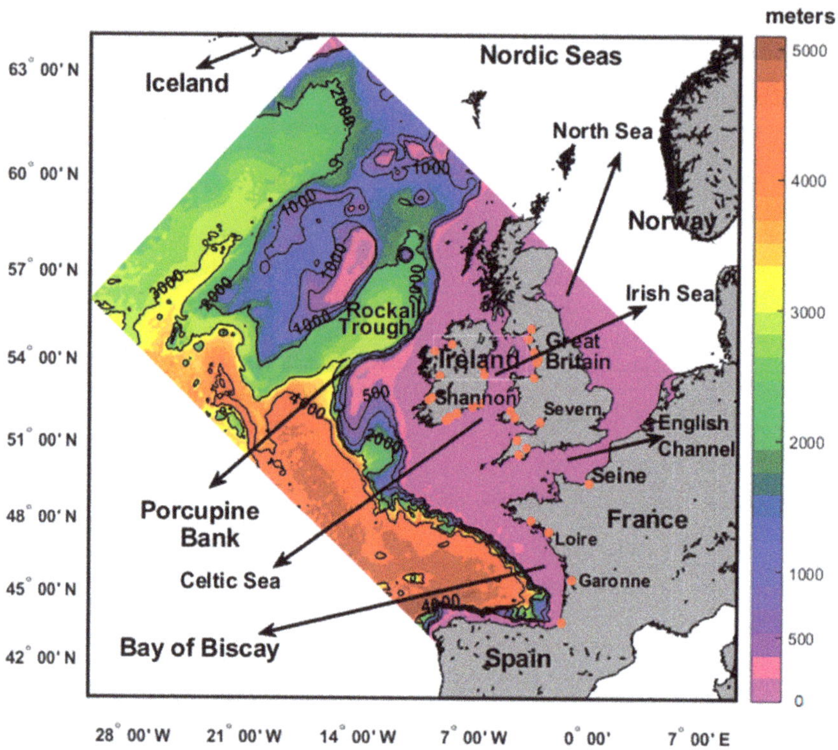

Figure 1. Bathymetric map of the area covered by the model. Major rivers included in the model are marked by red circles.

Operational modelling of physical ocean fields has developed rapidly over the last decade in response to the needs of the marine community. The availability of hindcasts and forecasts of the hydrodynamic properties for specific regions is of great benefit to decision makers in areas such as marine environmental management, and the aquaculture and fisheries industries. Operational models that forecast the physical state of the oceans exist for European oceans (e.g., [13–27]). The first operational models developed in Ireland and described by [15,16].

Ref. [24] described an operational biogeochemical model for the North East Atlantic Ocean region. The model domain covered a significant portion of the North-West European continental shelf with a horizontal and vertical resolution of ≈4 km and 40 sigma levels, respectively. It was one-way nested within the high resolution (1/12°) CMEMS Ocean (http://www.CMEMS-ocean.fr) PSY2V4R2 operational model of the North Atlantic whereby daily values for potential temperature, sea surface height and velocity are linearly interpolated at the open ocean boundaries. A simple nitrogen-based NPZD model developed described in [28] was dynamically coupled to ROMS. The model has successfully represented the intra-annual variability of surface chlorophyll and nitrate concentrations at monthly time scales.

There have already been several studies simulating the North Atlantic region including [27,29–36]. Overall, properly validated hydrodynamic models can provide information on the present and near-future states of ocean systems, which is of value to a wide range of users from both the private sector and the general public.

Objectives of the Study

This study presents the set-up and results from the validation of a high-resolution numerical operational model (hereafter called NEA_ROMS) developed at Irish Marine Institute. Validation of NEA_ROMS was carried out for the time period from January 2016 until December 2019 using observational data from various sources. The model has been compared with satellite SST, tide gauges, Argo, and CTD temperature and salinity profiles. In addition, the Geostrophic and Ekman Current Observatory GEKCO surface data was used to validate model velocity fields. Additionally, this study will focus on the Irish coastal waters and the authors will examine the representation of the Irish coastal current (ICC) in the model. Section 2 describes the model implementation, and nesting procedures, Section 3 presents the validation against observational data, Section 4 describes the model results related to the ICC, and Section 5 provides the discussion and conclusions.

2. Model Design and Implementation

The model used in this study is an implementation of the Regional Ocean-Modelling System (ROMS). ROMS has been proven to demonstrate substantial skill in forecasting [37] and has been used in a number of successful operational systems [38–40]. ROMS is a primitive-equation, free-surface, hydrostatic model as described in [41]. In the horizontal, it uses orthogonal curvilinear coordinates on an Arakawa "C" grid while utilizing a terrain-following (sigma) coordinate in the vertical. The momentum equations are solved using a split-explicit time-stepping scheme whereby a whole number of barotropic time steps are carried out within each baroclinic time step to solve the barotropic momentum equations. The prognostic variables of the model are surface elevation, potential temperature, salinity, and velocity. All the model equations are written in rectangular coordinates and contain spatially and temporally varying horizontal eddy viscosity and diffusion coefficients. Vertical diffusion is calculated using the K-profile parameterization (KPP) [42] and a modified Jacobian method [43] is used to calculate horizontal pressure-gradient forces. The associated parameters with KPP vertical turbulent closure scheme have been enhanced to tune the vertical profile of currents, temperature, and salinity by [44–47]. Recursive MPDATA is used for the advection of tracers [48]. Bottom stress is applied using the logarithmic "law of the wall" with a constant roughness length of 0.01 m.

The NEA_ROMS model domain covers a significant portion of the North-West European continental shelf and also the Porcupine and Rockall Banks and the Rockall Trough as shown in Figure 1. NEA_ROMS model has a horizontal resolution of 1.1 km and 40 vertical sigma levels with the highest concentration of levels at the surface and the bottom. The basin-scale daily average lateral boundary conditions are provided from the high resolution (1/12°) CMEMS_GLOBAL_ ANALYSIS_ FORECAST_ PHY _001_024 model [49]. The daily values are for potential temperature, salinity, sea surface height, and velocity components at the three open boundaries. The discharges from 29 rivers are included in the NEA_ROMS model. The rivers comprise the major rivers of Ireland, west Britain, and west France, as well as some more minor Irish rivers. The discharge rates are average daily values calculated from many years of historical discharge data for 29 rivers across Ireland, west Britain, and west France. Furthermore, the nested model temperature and salinity fields were nudged toward the GLOBAL_ ANALYSIS_ FORECAST_ PHY _001_024 operational model. Within the NEA_ROMS, several one-way nested shelf models have been implemented to downscale and forecast near the Irish coasts.

The model bathymetry utilizes data from a number of sources. Large-scale multi-beam mapping of Irish territorial waters has been carried out under the Irish National Seabed Survey (INSS) and the Integrated Mapping for the Sustainable Development of Ireland's Marine Resource (INFOMAR) programs. These datasets along with an extensive single-beam archive maintained by the Marine Institute were used to develop the model bathymetry. A 1 × 1 km Celtic Sea bathymetry [50] complements the bathymetry in the Irish and Celtic Seas while GEBCO data is used for the bathymetry in non-Irish waters. The model minimum depth is set at 10 m near the coastline. In order to reduce the pressure gradient errors, we smoothed the model bathymetry according to the rx_0 factor of Beckman and Haidvogiel [51,52].

2.1. Lateral Boundary Conditions

The model has three open boundaries in the North, East, and West. The NEA-ROMS lateral boundary conditions are taken from the CMEMS GLOBAL_ANALYSIS_FORECAST_ PHY _001_024 at 50 vertical levels are ranging from 0 to 5500 m. The CMEMS model is version 3.1 of NEMO [49]. The model data is freely available via the Copernicus Marine Environment Monitoring Service (CMEMS) web site marine.copernicus.eu. The analysis is updated weekly while a 10-day forecast is updated daily. The model boundary parameters are temperature, salinity, total velocity components, surface elevation, and barotropic velocity components. The nesting was designed to ensure that the volume transport across the open boundary of the nested model matches the volume transport of the CMEMS global model.

For the barotropic velocity, we used the scheme proposed by [53]. At the outflow, we impose

$$\overline{\varphi_{nnested}} = \left[\sqrt{\frac{g}{H_{nested}}}\left(\eta_{nested} - \eta_{nesting}\right)\right] - \overline{\varphi_{nnested}}\frac{H_{nesting} + \eta_{nesting}}{H_{nested} + \eta_{nested}} \tag{1}$$

where $\overline{\varphi_n} = \frac{1}{H+\eta}\int_H^\eta \varphi_n dz$ is the normal barotropic velocity at the open boundaries and g the gravity, while at the inflow will be $\overline{\varphi_{nnesting}} = \overline{\varphi_{nnested}}$

For the baroclinic velocity, we impose the normal velocity according to the following formula:

$$\int_{n_2}^{n_1}\int_{H_{nested}}^{\eta_{nested}} \varphi_{nnested} dzdn = \int_{n_2}^{n_1}\int_{H_{nesting}}^{\eta_{nesting}} \varphi_{nnesting} dzdn \tag{2}$$

where φ_n is the normal component to the three open boundaries in m/s. For the temperature and salinity, we used a third-order upstream scheme with implicit mixing for horizontal advection [54]. Whenever the normal velocity is directed outwards:

$$\frac{\partial T_{nested}}{\partial t} + \varphi_{nnested}\frac{\partial T_{nested}}{\partial y} = 0; \frac{\partial S_{nested}}{\partial t} + \varphi_{nnested}\frac{\partial S_{nested}}{\partial y} = 0 \tag{3}$$

While on the inflow the temperature and salinity are prescribed to be identical to that interpolated to CMEMS global model values. Where n1, n2 are the extremes of the open boundary section, ηnesting, and Hnesting are the surface elevation and the bathymetry of the nested model, while ηnested and Hnested are the surface elevation and bathymetry of the CMEMS global model and φ_n is the normal velocity component to the open boundaries. The total tangential velocity at the open boundaries is set to be equal between the CMEMS global model and the nested model.

The nested model temperature and salinity fields were nudged toward the CMEMS global model values in the right-hand side (r.h.s) of the prognostic fields following [55]. Indicating temperature or salinity with 'gamma' we add a relaxation term such as:

$$\frac{\partial \gamma}{\partial t} = \text{r.h.s}\ \frac{1}{\Gamma}\left(\gamma - \gamma^{\text{CMEMS}}\right) \tag{4}$$

Here Γ is the relaxation time that varies smoothly from 20 model grid internal points to the open boundaries with values of 240 s to 100 days. This nudging prevents a substantial model drift from the CMEMS global model values.

Finally, the tidal forcing is prescribed at the model boundaries by applying elevations and barotropic velocities for 10 major tide constituents (M_2, S_2, N_2, K_2, K_1, O_1, P_1, Q_1, M_f, M_m) which are obtained from the TPXO8 global inverse barotropic tide model [56]. Radiation conditions are specified at the boundaries for depth-averaged velocity after [53] and also for water level after [57]. Radiation conditions are prescribed for tracers and baroclinic momentum [55].

2.2. Surface Boundary Conditions

The atmospheric data for the computation of the surface forcing are taken from the global high resolution (0.125°) atmospheric model run by the European Centre for Medium Range Weather Forecasts (ECMWF) with 3-hourly temporal step. This has changed in 2019 to hourly fields with spatial resolution (0.1°). The atmospheric fields used are air temperature, relative humidity, wind velocity at 10 m above sea level, mean sea level pressure, cloud cover, total precipitation, and solar radiation (net surface and longwave radiation). In order to parametrize the air–sea interaction processes, the wind stress, heat fluxes, and evaporation rate are computed by means of interactive bulk formulae making use of atmospheric data and the model predicted sea surface temperature according to [58]. The NEA-ROMS heat fluxes are computed interactively according to [13] and described in [59] the heat flux boundary condition is given by

$$\rho_o C_P K_H \left.\frac{\partial T}{\partial z}\right|_{z=\eta} = Q_T + \frac{\partial Q}{\partial T}\ (SST_{CMEMS} - SST_m\Big) \tag{5}$$

where Q_T is the total net heat flux according to the formula:

$$Q_T = Q_S - Q_b - Q_e - Q_h \tag{6}$$

The last term in Equation (5) is the heat flux correction term. $\partial Q/\partial T$ is the heat flux correction. which is set at 80 $W{\cdot}m^{-2}{\cdot}{}^{\circ}C^{-1}$ for NEA-ROMS. SST_m is the model sea surface temperature and salinity and SST_{CMEMS} is the reference SST. Q_S, Q_e, Q_b, and Q_h are the short, latent, back, and sensible heat fluxes in W/m^2. (K_H) is the vertical mixing coefficients for tracers in ($m^2{\cdot}s^{-1}$). This heat flux correction, which was extracted from CMEMS, sea surface temperature, is to control the simulation from drifting away from the CMEMS values.

The momentum flux boundary condition for the surface takes the form:

$$\rho_o K_M \left.\frac{\partial(u,v)}{\partial z}\right|_{z=\eta} = (\tau_{wx}, \tau_{wy}) \tag{7}$$

In Equation (7), η is the free surface elevation in (m) and τ_{wx}, τ_{wy} are the wind stress components in Nm^{-2} calculated from the equation

$$\tau = C_D\ \rho_a\ |W|^*\ W \tag{8}$$

where ρ_a = 1.2 kg·m^{-3} is the density of air, C_D is the drag coefficient in Equation (8) computed according to [60]. While (W) is the wind speed component and (K_M) is the vertical mixing coefficient for momentum expressed in m^2·s^{-1}.

Daily averaged freshwater discharges were specified for a total of 29 rivers across Ireland, west Britain, and west France. The climatological mean river discharge values were obtained from various sources (Table 1), and the location of each river is shown in Figure 1. The NEA_ROMS rivers salinity was set to zero and the river temperatures were not prescribed. Discharge rates for the Irish rivers were obtained from the Office of Public Works (OPW) http://www.opw.ie/ and Electricity Supply Board (ESB) databases. The NEA_ROMS operational river discharge system is evolving. In 2019 our team managed to implement near real time flow rates for Corrib River obtained from OPW. The Corrib River is located inside Galway Bay on the west coast of Ireland. UK rivers data were downloaded from the Centre for Ecology and Hydrology website (https://www.ceh.ac.uk/our-science/projects/national-river-flow-archive). Data from the French rivers were provided from the national database "HYDRO", run by the French ministry of ecology and solidarity transition. The corresponding model salt flux boundary condition is given according to [61]:

$$K_H \frac{\partial S}{\partial z}\bigg|_{z=\eta} = SSS(E - P - (\frac{R}{A})) \quad (9)$$

where E is the evaporation, P the precipitation, R is the Rivers climatological daily discharge, A is mouth cross-section in m^2, and SSS in Equation (8) represents the sea surface model salinity.

Table 1. Mean annual climatological river discharge values (m^3/s) included in the NEA_ROMS.

Country	River Name	Mean Annual Discharge (m^3/s)	River Salinity
Ireland	Shannon_Maigue	195.57	0
	Erne	96.72	0
	Corrib	101.87	0
	Barrow-Nore-Suir_Anner	130.87	0
	Boyne	37.63	0
	Blackwater	55.72	0
	Moy	56.16	0
	Slaney	31.83	0
	Liffey	10.93	0
	Lee	27.03	0
	Feale_Galey	26.16	0
	Bandon	15.00	0
Great Britain	Severn_Wye	179.53	0
	Eden_Annan	81.34	0
	Ribble	32.96	0
	Tywi	39.33	0
	Dee	31.03	0
	Lune	35.36	0
	Tamar	22.39	0
	Teifi	28.60	0
	Taw_Torridge	33.74	0
	Derwent	26.01	0
	Leven	43.82	0
	Exe	16.02	0
France	Blavet	29.05	0
	Loire	873.49	0
	Seine	579.11	0
	Dordogne_Garonne	730.72	0
	Adour	44.82	0

2.3. Operational Setup

Figure 2 describes the schematic of the NEA_ROMS operational functioning during the simulation period. At the beginning of each year, a new NEA-ROMS run is initialized using interpolated fields from the CMEMS Ocean global model. Tide and river forcing files are created to provide forcing data for the full year ahead. The NEA_ROMS model is then run in hindcast/forecast mode for the year. This entails accessing the CMEMS Ocean weekly analysis run to provide boundary data for the NEA-ROMS hindcast run, while the latest CMEMS Ocean forecast data is used for our forecast boundary forcing. Similarly, the latest ECMWF forecasts are used for atmospheric forcing. The 7-day hindcast is initialized using a restart file from the end of the previous 7-day hindcast. The final output file of the hindcast is used to initialize the first forecast of the current week. A 3-day forecast is created every day using a restart file from the previous forecast to initialize the run. NEA_ROMS state variables (i.e., temperature, salinity, sea surface height, barotropic and baroclinic velocity fields) are saved to netCDF files in hourly snapshots. Selected points in the domain have data saved at the higher frequency of 10 min for use in model validation and also to provide boundary data for two high-resolution coastal models which are also part of the operational system. Model results are published on the Marine Institute website http://www.marine.ie/Home/site-area/data-services/marine-forecasts/marine-forecasts. A rolling month of model output is published to the Marine Institute THREDDS http://milas.marine.ie/thredds/catalog.html and ERDDAP https://erddap.marine.ie/erddap/index.html servers. Monthly mean fields of some parameters (e.g., surface and bottom temperature) are also published to these servers.

2.4. Observations Data and Methods of Analysis

Data used to validate the NEA-ROMS include Sea Surface Temperature (SST) from satellites, tide gauges, ARGO floats, and CTD salinity and temperature profiles.

The satellite SST product used is the European North West Shelf/Iberia Biscay Irish Seas High Resolution ODYSSEA L4 Sea Surface Temperature Analysis dataset which is available for download from the CMEMS website (data product, SST_ATL_SST_L4_NRT_OBSERVATIONS_010_025). The product consists of daily averaged SST values. The data valid for a particular day (from previous day 12:00 to current day 12:00) at a horizontal resolution of 0.02° × 0.02° for a domain covering the European North West Shelf and Iberia-Biscay-Irish Seas. It is a multi-sensor Level 4 analysis which aims to provide an estimate of the night time SST based on original SST observations http://resources.marine.copernicus.eu/documents/QUID/CMEMS-SST-QUID-010-025.pdf. The data from 1 June 2019 were replaced by ODYSSEA product, SST_NWS_SST_L4_NRT_OBSERVATIONS_010_003, https://cmems-resources.cls.fr/documents/PUM/CMEMS-OSI-PUM-010-003-010.pdf.

The authors validated NEA_ROMS Sea Surface Height (SSH) using observed time series from tide gauges around the Irish coast (Irish National Tide Gauge Network) https://data.gov.ie/dataset/irish-national-tide-gauge-network. The sampling frequency of the measured tide data ranged from 5 to 15 min while the model output is produced at a frequency of 10 min. Harmonic analysis was carried out on the observed and modelled time series using the T-TIDE software in MATLAB [62]. In addition, the surge component was calculated following [62], for observed and modelled data at the locations of Irish tide gauges, and the difference statistics are presented. According to [63], the surge (residual) component of sea level is defined as the total water level minus the tide.

$$\text{Sea level elevation} = \text{predicted tide level} + \text{storm surge height} \tag{10}$$

In this study, ARGO float profiles were used to validate the NEA_ROMS temperature and salinity fields. Drifting ARGO floats record vertical profiles of temperature and salinity in the upper 2000 m of the ocean approximately every 10 days and transmit the data via the ARGO satellite system to a number of data processing centers [64–67]. The data is then made freely available via public access site http://www.argo.ucsd.edu/. Irish Marine Institute (IMI) have used only ARGO data in delayed mode. For the corresponding modelled profile, the authors used the hourly snapshots model output file corresponding to the date the ARGO profile was acquired, and the grid point closest in location to the ARGO profile. The ARGO temperature and salinity were then interpolated onto the modelled vertical profile for that grid point. The authors chose the year 2019 as one that has the best ARGO tracks coverage for the south of the Porcupine Bank, the southern and northern entrance to the Rockall Trough area.

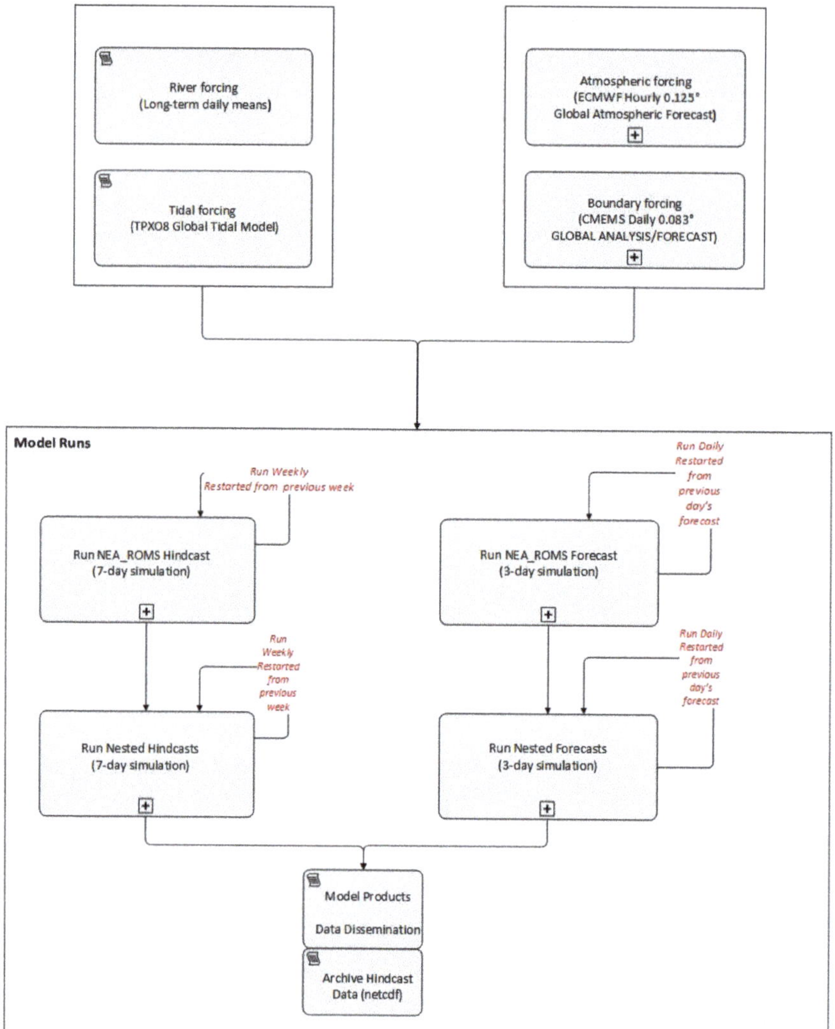

Figure 2. Schematic of the NEA_ROMS operational functioning during the simulation period.

The IMI maintains an archive of all Conductivity Temperature Depth (CTD) measurements taken from the research vessels operated by the Institute. During the period of January 2016 to December 2019, CTDs were acquired from the research vessels in the shelf seas around Ireland with the highest frequency in the months of spring and summer. The authors chose the year 2017 as one that provides the best Irish shelf waters coverage. These data are analyzed to assess the model performance in the shallow, coastal waters around Ireland.

In the study, the authors used the Geostrophic and Ekman Current Observatory GEKCO surface data to validate model velocity fields and Eddy Kinetic Energy (EKE), then compare it with CMEMS Ocean global model (EKE). The data are daily averages 0.25° spatial resolution obtained from Mean Absolute Dynamic Topography (MADT) at the gridded DUACS products. The DUACS system has been producing, as part of the CNES/SALP project, and the Copernicus Marine Environment and Monitoring Service (CMEMS), high quality multi-mission altimetry Sea Level products for oceanographic applications, climate forecasting centers, geophysics and biology communities http://www.aviso.altimetry.fr [68,69]. The geostrophic velocity anomalies presented in this study were deduced from MADT maps with geostrophic approximation as follows:

For NEA_ROMS domain, the mean velocity $\overline{U}_g, \overline{V}_g$ in the x and y directions and the anomalies U'_g , V'_g from the mean were computed

$$U'_g = -\frac{g}{f}\frac{\partial h'}{\partial y}, V'_g = \frac{g}{f}\frac{\partial h'}{\partial x} \tag{11}$$

where h' is the absolute dynamic topography anomaly in meters, g is the gravity acceleration in m/s^2 and $f = 2\Omega \sin \phi$, where $\Omega = 2\pi/T$ is the earth angular velocity in s^{-1}, T is the earth periodic time = 86,400 s (1 day in seconds), and ϕ is the latitude in degrees. The variance of these velocity anomalies was considered as Eddy Kinetic Energy (EKE), representative of mesoscale activity.

$$EKE = \frac{1}{2}\left(U'^2_g + {}'V'^2_g\right) \tag{12}$$

Model velocities were taken daily at the first sigma layer to compute EKE_{NEA_ROMS} to be exactly like EKE_{GEKCO} and EKE_{CMEMS} between 1 January 2017 and 31 December 2017. Velocity time series were subsampled at a daily period of GEKCO MADT as described in [70]. Space and time dependent EKE_{GEKCO} and EKE_{NEA_ROMS} fields were eventually averaged temporally to produce mean EKE fields.

3. Validation of NEA_ROMS Results against Observations

In this part, the authors will validate and discuss the NEA_ROMS results against available observations described previously in Section 2.4.

3.1. Comparison between NEA_ROMS and ODYSSEA SST during the Simulation Period

Figure 3a–c shows the temporal evolution of the daily mean SST basin averaged statistics for the NEA_ROMS Model-ODYSSEA satellite data for the whole model simulation period from 1 January 2016 until 31 December 2019 in °C. The daily SST bias between the NEA_ROMS and ODYSSEA satellite is shown in Figure 3a. The ODYSSEA SST product is based on night-time data, so for better validation, we rejected all NEA_ROMS daytime SST data. The highest daily bias mean was about 1.8 °C and recorded in mid-June 2018. While the lowest daily bias mean was −0.8 °C in mid-November 2019. The SST for the NEA_ROMS was greater than satellite during the summer months of June and July over the whole simulation period. This may be due to model error as a result of air–sea physics parametrizations during these months. This error has been discussed by [71–73]. The lowest bias differences between the model and satellite SST were found from mid-February till the end of April and during October over the whole simulation period (see Figure 3a). The root mean square error (RMSE) between the model and the satellite is described in Figure 3b. This figure suggests that the RMSE between our model SST and the satellite is around 0.5 °C for most of the simulation period. The maximum RMSE was found to be more than 1.5 °C during the summer season for all years except 2019. This may be attributed to problems mentioned above of the model air–sea physics parametrizations during the summer season. Regarding Figure 3c, the authors found that the NEA_ROMS SST is significantly correlated to satellite SST in all simulation period at 95% confidence limit. The highest correlation coefficient was more than 0.97 and recorded in March and April 2018. The minimum correlation coefficient was 0.85 and recorded in mid-June 2018 and December 2019. The annual bias maps between the model and satellite SST are made for the whole simulation period as shown in Figure 4a–d. The annual bias SST maps pattern shows a consistent improvement from the year 2016 until 2019. The maps suggest that minimum bias patterns found in 2019, while the maximum one recorded in 2016. During the simulation period, an effort was made to improve the temperature field by nudging of the surface temperatures towards a reference solution and nudging of the open ocean boundary temperatures and salinity in the relaxation zone towards a reference solution. Referring to Equations (4) and (5), we tuned best values for Γ the relaxation time and heat flux correction terms, $\partial Q/\partial T$ to improve our temperature and salinity fields. The NEA_ROMS model appears to overestimate SST by about 1 °C more than ODYSSEA off the European Northwest Shelf as shown in Figure 4a–d but this overestimation pattern has improved in 2019 (see Figure 4d). The authors observed a persistent positive SST bias of more than 1 °C in the Bay of Biscay and along the Iberian Coast except in 2019 as present in Figure 4d. The SST deficiencies in the Bay of Biscay could be due to the model spatial resolution or inaccuracies in the bathymetry data. This region features very deep waters and steep topography. Additionally, the north Atlantic is affected by large surface waves throughout the year, especially during winter time as mentioned by [74]. The surface wave is not included in our model and nonbreaking wave-induced mixing effect could be important in simulating SST as described in [75]. This may lead to significant cooling of the simulated SST [75]. In summary, the spatial distribution of the bias value clearly shows that there are improvements at the model boundary as well as over the whole domain which demonstrates that both changes in the model configuration are contributing to the improvement in model skill.

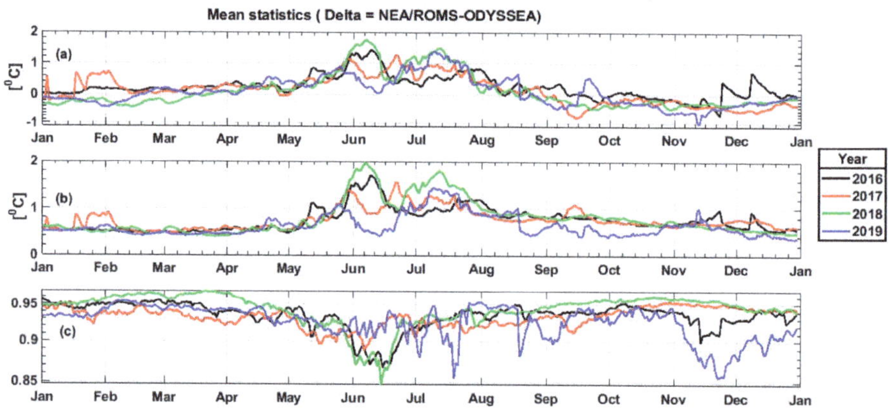

Figure 3. Temporal evolution of the mean Sea Surface Temperature (SST) basin averaged statistics for the (NEA_ROMS Model-ODYSSEA satellite data) in °C for the model simulation period from 1 January 2016 until 31 December 2019. (**a**) Bias, (**b**) root mean square error (RMSE), (**c**) correlation coefficient, the critical correlation R = 0.1 at 95% confidence limit. The continuous black, red, green, and blue lines present the years 2016, 2017, 2018, and 2019, respectively. Red, green, and blue lines present the years 2016, 2017, 2018, and 2019, respectively.

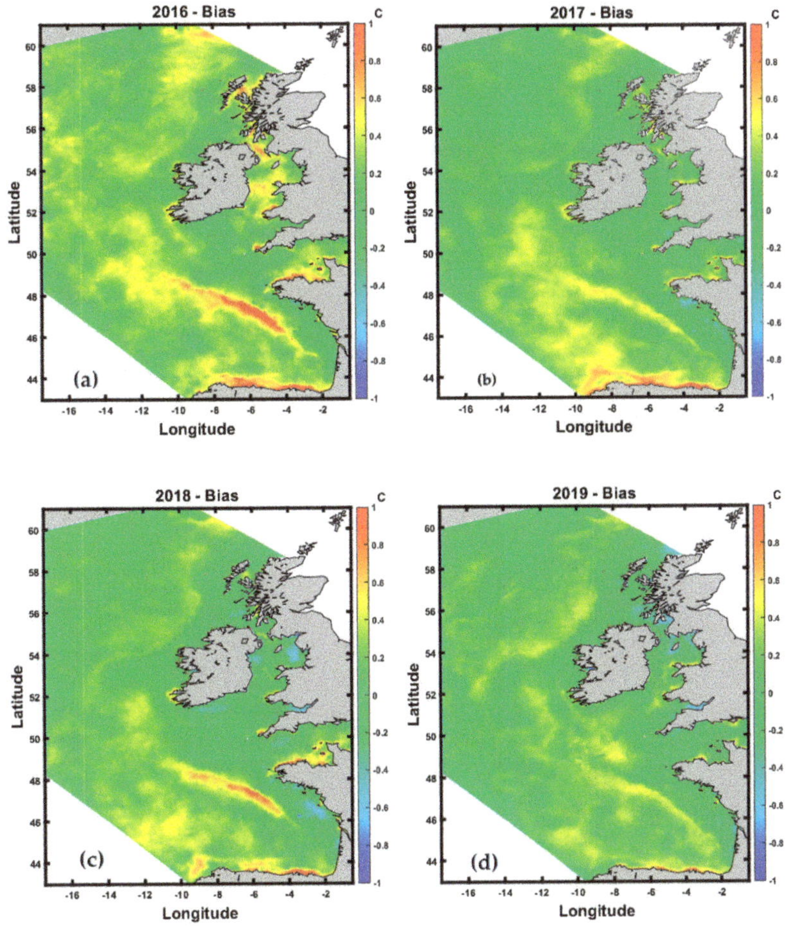

Figure 4. Bias maps between (NEA_ROMS Model-ODYSSEA) satellite data in °C for the model simulation period from 1 January 2016 until 31 December 2019. **(a)** 2016, **(b)** 2017, **(c)** 2018, **(d)** 2019.

3.2. Validation of NEA_ROMS Sea Surface Height against Tide Gauge Stations

Observed time series from tide gauges around the Irish coast are compared with model sea surface heights output at the same locations as shown in Figure 5, Tables 2 and 3. The tide gauge stations are distributed around the whole of the Republic of Ireland. Harmonic analysis was carried out on the eight locations using the T-TIDE software in MATLAB [62]. The aim of the analysis was to compare the modelled and measured values of the main tidal constituents (magnitude and phase angle). This analysis demonstrated that the tidal signal in the Sea Surface Height (SSH) data was dominated by three semi-diurnal constituents (M_2, S_2, N_2) and three diurnal constituents (K_1, O_1, and Q_1). Regarding our tidal analysis the authors found that M_2 and S_2 are responsible for most of the tide in our domain in agreement with [76–79]. The spatial variability in the mean error between model and observations for the two main constituents, M_2 and S_2, is shown in Figure 5a–d. The model shows good skills on the west coast of Ireland and also in the western Irish Sea. The maximum M_2 magnitude error for the west Irish coast was about 0.07 m for Malin Head, Aranmore, and Galway tide gauge stations. However, the model has a higher negative error of about 5% of the total M_2 amplitude in the Celtic Sea recorded in Dunmore East tide gauge station where the M_2 amplitude is consistently underestimated. The S_2 amplitude error shows a broadly similar pattern of M_2. The maximum S_2 error of about (-) 0.03 m (about 6% of the total S_2 amplitude) was recorded in the Celtic Sea at the Ballycotton tide gauge station. One explanation for the poorer skill in the Celtic Sea is an excessive dissipation of tidal energy over the broad expanse of shelf south of Ireland, leading to a reduction in the amplitude of the tidal wave in agreement with [80]. The phase angle errors for both components M_2 and S_2 are negative at most tide gauges stations except in Dunmore East station as shown in Figure 5c,d. The highest negative phase angle error for both components was found to be (-) 7° in the Celtic Sea and also recorded at the Howth gauge station. The aggregated results for the most significant tidal constituents are presented in Table 2. The magnitude difference was very small for all tidal constituents. Ratios of the modelled to the observed amplitudes of the constituents show that the model has good skill at N_2, O_1, and Q_1 though these constituents are much less significant to the total tide than the other three.

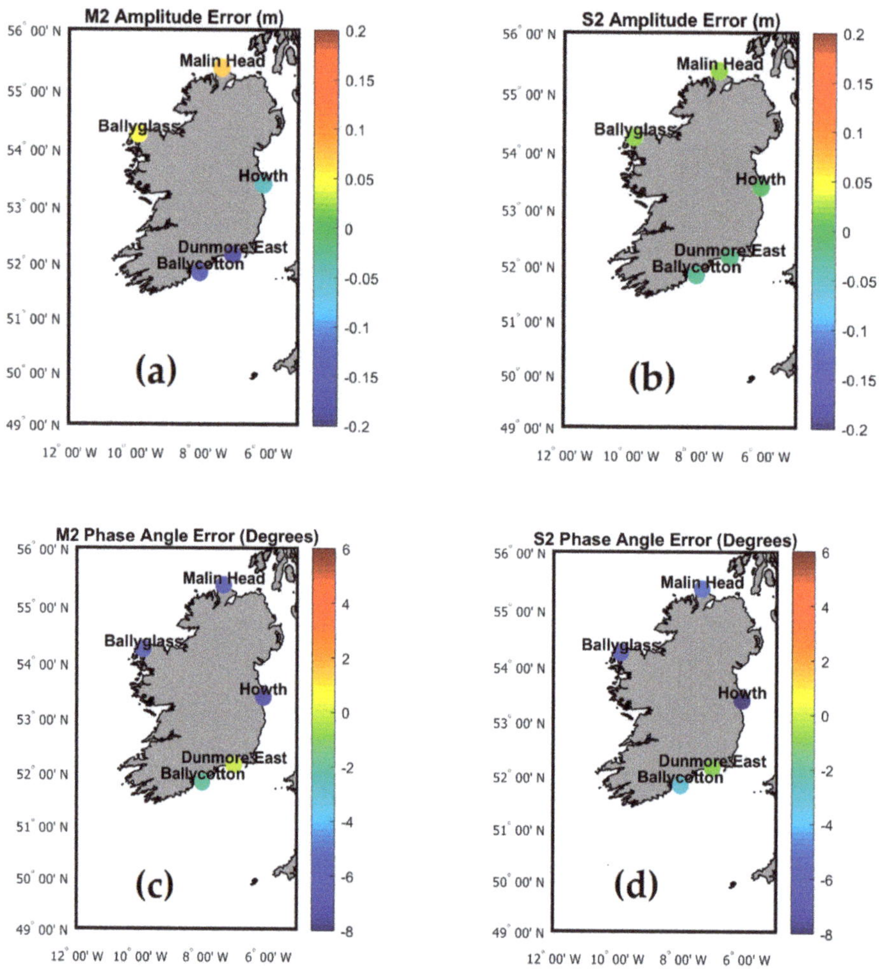

Figure 5. Error maps for NEA_ROMS minus observed amplitudes and phases for M_2 and S_2 tidal constituents at five Irish tide gauge locations, Malin Head, Ballyglass, Ballycotton, Dunmore East, and Howth. **(a)** M2 Amplitude Error (m) **(b)** S2 Amplitude Error (m) **(c)** M2 phase Angle Error (Degrees) **(d)** S2 Angle Error (Degrees).

Table 2. The amplitudes in meters and phases in degrees for six of the principal tidal constituents calculated, for the measured and modelled data.

Tidal Constituent		Gauge		NEA_ROMS	
		Amp	Phase	Amp	Phase
Aranmore	M2	1.22	161	1.29	156
	S2	0.44	194	0.46	189
	N2	0.25	139	0.26	135
	K1	0.12	154	0.14	142
	O1	0.08	359	0.08	5
	Q1	0.03	299	0.03	292
Ballycotton	M2	1.42	148	1.26	146
	S2	0.44	193	0.41	190
	N2	0.26	128	0.23	128
	K1	0.02	178	0.02	178
	O1	0.03	35	0.02	55
	Q1	0.01	340	0.01	348
Ballyglass	M2	1.16	158	1.21	152
	S2	0.41	191	0.43	185
	N2	0.23	136	0.25	131
	K1	0.14	122	0.16	116
	O1	0.09	337	0.08	342
	Q1	0.03	277	0.03	268
Castletownbere	M2	1.12	131	1.09	124
	S2	0.36	161	0.36	155
	N2	0.23	108	0.22	101
	K1	0.04	50	0.05	49
	O1	0.01	279	0.02	277
	Q1	<0.01	185	<0.01	184
Dunmore East	M2	1.38	150	1.20	150
	S2	0.45	199	0.43	198
	N2	0.25	133	0.22	136
	K1	0.04	178	0.04	177
	O1	0.04	29	0.03	49
	Q1	0.01	344	0.01	341
Galway	M2	1.57	141	1.64	137
	S2	0.55	172	0.58	170
	N2	0.32	119	0.34	117
	K1	0.09	76	0.10	79
	O1	0.06	311	0.05	311
	Q1	0.02	261	0.02	243
Howth	M2	1.44	325	1.39	318
	S2	0.41	357	0.41	349
	N2	0.28	297	0.28	289
	K1	0.10	198	0.11	187
	O1	0.08	37	0.07	54
	Q1	0.03	343	0.03	340
Malin Head	M2	1.12	178	1.19	172
	S2	0.42	206	0.44	201
	N2	0.23	156	0.26	151
	K1	0.09	167	0.11	154
	O1	0.07	7	0.07	18
	Q1	0.03	310	0.03	306

Accurate prediction of the surge is vital if a model is to be effective as part of a coastal flooding early warning system. The surge component was calculated following [63], for observed and modelled data at the locations of Irish tide gauges and the different statistics are presented in Table 3. Surge is caused by atmospheric pressure and wind. In order to obtain the surge signal we removed tides. The tidal signal in the Sea Surface Height (SSH) data was dominated by three semi-diurnal constituents (M_2, S_2, N_2,) and three diurnal constituents (K_1, O_1, and Q_1). The bias results were approximately zero for most gauge stations except in Ballyglass, Dunmore East, and Malin Head that were −0.01, 0.01, and 0.01 m respectively. The results show that the RMSE is around 0.15 m for all gauge stations. There are a number of possible contributors to this error, including the slight phase difference in the tidal signal and inaccuracies in the modelled weather fields both in space and time.

Table 3. Bias and RMSE of error for modelled minus observed at Irish tide gauge locations in meters.

Tide Gauge Station	NEA_ROMS Bias (m)	NEA_ROMS RMSE (m)
Aranmore	0	0.14
Ballycotton	0	0.16
Ballyglass	−0.01	0.14
Castletownbere	0	0.14
Dunmore East	0.01	0.17
Galway	0	0.15
Howth	0	0.15
Malin Head	0.01	0.15

3.3. Validation of NEA_ROMS with ARGO Floats Temperature and Salinity Profiles

The locations of ARGO float profiles in the NEA_ROMS domain are shown in Figure 6a–e. The authors used 304 quality checked ARGO profiles data for the year 2019. The bias between modelled and measured ARGO float temperature and salinity profiles are presented in Figure 6a,b. The bias results reveal an agreement between NEA_ROMS and ARGO temperature and salinity profiles for most of the locations. The maximum bias between the NEA_ROMS and Argo floats temperature was 0.8 °C and located at the western open boundary (see Figure 6a). The same locations recorded maximum salinity bias of around 0.15. This may be due to the fact that some of the Argo floats are located too close to the boundary where they are influenced by boundary nudging effects. Additionally, the absence of the wave effect in our model could be an important reason for poor representation of the salinity and temperature across the water column as described in [74]. Biases were used to calculate the RMSE as shown in Figure 6c,d. The maximum RMSE between the NEA_ROMS and Argo floats temperature was 1 °C at the same locations described above. The RMSE magnitude of the temperatures is warmer in the model than measured. While the maximum RMSE for the salinity was about 0.15 and recorded again at the northwestern open boundary edge as seen in Figure 6d. This can be attributed to the reason mentioned above. The NEA model correctly reproduces salinity/temperature profiles in the inner domain and high latitude areas. Figure 6e shows the T-S diagram for both model and ARGO float profiles combined with σ_θ contours, while, the water masses description for the Rockall Trough area according to [24,81] is shown in Figure 7. Figure 6e suggests that ARGO and model output broadly match. The model could describe the high saline dense Mediterranean Water (MW). This water mass is located to the south of the Porcupine Bank at 1000 m depth with a maximum salinity of about 35.8, temperature range from 8 to 10 °C, and a σ_θ of approximately 27.6 in agreement with [82]. In addition, the deep water mass of Labrador Sea Water (LSW) is well developed by the model and ARGO floats T-S diagram. This cold deep water mass is generated over the Labrador basin and slowly makes its way east across the Atlantic Ocean [83]. This water mass is found at 2000 m depth with a salinity of about 34.8, temperature 3 °C, and σ_θ of approximately 27.8 in agreement with [83–85] and shown by Figures 6e and 7. The water mass Subarctic Intermediate Water (SAIW) signal is presented by Model and ARGO floats T-S diagram. This water mass is located at the northern entrance to the Rockall Trough at 900 m depth with a maximum salinity of about 35.2 and σ_θ of approximately 27 as shown in Figures 6e and 7, and described by [86]. There was a problem in presenting the surface water mass Eastern North Atlantic Water ENAW at the model T-S diagram. The model does not properly recreate this signal maybe because it mixed with the underneath SAIW water mass. This may be as indirect effect of model excess vertical mixing due to the use of associated parameters with KPP vertical turbulent closure scheme as described in [46,47]. This parameters were used to tune the vertical profiles of currents, temperature, and salinity [46,47]. In conclusion, the comparison with ARGO floats profiles has shown a good capability of the model to reproduce the main structures of the field in the area.

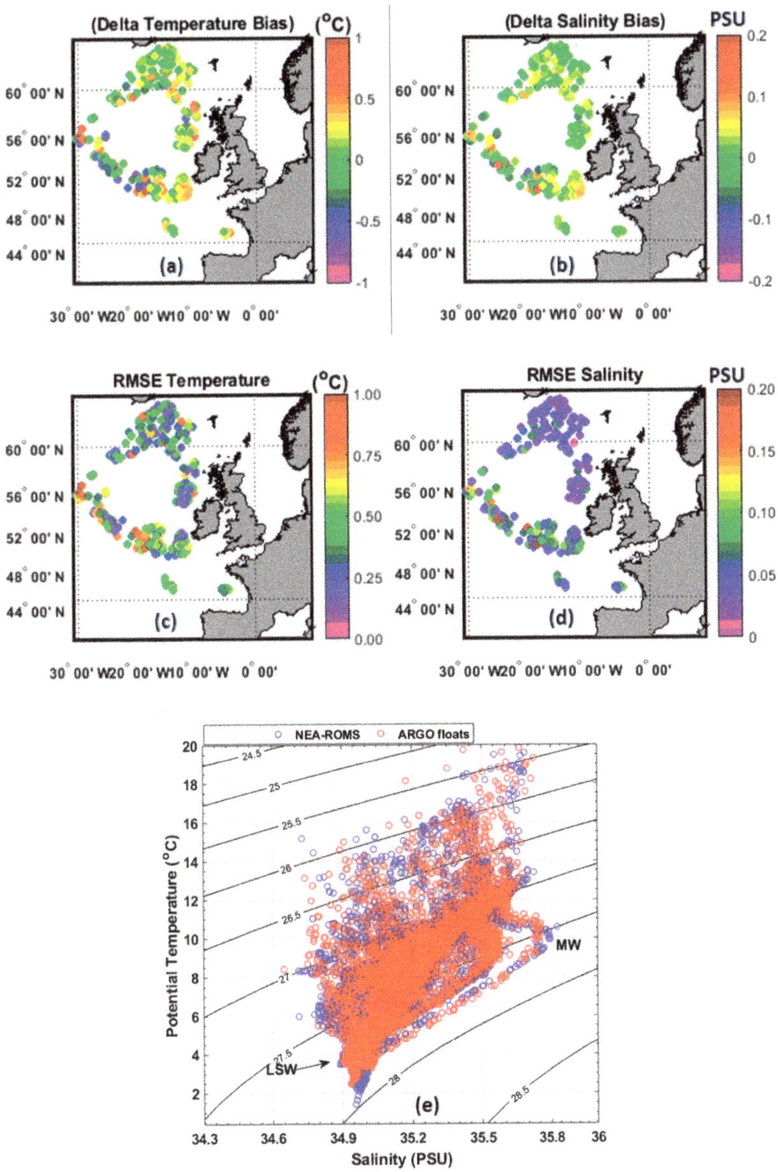

Figure 6. The bias and RMSE between NEA_ROMS and 304 ARGO float profiles locations in 2019. (**a**) Bias temperature, (**b**) bias salinity, (**c**) RMSE temperature, (**d**) RMSE salinity, (**e**) T-S diagram for NEA_ROMS and ARGO floats profiles.

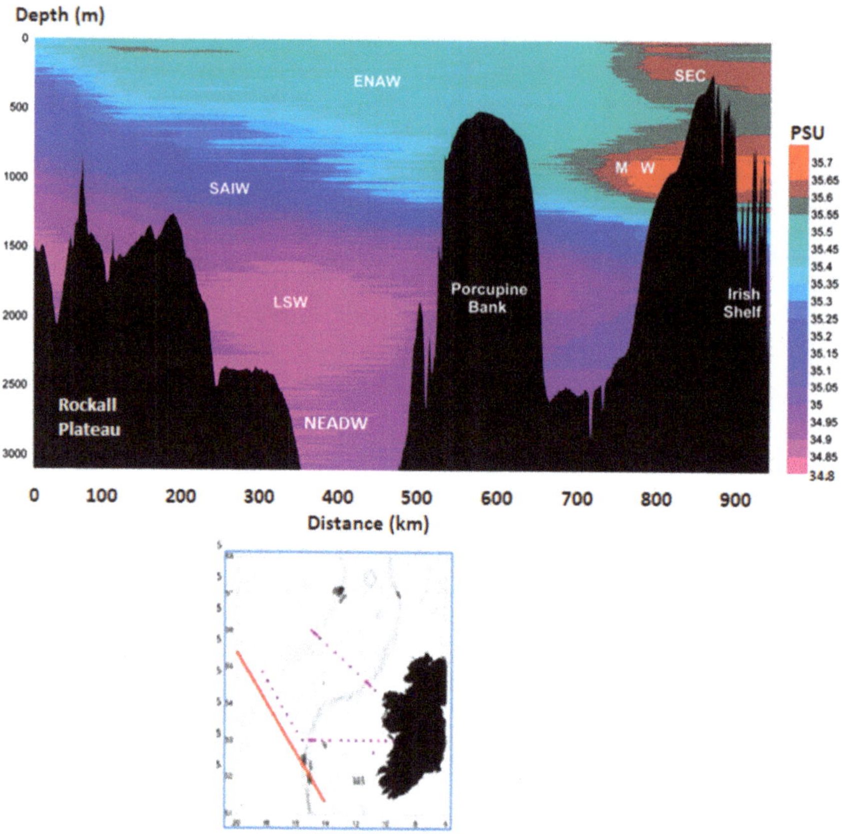

Figure 7. Schematic map of the main water masses across the southern Rockall Trough, adapted from [24,81] with permission of Tomasz Dabrowski & in 2019, where the transect is along the Conductivity Temperature Depth (CTD) points (the purple points are locations of repeat CTD profiles taken as part of an annual ocean climate cruise carried out by the Marine Institute), and the water masses in the plot are ENAW = Eastern North Atlantic Water; SAIW = Subarctic Intermediate Water; MW = Mediterranean Water; LSW = Labrador Sea Water; NEADW = Northeast Atlantic Deep Water; SEC = Shelf Edge Current. This Figure adapted from [24,81] with permission from Tomasz Dabrowski & McGrath in 2019.

3.4. Validation of NEA_ROMS with CTDs Temperature and Salinity Profiles

To measure the model performance in shelf waters IMI used Conductivity Temperature Depth (CTD) profiles. The method used to compare the model and CTD data is the same as that used for the ARGO data. The authors analyzed 223 CTD profiles From January 2017 to December 2017, CTDs were acquired from the IMI research vessels in the shelf seas around Ireland as shown in Figure 8a–d. This data was analyzed to assess the model performance in the shallow, coastal waters around Ireland. Figure 8a,b presents the bias between modelled and measured CTD temperature and salinity profiles. The temperature bias results propose a good agreement between NEA_ROMS and CTD profiles as shown in Figure 8a. The temperature bias results are almost zero for most locations except one profile situated to the west of the Irish coast. The temperature bias in this single location was found to be 0.9 °C. The salinity bias results suggest an overestimation by the NEA_ROMS of about 0.15 for most profiles located at the southern entrance to the Irish Sea (see Figure 8b). The reason behind this could be related to the lower accuracy of the salinity field for the CMEMS model in this area. The CMEMS salinity field has improved a lot since 2017 as described in [87]. This improvement would obviously have a big impact on our model salinity field. The maximum RMSE between the model and the CTD temperature profiles was found to be 0.8 °C while it was 0.17 for salinity as described in Figure 8c,d. The distribution patterns of the RMSE values clearly show the model robustness in presenting the main water masses at the Irish shelf waters.

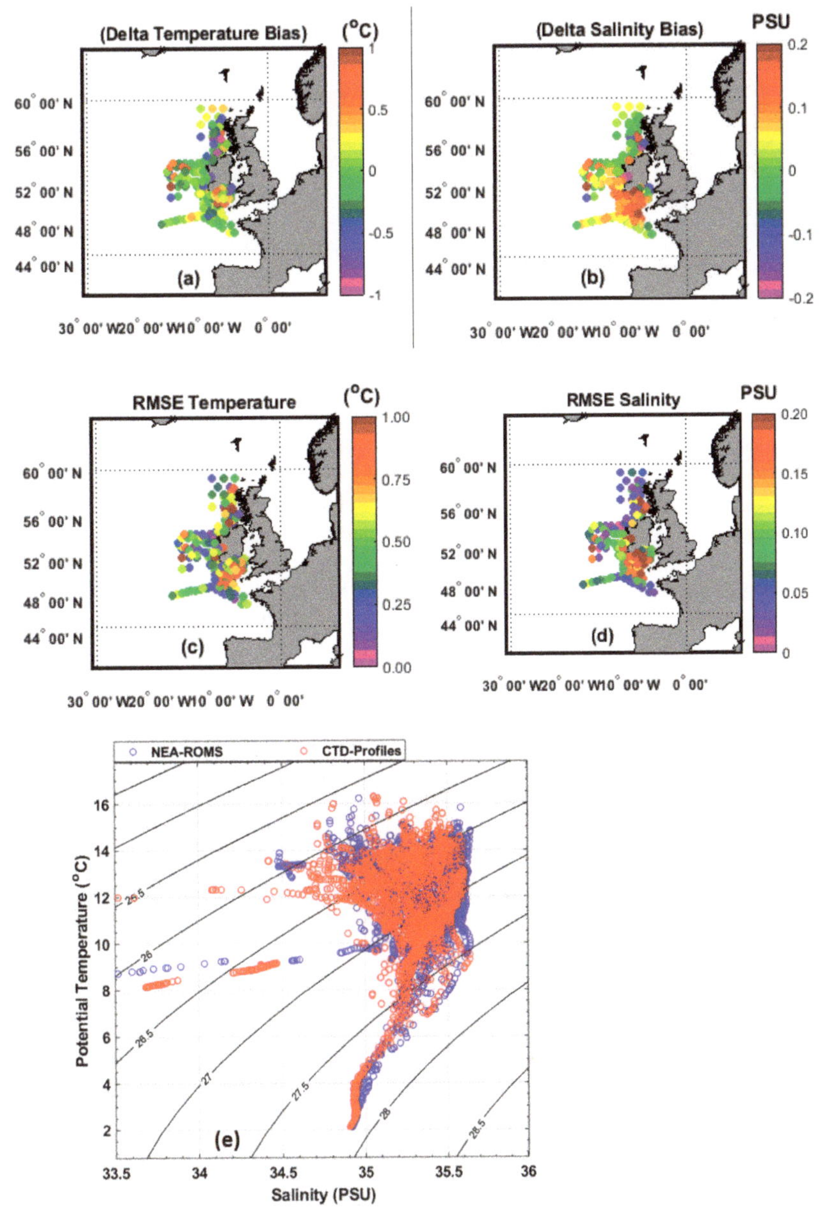

Figure 8. The bias and RMSE between NEA_ROMS and 223 CTD profiles locations in 2017. (**a**) Bias temperature, (**b**) bias salinity, (**c**) RMSE temperature, (**d**) RMSE salinity, (**e**) T-S diagram.

The T-S diagram for NEA_ROMS and 223 CTDs profiles describes the Irish coastal waters around Ireland as shown in Figure 8e. The comparison between the CTDs T-S diagram and the model T-S shows a very good agreement. We observed a significant resemblance in the T-S diagram patterns between the CTD data and the model in the whole water column. The NEA_ROMS T-S diagram represents deep water masses well, while it overestimates salinities in some surface shallow coastal waters with maximum centered salinity of about 35.5 and σ_θ of approximately 26.5 as shown in Figure 8b. The reason for the overestimation model salinity is probably a significant underestimation of the discharge of fresh water into the coastal waters of Ireland and west Britain. Another possibility, as previously mentioned, may be due to the use of KPP vertical turbulent closure scheme inside the ROMS model which can affect the vertical mixing as described in [46,47]. This issue is a potential source of deficiencies between model output and observations.

In conclusion, the NEA model correctly reproduces the shelf water masses around Ireland.

3.5. *Validation of NEA_ROMS with GEKCO Surface Data*

Mean Kinetic Energy (MKE) and EKE calculated from geostrophic current from altimetry GEKCO data and surface current from NEA and CMEMS global models are shown in Figure 9 which was previously described in Section 2.4. The authors combined surface current with MKE maps in Figure 9a–c to validate NEA_ROMS current pattern. The GEKCO data and both models are characterized by the North Atlantic Current and European Slope Current which transports heat and salt from the north-east Atlantic, interacts with the continental shelf slope and forms branches that flow into the North Sea in agreement with [23,35]. Both NEA_ROMS and CMEMS Global models produce similar mean circulation to the GEKCO currents. NEA_ROMS has well-resolved meanders and eddies, which are not well-resolved by CMEMS global (see Figure 9b). For example, the eddies and meanders associated with the North Atlantic Current at the middle of NEA_ROMS domain seem to be realistic as shown before by [32,35,88]. The NEA_ROMS has a higher resolution and more detailed bathymetry in this area than the CMEMS model, and this is probably the reason that it is better able to resolve the eddy field in the area. Figure 9a–c suggests that CMEMS global model generally has the largest MKE field, except for the North Ireland region. However, CMEMS global assimilates observations such as altimetric sea level, temperature and salinity profiles, and corrects toward satellite SST at the surface, as described in [87], so that MKE could be larger at times due to the correction of physics uncertainties [73,89]. Figure 9e–f represents the mean EKE fields for GEKCO data, NEA_ROMS, and CMEMS global, respectively. The basin averaged EKE for GEKCO data, NEA_ROMS, and CMEMS were 0.0063, 0.0153, and 0.0147 $m^2 s^{-2}$, respectively. The EKE comparison suggests the NEA_ROMS has the highest EKE field. The finer model resolution and the better river discharges of NEA_ROMS may produce this change according to [73,90]. Overall, NEA_ROMS reproduces the main regional circulation patterns, improving the dynamics forced by finer resolution and more detailed bathymetry in this area.

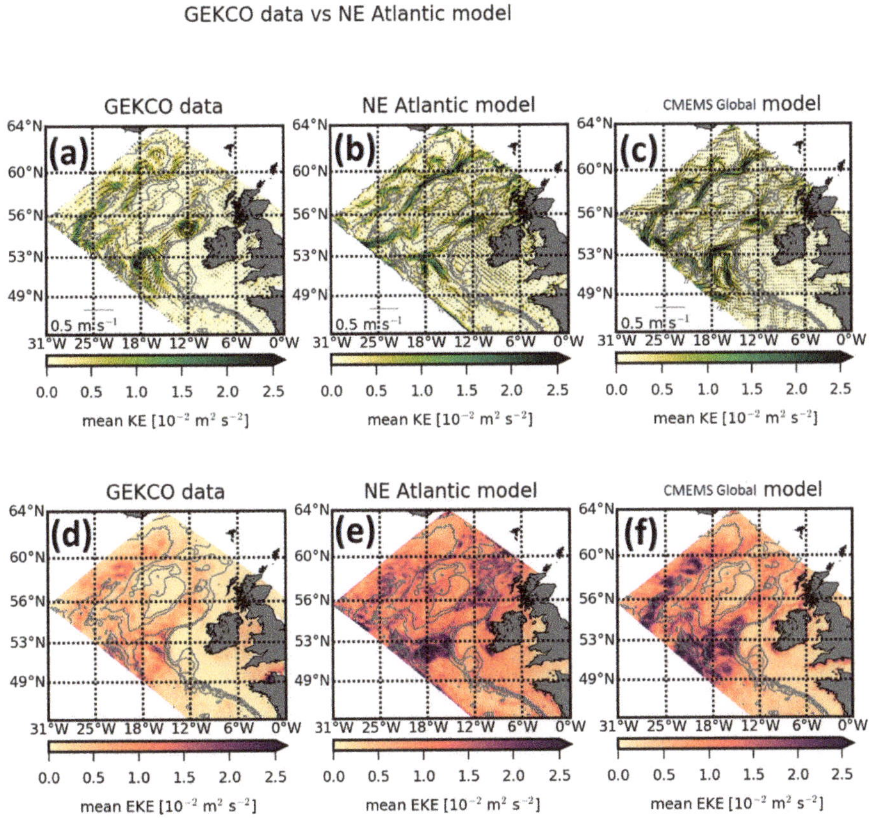

Figure 9. Mean Kinetic Energy fields combined with surface current from (**a**) GEKCO data, (**b**) NEA_ROMS model, (**c**) CMEMS global model, and the mean EKE from (**d**) GEKCO data, (**e**) NEA_ROMS model, and (**f**) CMEMS global model.

4. The Irish Coast Current (ICC)

Whilst it is fundamental that a hydrodynamic model is quantitatively validated, it is also necessary that it reproduces the important hydrographical features within its domain. To that end, an assessment is made of the model's ability to reproduce the summer Irish Coastal Current (ICC) [10,11,91]. The ICC flows northward parallel to the western coast of Ireland [92]. The flow of ICC depends on the local wind forcing, so the Atlantic Water then either enters the Irish Sea through the North Channel or, more likely, continues flowing northward producing the ICC until the Hebrides [92]. Additionally, the ICC shows a shelf-located meandering surface current of 10–20 cm·s^{-1} which, has a full depth presence over the 100–120 m contours [93,94]. The authors followed the methodology of [11] to produce the annual maps of the summer ICC simply by averaging two months (July and August) of the mid-depth residual NEA_ROMS velocity field for the Celtic Sea and Irish Shelf. We used the model output hourly snapshots velocity field. To produce annual maps, we get velocities away from ICC region out of the picture.

Figure 10 displays the annual results for this analysis over the whole simulation period from 2016 until 2019. The velocity of the current is found to be in the range of 6–20 cm s^{-1} which is in agreement with reported values in the published literature [10,11]. The localized "hotspots" of increased velocities near headlands on the western coast are presumably induced by the tidal current while the weakening of the current north of the Shannon may be indicative of tidal mixing being more dominant than stratification [11,93]. The map patterns suggest a well-developed ICC for all years except for 2016. The highest velocity for the ICC was about 19 cm s^{-1} and recorded in 2017, 2018, and 2019 as shown in Figure 10b–d. There is an interesting difference in 2019 where the ICC is very strong and well defined along most of the western Irish coast (see Figure 10d). In conclusion, the results show that the NEA_ROMS has reproduced the ICC front for the whole simulation period.

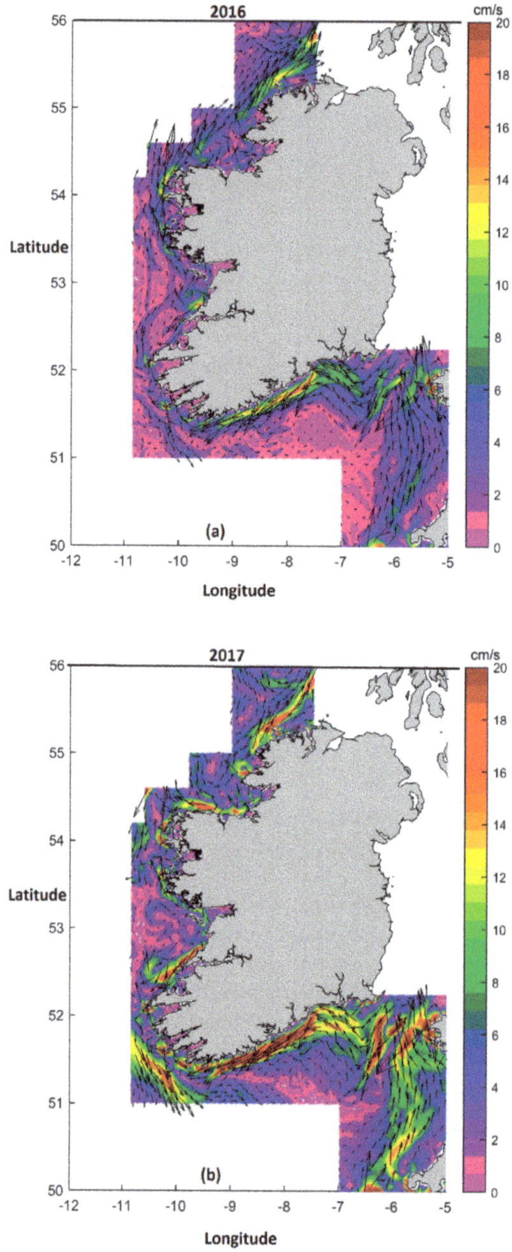

Figure 10. *Cont.*

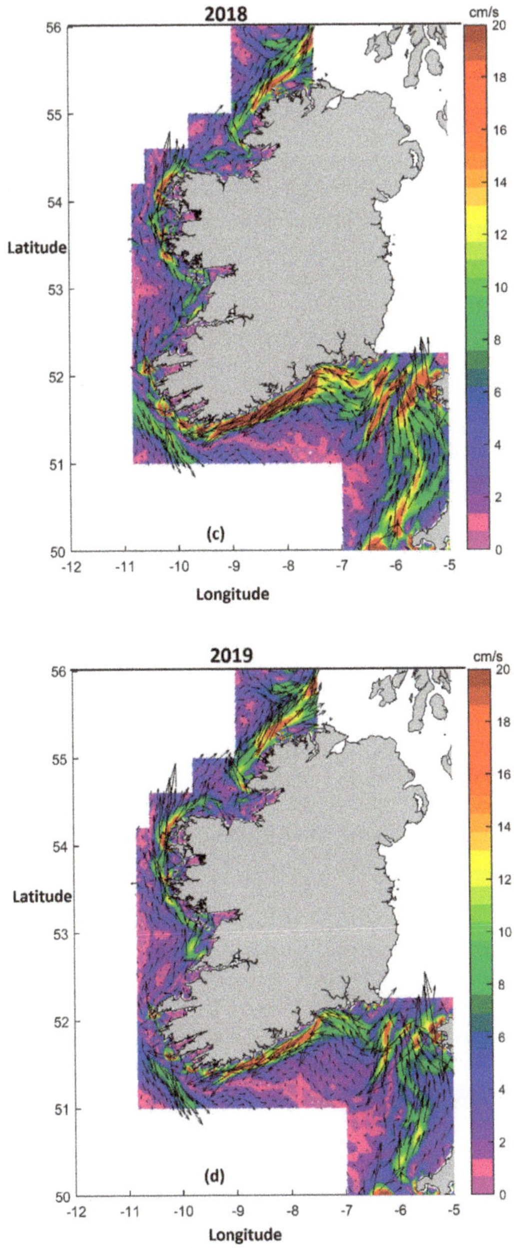

Figure 10. Magnitude current maps for NEA_ROMS Model describing the Irish Coastal Current in [cm/sec] for the model simulation period from 2016 until 2019 where; (**a**) 2016 (**b**) 2017 (**c**) 2018 (**d**) 2019.

5. Conclusions and Further Discussion

This paper presents an operational model for Irish waters NEA_ROMS. The model is based on ROMS and is forced by operationally available atmospheric and boundary data. The system is robust following several years in operation and model data is made freely available via THREDDS server.

The authors validated and calibrated the NEA_ROMS with available observations in our domain area. The observations were the SST obtained from satellite covering the simulation period, time series of eight tide gauges around the Irish coast obtained from the Irish National Tide Gauge Network, Argo, and CTD temperature and salinity profiles.

The SST for the NEA_ROMS bias was greater than the satellite during the summer months of June and July over the whole simulation period. The highest daily bias mean was about 1.8 °C and recorded in mid-June 2018. While the lowest daily bias mean was −0.8 °C in mid-November 2019. The authors attribute this to the model error as a result of air–sea physics parametrizations as discussed by [71–73]. The highest correlation coefficient was found to be more than 0.97 and recorded in March and April 2018. While, the minimum correlation coefficient was 0.85 and recorded in mid-June 2018 and December 2019.

The model has a higher negative error of about 5% of the total M_2 amplitude in the Celtic Sea recorded in Dunmore East tide gauge station. The S_2 amplitude error showed a broadly similar pattern to M_2. The maximum S_2 error was about 6% of the total S_2 amplitude recorded in the Celtic Sea at the Ballycotton tide gauge station. That was due to the poorer skill in the Celtic Sea due to excessive dissipation of tidal energy over the broad expanse of shelf south of Ireland, leading to a reduction in the amplitude of the tidal wave in agreement with [80].

The validation of NEA_ROMS with ARGO floats revealed the model skills in representing the main water masses in particular MW and LSW for the area located to the south of the Porcupine Bank and the entrance to the Rockall Trough. The location of some Argo floats too close to the boundary has increased the temperature and salinity RMSE because they may have been influenced by model boundary nudging effects.

The maximum RMSE between the model and the CTD temperature profiles was found to be 0.8 °C while it was 0.17 for salinity. The NEA model correctly reproduces the shelf water masses around Ireland.

The use of KPP vertical turbulent closure scheme in the ROMS model can affect the vertical mixing of the salinity and temperature profiles as described in [46,47]. This issue was a potential source of deficiencies between model predictions and observations.

The meanders and eddies at the model domain were well resolved by NEA_ROMS which were not well resolved by CMEMS global. The NEA_ROMS has a higher resolution and more detailed bathymetry in this area, which may be the reasons why the model resolved these meanders and eddies well.

In 2019 the ICC was very strong and well defined along most of the western Irish coast. In conclusion, the results showed that the NEA_ROMS well reproduced the ICC front for the whole simulation period.

Future work requires the use of high frequency near real time river discharges. The inclusion of data assimilation in the NEA_ROMS operational system is considered to be the next logical step in its ongoing development. Recent applications also include the modelling studies of the biogeochemical cycling. Additionally, indices supporting the implementation of the Marine Strategy Framework Directive (MSFD) are usually obtained from in-situ data. This is a major difficulty in oceanic areas where data are scarce. Validated numerical models can fill this gap. Tools will be developed to obtain indices of interest directly from model results, e.g., areas of upwelling, fronts, eddy index, primary production, trophic status, or even conditions for propagation of noise.

Author Contributions: Conceptualization, H.N., K.L., G.N., T.D. and M.C.; methodology, H.N. and K.L.; The developer of the first version of the North East Atlantic model, M.C.; validation, H.N., K.L., G.N. and T.D.; formal analysis, H.N., K.L., G.N. and T.D., investigation, H.N., K.L., G.N. and T.D.; resources, H.N., K.L., G.N., T.D. and M.C.; data curation, H.N., K.L., G.N., T.D. and M.C.; writing—original draft preparation, H.N. K.L., G.N., T.D. and M.C.; writing—review and editing, K.L., G.N., T.D. and M.C.; visualization, H.N., K.L., G.N., and T.D.; supervision, G.N. and T.D.; project administration, G.N. and T.D.; funding acquisition, INTERREG Atlantic Area Cross-border Cooperation Programme project "Innovation in the Framework of the Atlantic Deep Ocean" (iFADO, under contract EAPA 165/2016. All authors have read and agreed to the published version of the manuscript.

Funding: The validation of the model was funded by the INTERREG Atlantic Area Cross-border Cooperation Programme project "Innovation in the Framework of the Atlantic Deep Ocean" (iFADO, under contract EAPA 165/2016).

Acknowledgments: The authors are grateful to INTERREG Atlantic Area Cross-border Cooperation Programme project "Innovation in the Framework of the Atlantic Deep Ocean" (iFADO, under contract EAPA 165/2016) for supporting this study. We wish to thank Maximo Garcia-Jove Navarro for processing GEKCO and CMEMS data and developing Figure 9. Finally, we would like to thank Achref Othmani for preparing Table 1 the mean annual climatological river discharge values that were included in the NEA_ROMS. Special thanks to the two anonymous reviewers for helpful comments on the manuscript.

Conflicts of Interest: The authors declare no conflict of interest.

References

1. McCartney, M.S.; Mauritzen, C.M. On the origin of the warm inflow to the Nordic Seas. *Prog. Oceanogr.* **2001**, *51*, 125–214. [CrossRef]
2. Holliday, N.P.; Pollard, R.T.; Read, J.F.; Leach, H. Water mass properties and fluxes in the Rockall Trough, 1975–1998. *Deep Sea Res.* **2000**, *47*, 1303–1332. [CrossRef]
3. Holliday, N.P.; Bacon, S.; Cunningham, S.A.; Gary, S.F.; Karstensen, J.; King, B.A.; Li, F.; Mcdonagh, E.L. Subpolar North Atlantic Overturning and Gyre-Scale Circulation in the summers of 2014 and 2016. *J. Geophys. Res. Ocean.* **2018**, *123*, 4538–4559. [CrossRef]
4. Pingree, R.D.; Le Cann, B. 1989. Celtic and Armorican slope and shelf residual currents. *Prog. Oceanogr.* **1989**, *23*, 303–338. [CrossRef]
5. Turrell, W.R.; Jorge da Silva, A.; Mohn, C.; Griffiths, C.; Lavin, A.; Svendsen, E.; White, M. *A Quasi-Synoptic Survey of the European Continental Shelf Edge during the SEFOS Project*; ICES CM 1995, 1995/C:10; Hydrography Committee: Copenhagen, Denmark, 1995.
6. Pingree, R.D.; Sinha, B.; New, A.L.; Waddington, I.; Head, R.N.; Nechvolodov, L.V. Will deep subtropical ring 'Storm Physalia' cross the Mid Atlantic Ridge and reach America? *Mar. Biol. Assoc. UK* **1996**, *76*, 553–567. [CrossRef]
7. Huthnance, J.M.; Holt, J.T.; Wakelin, S.L. Deep ocean exchange with west-European shelf seas. *Ocean Sci.* **2009**, *5*, 621–634. [CrossRef]
8. Huang, W.G.; Cracknell, A.P.; Vaughan, R.A.; Davies, P.A. A satellite and field view of the Irish Shelf Front. *Cont. Shelf Res.* **1991**, *11*, 543–562. [CrossRef]
9. McCarthy, T.K.; Frankiewicz, P.; Cullen, P.; Blaszkowski, M.; O'Connor, W.; Doherty, D. Long term effects of hydropower installations and associated river regulation on River Shannon eel populations: Mitigation and management. *Hydrobiologia* **2008**, *609*, 109–124. [CrossRef]
10. Brown, J.; Carillo, L.; Fernand, L.; Horsburgh, K.J.; Hill, A.E.; Young, E.F. Observations of the physical structure and seasonal jet-like circulation of the Celtic Sea and St. George's Channel of the Irish Sea. *Cont. Shelf Res.* **2003**, *23*, 533–561. [CrossRef]
11. Fernand, L.; Nolan, G.D.; Raine, R.; Chambers, C.E.; Dye, S.R.; White, M.; Brown, J. The Irish coastal current: A seasonal jet-like circulation. *Cont. Shelf Res.* **2006**, *26*, 1775–1793. [CrossRef]
12. Hill, A.E.; Simpson, J.H. Low frequency variability of the Scottish Coastal Current induced by alongshore pressure gradients. *Estuar. Coast. Shelf Sci.* **1988**, *27*, 163–180. [CrossRef]
13. Pinardi, N.; Allen, I.; Demirov, E.; De Mey, P.; Korres, G.; Lascratos, A.; Le Traon, P.Y.; Maillard, C.; Manzella, G.; Tziavos, C. Mediterranean Forecasting System: First Phase of Implementation (1998–2001). *J. Ann. Geophys.* **2003**, *21*, 1–21. [CrossRef]
14. Holt, M.; Li, Z.; Osborne, J. Real time forecast modeling for the NW European Shelf Seas. In *Building the European Capacity in Operational Oceanography*; Elsevier Oceanography Series; Dahlin, H., Flemming, N.C., Nittis, K., Petersson, S.E., Eds.; Elsevier Science: Amsterdam, The Netherlands, 2003; Volume 69, pp. 484–489.
15. Cure, M.; Lyons, K.; Nolan, G. Operational Forecasting in the IBIROOS Region. In Proceedings of the Adjoint Modeling and Applications, La Jolla, CA, USA, 24–26 October 2005.
16. Elliott, A.; Hartnett, M.; O'Riain, G.; Dollard, B. *The PRISM Project: Predictive Irish Sea Models*; Final Report; Catchment to Coast Research Centre University of Wales Aberystwyth and Bangor Ceredigion: Cardiff, UK, 2007.

17. Siddorn, J.R.; Allen, J.I.; Blackford, J.C.; Gilbert, F.J.; Holt, J.T.; Holt, M.W.; Osborne, J.P.; Proctor, R.; Mills, D.K. Modelling the hydrodynamics and ecosystem of the North-West European continental shelf for operational oceanography. *Mar. Syst.* **2007**, *65*, 417–429. [CrossRef]
18. Kordzadze, A.A.; Demetrashvili, D.I. Operational forecast of hydrophysical fields in the Georgian Black Sea coastal zone within the ECOOP. *Ocean Sci.* **2011**, *7*, 793–803. [CrossRef]
19. Korotaev, G.K.; Oguz, T.; Dorofeyev, V.L.; Demyshev, S.G.; Kubryakov, A.I.; Ratner, Y.B. Development of Black Sea nowcasting and forecasting system. *Ocean Sci.* **2011**, *7*, 629–649. [CrossRef]
20. Stanev, E.V.; Schulz-Stellenfleth, J.; Staneva, J.; Grayek, S.; Seemann, J.; Petersen, W. Coastal observing and forecasting system for the German Bight estimates of hydrophysical states. *Ocean Sci.* **2011**, *7*, 569–583. [CrossRef]
21. Zhuang, S.Y.; Fu, W.W.; She, J. A pre-operational three dimensional variational data assimilation system in the North/Baltic Sea. *Ocean Sci.* **2011**, *7*, 771–781. [CrossRef]
22. Mateus, M.; Riflet, G.; Chambel, P.; Fernandes, L.; Fernandes, R.; Juliano, M.; Campuzano, F.; de Pablo, H.; Neves, R. An operational model for the West Iberian coast: Products and services. *Ocean Sci.* **2012**, *8*, 713–732. [CrossRef]
23. Siddorn, J.R.; Furner, R. An analytical stretching function that combines the best attributes of geopotential and terrain-following vertical coordinates. *Ocean Model.* **2013**, *66*, 1–13. [CrossRef]
24. Dabrowski, T.; Lyons, K.; Berry, A.; Cusack, C.; Nolan, G. An operational biogeochemical model of the North-East Atlantic: Model description and skill assessment. *J. Mar. Syst.* **2014**, *129*, 350–367. [CrossRef]
25. Danabasoglu, G.; Yeager, S.G.; Bailey, D.; Behrens, E.; Bentsen, M.; Bi, D.; Biastoch, A.; Böning, C.; Bozec, A.; Canuto, V.M.; et al. North Atlantic simulations in Coordinated Ocean-ice Reference Experiments phase II (CORE-II). Part I: Mean states. *Ocean Model.* **2014**, *73*, 76–107. [CrossRef]
26. Storkey, D.; Blaker, A.T.; Mathiot, P.; Megann, A.; Aksenov, Y.; Blockley, E.W.; Calvert, D.; Graham, T.; Hewitt, H.T.; Hyder, P.; et al. UK Global Ocean GO6 and GO7: A traceable hierarchy of model resolutions. *Geosci. Model Dev.* **2018**, *11*, 3187–3213. [CrossRef]
27. Gutknecht, E.; Reffray, G.; Mignot, A.; Dabrowski, T.; Sotillo, M.G. Modelling the marine ecosystem of Iberia–Biscay–Ireland (IBI) European waters for CMEMS operational applications. *Ocean Sci.* **2019**, *15*, 1489–1516. [CrossRef]
28. Fennel, K.; Wilkin, J.; Levin, L.; Moisan, J.; O'Reilly, J.; Haidvogel, D. Nitrogen cycling in the Middle Atlantic Bight: Results from a three-dimensional model and implications for the North Atlantic nitrogen budget. *Glob. Biogeochem. Cycles* **2006**, *20*, GB3007. [CrossRef]
29. O'Dea, E.J.; Arnold, A.K.; Edwards, K.P.; Furner, R.; Hyder, P.; Martin, M.J.; Siddorn, J.R.; Storkey, D.; While, J.; Holt, J.T.; et al. An operational ocean forecast system incorporating NEMO and SST data assimilation for the tidally driven European North-West shelf. *Oper. Oceanogr.* **2012**, *5*, 3–17. [CrossRef]
30. Sotillo, M.G.; Cailleau, S.; Lorente, P.; Levier, B.; Aznar, R.; Reffray, G.; Amo-Baladrón, A.; Chanut, J.; Benkiran, M.; Alvarez-Fanjul, E. The My Ocean IBI Ocean Forecast and Reanalysis Systems: Operational products and roadmap to the future Copernicus Service. *Oper. Oceanogr.* **2015**, *8*, 63–79. [CrossRef]
31. Guihou, K.; Polton, J.; Wakelin, S.; O'Dea, E.; Holt, J. Kilometric Scale modeling of the North West European Shelf Seas: Exploring the spatial and temporal variability of internal tides. *Geophys. Res.* **2017**, *123*, 688–707. [CrossRef]
32. Graham, J.A.; O'Dea, E.; Holt, J.; Polton, J.; Hewitt, H.T.; Furner, R.; Guihou, K.; Brereton, A.; Arnold, A.; Wakelin, S.; et al. AMM15: A new high-resolution NEMO configuration for operational simulation of the European north-west shelf. *Geosci. Model Dev.* **2018**, *11*, 681–696. [CrossRef]
33. King, R.; While, J.; Martin, M.J.; Lea, D.J.; Lemieux-Dudon, B.; Waters, J.; O'Dea, E. Improving the initialisation of the Met Office operational shelf-seas model. *Ocean Model.* **2018**, *130*, 1–14. [CrossRef]
34. Tinker, J.; Krijnen, J.; Wood, R.; Barciela, R.; Dye, S.R. What are the prospects for seasonal prediction of the marine environment of the North-west European Shelf? *Ocean Sci.* **2018**, *14*, 887–909. [CrossRef]
35. Tonani, M.; Sykes, P.; King, R.R.; McConnell, N.; Péquignet, A.C.; O'Dea, E.; Graham, J.A.; Polton, J.; Siddorn, J. The impact of a new high-resolution ocean model on the Met Office North-West European Shelf forecasting system. *Ocean Sci.* **2019**, *15*, 1133–1158. [CrossRef]
36. Ghantous, M.; Ayoub, N.; De Mey-Frémaux, P.; Vervatis, V.; Marsaleix, P. Ensemble Downscaling of a Regional Ocean Model. *Ocean Model.* **2020**, *145*, 101511. [CrossRef]

37. Haidvogel, D.B.; Arango, H.; Budgell, W.P.; Cornuelle, B.D.; Curchitser, E.; Di Lorenzo, E.; Fennel, K.; Geyer, W.R.; Hermann, A.J.; Lanerolle, L.; et al. Ocean forecasting in terrain-following coordinates: Formulation and skill assessment of the Regional Ocean Modeling System. *Comput. Phys.* **2008**, *227*, 3595–3624. [CrossRef]
38. Chiggiato, J.; Oddo, P. Operational ocean models in the Adriatic Sea: A skill assessment. *Ocean Sci.* **2008**, *4*, 61–71. [CrossRef]
39. Chao, Y.; Li, Z.; Farrara, J.; McWilliams, J.C.; Bellingham, J.; Capet, X.; Chavez, F.; Choi, J.K.; Davis, R.; Doyle, J.; et al. Development, implementation and evaluation of a data-assimilative ocean forecasting system off the central California coast. *Deep Sea Res. II* **2009**, *56*, 100–126. [CrossRef]
40. Wilkin, J.; Zhang, W.G.; Cahill, B.; Chant, R.C. Integrating coastal models and observations for studies of ocean dynamics, observing systems and forecasting. In *Operational Oceanography in the 21st Century*; Schiller, A., Brassington, G., Eds.; Springer: Dordrecht, The Netherlands, 2011; pp. 487–512. [CrossRef]
41. Shchepetkin, A.F.; McWilliams, J.C. The Regional Ocean Modeling System: A split-explicit, free-surface, topography following coordinates ocean model. *Ocean Model.* **2005**, *9*, 347–404. [CrossRef]
42. Large, W.G.; McWilliams, J.C.; Doney, S.C. Oceanic vertical mixing: A review and a model with a nonlocal boundary layer parameterization. *Rev. Geophys.* **1994**, *32*, 363–403. [CrossRef]
43. Shchepetkin, A.F.; McWilliams, J.C. A method for Computing Horizontal Pressure-Gradient Force in an Oceanic Model with a Non- Aligned Vertical Coordinate. *Geophys. Res.* **2003**, *108*, 3090. [CrossRef]
44. Keen, T.R.; Glenn, S.M. Factors influencing hindcast skill for modelling shallow water currents during Hurricane Andrew. *J. Atmos. Ocean. Technol.* **1998**, *15*, 221–236. [CrossRef]
45. Keen, T.R.; Glenn, S.M. Shallow water currents during Hurricane Andrew. *J. Geophys. Res.* **1999**, *104*, 23443–23458. [CrossRef]
46. Allahdadi, M.N.; Li, C. Numerical Simulation of Louisiana Shelf Circulation under Hurricane Katrina. *J. Coast. Res.* **2017**, *34*, 67–80. [CrossRef]
47. Allahdadi, M.N.; Li, C. Effect of Stratification on Current Hydrodynamics over the Louisiana shelf during Hurricane Katrina. *Water Sci. Eng.* **2017**, *10*, 154–165. [CrossRef]
48. Margolin, L.; Smolarkiewicz, P.K. Antidiffusive velocities for multipass donor cell advection. *Siam J. Sci. Comput.* **1988**, *20*, 907–929. [CrossRef]
49. Madec, G.; the NEMO team. *NEMO Ocean Engine, Note du Pôle de Modélisation 27*; Institut Pierre-Simon Laplace (IPSL): Paris, France, 2008; ISSN 1288–1619.
50. Brown, J.; Joyce, A.E.; Aldridge, J.N.; Young, E.F.; Fernand, L.; Gurbutt, P.A. Further Identification and Acquisition of Bathymetric Data for Irish Sea Modelling; DETR research contract CW075; 1999. Available online: https://www.bodc.ac.uk/resources/products/data/external_products/celtic_seas/documents/bathy_doc.pdf (accessed on 28 August 2020)
51. Haney, R.L. On the pressure gradient force over steep bathymetry in sigma coordinates ocean models. *Phys. Oceanogr.* **1991**, *21*, 610–619. [CrossRef]
52. Shapiro, R. Linear filtering. *Math. Comput.* **1975**, *29*, 1094–1097. [CrossRef]
53. Flather, R.A. A tidal model of the northwest European continental shelf. *Mem. Soc. R. Sci. Liege* **1976**, *6*, 141–164.
54. Shchepetkin, A.; McWilliams, J. Quasi-monotone advection schemes based on explicit locally adaptive dissipation. *Mon. Weather Rev.* **1998**, *126*, 1541–1580. [CrossRef]
55. Marchesiello, P.; McWilliams, J.C.; Shchepetkin, A. Open boundary conditions for long term integration of regional oceanic models. *Ocean Model.* **2001**, *3*, 1–20. [CrossRef]
56. Egbert, G.D.; Erofeeva, S.Y. Efficient inverse modeling of barotropic ocean tides. *Atmos. Ocean. Technol.* **2002**, *19*, 183–204. [CrossRef]
57. Chapman, D.C. Numerical treatment of cross-shelf open boundaries in a barotropic coastal ocean model. *Phys. Oceanogr.* **1985**, *15*, 1060–1075. [CrossRef]
58. Zavatarelli, M.; Pinardi, N.; Kourafalou, V.H.; Maggiore, A. Diagnostic and prognostic model studies of the Adriatic Sea circulation. Seasonal variability. *Geophys. Res.* **2002**, *107*, 3004. [CrossRef]
59. Nagy, H.; Di-Lorenzo, E.; El-Gindy, E. The Impact of Climate Change on Circulation Patterns in the Eastern Mediterranean Sea Upper Layer Using Med-ROMS Model. *Prog. Oceanogr.* **2019**, *175*, 244. [CrossRef]
60. Hellerman, S.; Rosenstein, M. Normal monthly wind stress over the world ocean with error estimates. *Phys. Oceanogr.* **1983**, *13*, 1093–1104. [CrossRef]

61. Beron-Vera, F.J.; Ochoa, J.; Ripa, P. A note on the boundary conditions for salt and freshwater balances. *Ocean Model.* **1999**, *1*, 111–118. [CrossRef]
62. Pawlowicz, R.; Beardsley, B.; Lentz, S. Classical tidal harmonic analysis including error estimates in MATLAB using T_TIDE. *Comput. Geosci.* **2002**, *28*, 929–937. [CrossRef]
63. Flather, R.A.; Williams, J.A. Climate change effects on the storm surge: Methodologies and results. In *Climate Scenarios for Water-Related and Coastal Impact*; ECLAT-2 Workshop Report; Beersma, J., Agnew, M., Viner, D., Hulme, M., Eds.; CRU: Norwich, UK, 2000; pp. 66–78.
64. Argo. Argo float data and metadata from Global Data Assembly Centre (Argo GDAC). *SEANOE* **2020**. [CrossRef]
65. Roemmich, D.; Johnson, G.C.; Riser, S.; Davis, R.; Gilson, J.; Owens, W.B.; Garzoli, S.L.; Schmid, C.; Ignaszewski, M. The Argo program observing the global ocean with profiling floats. *Oceanography* **2009**, *22*, 34–43. [CrossRef]
66. Lumpkin, R.; Centurioni, L.; Perez, R. Fulfilling observing system implementation requirements with the global drifter array. *Atmos. Ocean. Technol.* **2016**, *33*, 685–695. [CrossRef]
67. Gasparin, F.; Guinehut, S.; Mao, C.; Mirouze, I.; Rémy, E.; King, R.R.; Hamon, M.; Reid, R.; Storto, A.; Le Traon, P.Y.; et al. Requirements for an Integrated in situ Atlantic Ocean Observing System From Coordinated Observing System Simulation Experiments. *Front. Mar. Sci.* **2019**, *6*, 83. [CrossRef]
68. Sudre, J.; Maes, C.; Garçon, V. On the global estimates of geostrophic and Ekman surface currents. *Limnol. Oceanogr. Fluids Environ.* **2013**, *3*, 21–39. [CrossRef]
69. Ubelmann, C.; Klein, P.; Fu, L.L. Dynamic interpolation of sea surface height and potential applications for future high-resolution altimetry mapping. *J. Atmos. Ocean. Technol.* **2015**, *32*, 177–184. [CrossRef]
70. Penduff, T.; Barnier, B.; Dewar, W.K.; O'Brien, J.J. Dynamical response of the oceanic eddy field to the North Atlantic Oscillation: A model-data comparison. *J. Phys. Oceanogr.* **2004**, *34*, 2615–2629. [CrossRef]
71. Tonani, M.; Pinardi, N.; Fratianni, C.; Pistoia, J.; Dobricic, S.; Pensieri, S.; Alfonso, M.; Nittis, K. Mediterranean Forecasting System: Forecast andanalysis assessment through skill score. *Ocean Sci.* **2009**, *5*, 649–660. [CrossRef]
72. Pettenuzzo, D.; Large, W.G.; Pinardi, N. On the corrections of ERA-40 surface flux products consistent with the Mediterranean heat and water budgets and the connection between basins surface total heat flux and NAO. *Geophys. Res.* **2010**, *115*, C06022. [CrossRef]
73. Nagy, H.; Elgindy, A.; Pinardi, N.; Zavatarelli, M.; Oddo, P. A nested pre-operational model for the Egyptian shelf zone: Model configuration and validation/calibration. *Dyn. Atmos. Ocean.* **2017**, *80*, 75–96. [CrossRef]
74. Allahdad, M.A.; He, R.; Neary, V.S. Predicting ocean waves along the US east coast during energetic winter storms: Sensitivity to whitecapping parameterizations. *Ocean Sci.* **2019**, *15*, 691–715. [CrossRef]
75. Aijaz, S.; Ghantous, M.; Babanin, A.V.; Ginis, I.; Thomas, B.; Wake, G. Nonbreaking wave-induced mixing in upper ocean during tropical cyclones using coupled hurricane-ocean-wave modeling. *J. Geophys. Res. Ocean.* **2017**, *122*, 3939–3963. [CrossRef]
76. Mungall, J.C.H.; Matthews, J.B. The M_2 tide of the Irish Sea: Hourly configurations of the sea surface and of the depth-mean currents. *Estuar. Coast. Mar. Sci.* **1978**, *6*, 55–74. [CrossRef]
77. Robinson, I.S. The tidal dynamics of the Irish and Celtic Seas. *Geophys. J. Int.* **1979**, *56*, 159–197. [CrossRef]
78. O'Rourke, F.; Boyle, F.; Reynolds, A. Tidal energy update. *Appl. Energy* **2009**, *87*, 398–409. [CrossRef]
79. O'Rourke, F.; Boyle, F.; Reynolds, A. Tidal current energy resource assessment in Ireland: Current status and future update. *Renew. Sustain. Energy Rev.* **2010**, *14*, 3206–3212. [CrossRef]
80. Pugh, D.T. Tidal amphidrome movement and energy dissipation in the Irish Sea. *Geophys. J. Int.* **1981**, *67*, 515–527. [CrossRef]
81. McGrath, T.; Nolan, G.; McGovern, E. Chemical characteristics of water masses in the Rockall Trough. *Deep Sea Res. I Oceanogr. Res. Pap.* **2011**, *61*, 57–73. [CrossRef]
82. Lozier, M.S.; Stewart, N.M. On the temporally varying northward penetration of Mediterranean Overflow Water and eastward penetration of Labrador Sea Water. *J. Phys. Oceanogr.* **2008**, *38*, 2097–2103. [CrossRef]
83. McCartney, M.S.; Talley, L.D. The subpolar mode water of the North Atlantic Ocean. *J. Phys. Oceanogr.* **1982**, *12*, 1169–1188. [CrossRef]
84. Pickart, R.S.; Spall, M.A. Impact of Labrador Sea convection on the North Atlantic meridional overturning circulation. *J. Phys. Oceanogr.* **2007**, *37*, 2207–2227. [CrossRef]

85. LeBel, D.A.; Smethie, W.M.; Rhein, M.; Kieke, D.; Fine, R.A.; Bullister, J.L.; Min, D.H.; Roether, W.; Weiss, R.F.; Andrié, C.; et al. The formation rate of North Atlantic Deep Water and eighteen degree water calculate from CFC-11 inventories observed during WOCE. *Deep Sea Res. I* **2008**, *55*, 891–910. [CrossRef]
86. Xu, X.; Rhines, P.B.; Chassignet, E.P. On mapping the diapycnal water mass transformation of the upper North Atlantic Ocean. *J. Phys. Oceanogr.* **2018**, *48*, 2233–2258. [CrossRef]
87. Lellouche, J.M.; Greiner, E.; Le Galloudec, O.; Garric, G.; Regnier, C.; Drevillon, M.; Benkiran, M.; Testut, C.E.; Bourdalle-Badie, R.; Gasparin, F.; et al. Recent updates to the Copernicus Marine Service global ocean monitoring and forecasting real-time 1/12° high-resolution system. *Ocean Sci.* **2018**, *14*, 1093–1126. [CrossRef]
88. Graham, J.; Rosser, J.P.; O'dea, E.; Hewitt, H.T. Resolving the Shelf Break Exchange around the European Northwest Shelf. *Geophys. Res. Lett.* **2018**, *45*, 12–386. [CrossRef]
89. Dobricic, S.; Pinardi, N.; Adani, M.; Tonani, M.; Fratianni, C.; Bonazzi, A.; Fernandez, V. Daily oceanographic analyses by Mediterranean Forecasting System at the basin scale. *Ocean Sci.* **2007**, *3*, 149–157. [CrossRef]
90. Tonani, M.; Pinardi, N.; Dobricic, S.; Pujol, I.; Fratianni, C. A high resolution free surface model on the Mediterranean sea. *J. Ocean Sci.* **2008**, *4*, 1–14. [CrossRef]
91. Nolan, G.D.; Lyons, K. Ocean climate variability on the western Irish Shelf, an emerging time series. In Proceedings of the ICES Annual Science Conference, Gothenburg, Sweden, 9–12 September 2006; p. 28.
92. Xing, J.; Davies, A.M. The influence of shelf edge flows and wind upon the circulation on the Malin Shelf and in the Irish Sea. *Cont. Shelf Res.* **2001**, *21*, 21–45. [CrossRef]
93. Lynch, D.R.; Smith, K.W.; Cahill, B. Seasonal mean circulation on the Irish shelf—A model-generated climatology. *Cont. Shelf Res.* **2004**, *24*, 2215–2244. [CrossRef]
94. Porter, M.; Dale, A.C.; Jones, S.; Siemering, B.; Inall, M.E. Cross-slope flow in the Atlantic Inflow Current driven by the on-shelf deflection of a slope current. *Deep Sea Res. Part I* **2018**, *140*, 173–185. [CrossRef]

Journal of
Marine Science and Engineering

MDPI

Article

Improving Operational Ocean Models for the Spanish Port Authorities: Assessment of the SAMOA Coastal Forecasting Service Upgrades

Manuel García-León [1,2,*], Marcos G. Sotillo [2,3], Marc Mestres [4], Manuel Espino [4] and Enrique Álvarez Fanjul [2]

1 Nologin Consulting SL, NOW Systems, Avenida de las Ranillas, 1D, 50018 Zaragoza, Spain
2 Puertos del Estado, Área del Medio Físico, Avenida del Partenón, 10, 28042 Madrid, Spain; marcos@puertos.es (M.G.S.); enrique@puertos.es (E.Á.F.)
3 Mercator Ocean International, 31520 Ramonville-Saint-Agne, France
4 Laboratori d'Enginyeria Marítima, Universitat Politècnica de Catalunya BarcelonaTech, Jordi Girona 1-3, Mòdul D1, Campus Nord, 08034 Barcelona, Spain; marc.mestres@upc.edu (M.M.); manuel.espino@upc.edu (M.E.)
* Correspondence: manuel.garcia@nologin.es

Abstract: The Puertos del Estado SAMOA coastal and port ocean forecast service delivers operational ocean forecasts to the Spanish Port Authorities since 01/2017 (originally set-up for 9 ports). In its second development phase (2019–2021), the SAMOA service has been extended to 31 ports (practically, the whole Spanish Port System). Besides, the next generation of the SAMOA service is being developed. Research is being focused on (1) updating atmospheric forcing (by combining the AEMET HARMONIE 2.5 Km forecasts and the IFS-ECMWF ones), (2) upgrading the circulation model (ROMS), and (3) testing new methodologies to nest SAMOA systems in the Copernicus IBI-MFC regional solution (with emphasis on its 3D hourly dataset). Evaluation of specific model upgrades is here presented. Model sensitivity tests have been assessed using the available in-situ and remoted sensed (i.e., RadarHF) observations. The results show that SAMOA outperforms IBI-MFC in sea level forecasting at meso- and macro-tidal environments. Improvements by the herein proposed upgrades are incremental: some of these set-ups were used in the last SAMOA operational releases (i.e., the SAM_INI and the SAM_ADV ones; the later currently in operations), whereas the latest test (SAM_H3D) ensures more nesting consistency with the IBI-MFC and improves significantly surface currents and sea-surface temperature simulations.

Keywords: SAMOA; coastal models; operational forecasting; port circulation; dynamical downscaling; nesting techniques in operational applications; ocean model validation

Citation: García-León, M.; Sotillo, M.G.; Mestres, M.; Espino, M.; Fanjul, E.Á. Improving Operational Ocean Models for the Spanish Port Authorities: Assessment of the SAMOA Coastal Forecasting Service Upgrades. *J. Mar. Sci. Eng.* **2022**, *10*, 149. https://doi.org/10.3390/jmse10020149

Academic Editor: Christos Tsabaris

Received: 3 December 2021
Accepted: 19 January 2022
Published: 24 January 2022

Publisher's Note: MDPI stays neutral with regard to jurisdictional claims in published maps and institutional affiliations.

1. Introduction

Operational oceanography, as defined by EuroGOOS (European Global Ocean Observing System [1]) is the activity of systematic and long-term routine measurements of the ocean and atmosphere, and their fast interpretation and dissemination. Operational Oceanography is rapidly maturing, and its capabilities are being enhanced, mostly in response to a growing demand of regularly-updated ocean information. End-users and stakeholders are key to sustain operational oceanographic systems, as well as to foster new ones.

A complete list of the range of services and a description of how end-users and stakeholders use these operational services goes beyond the scope of this introduction (see Schiller et al., 2018 and Davidson et al., 2019 for more extensive reviews [2,3]). However, in general, primary sectors supported by operational oceanographic services are those related to improve the safety and efficiency of marine activities. Many of the existing operational oceanographic services focuses on regional coastal waters, where most of the protection of marine environments in sensitive habitats and human activities occurs (De Mey-Frémaux et al., 2019 [4]).

Among coastal activities, the ones related to port activity are certainly important due to the role that ports play in global and national economies. Note that the total gross weight of goods handled in EU (European Union) ports was estimated above 3.8 billion tonnes in 2015. The Spanish Port System contributes to this activity, handling 447 million tonnes during 2015, with a 4.5% of annual growth in the 2010–2015 period (EU EuroStat 2017 [5]). Approximately, 85% of total Spanish imports and 60% of exports are channelled through ports.

However, the ports are highly impacted by extreme met-ocean drivers (especially wind, waves, surface currents and sea level). Physical environment constrains Port equipment and their activities during all phases of Port life (from its design phase, in the planning tasks and during daily operations). Thus, oceanographic historical or reanalysis data are needed to be used in design criteria for offshore structures. Coastal infrastructure and maritime industries also require short-term high-resolution forecasts of wind, wave and current to manage their operational decisions. At a wider temporal horizon, longer-term forecasts and climate projections are needed for strategic decisions, port adaptation and planning. In order to answer to these complex challenges, the Spanish Port System launched SAMOA (Sistema de Apoyo Meteorológico y Oceanográfico a las Autoridades portuarias—System of Meteorological and Oceanographic Support for Port Authorities; Álvarez Fanjul et al., 2018 [6]). From 2014, Puertos del Estado (PdE) has led this SAMOA initiative developing and providing high-resolution met-ocean operational products to feed Decision Support Systems of the Spanish Port Authorities.

This SAMOA initiative combines met-ocean observation, ocean modelling and end-user service tools. One of the SAMOA components (the one referred in this work) is focused on the design and implementation of very high-resolution ocean circulation forecast systems, able to solve from coastal to very local scales within the ports. Sotillo et al., (2019) [7] provides an extensive description of this SAMOA integrated coastal/port forecast service. This system solves governing equations for ocean currents, sea level, temperature, salinity and concentrations of tracers; as one of the building blocks of any operational oceanographic systems aimed to forecast physical coastal processes.

In that sense, in the last decade, a great variety of operational ocean models were used, mostly at global (Lellouche et al., (2013, 2018), Liu et al., 2021 [8–10]) and basin scales on European Seas, based on different model codes (with different physical parameterization sets), spanning a wide range of spatial and temporal scales, using different forcing data sets and relying on data assimilation methods (Bell et al., 2009; Bell et al., 2015; Tonani et al., 2015 [11–13]). Today, the Copernicus Marine Service (CMEMS; https://marine.copernicus.eu/ accessed on 18 January 2022 [14]), focused on global and regional scales in European basins, is well-established and the sustained availability of its "core" operational ocean products has favoured the proliferation of "downstream" services devoted to coastal monitoring and forecasting (Letraon et al.; 2019 [15]). Capet et al., (2019) [16] in their review of current European capacity in terms of operational marine and coastal modelling systems, map 49 organizations around Europe delivering 104 operational model systems simulating mostly hydrodynamics, biogeochemistry and sea waves at regional and coastal scales. The PdE SAMOA coastal forecast model applications for Spanish ports, are nested into the CMEMS forecast solution for the Iberian-Biscay-Ireland (IBI) regional seas (Sotillo et al., 2015, Aznar et al., 2016 [17,18]), thus contributing to this European coastal model capacity.

In their detailed review on the basis of coastal modelling, Kourafalou et al., (2015) [19] pointed out that advancement of coastal ocean forecasting systems requires continuous scientific progress in several key topical areas. Among them, the following research lines were identified as priority ones to evolve the SAMOA service: (i) to understand primary mechanisms driving coastal circulation; (ii) to develop adequate methods to dynamically embed coastal systems in larger scale systems; and (iii) to include in downscaled solutions methods to adequately represent air–sea and land interactions, involving atmosphere-wave-ocean coupling. Dealing with these primary science topics was fundamental to face the initial SAMOA development phase (2014–2017) that resulted in nine new SAMOA coastal model applications (running operationally since 2017), as well as to align the research needed to improve the SAMOA circulation models. This second development phase, the SAMOA-2 Project (2019–2021), allowed to extend the SAMOA forecast service to a higher number of Spanish Ports (31) and to upgrade the ocean model applications in which the SAMOA service is based.

This contribution presents the research conducted to develop the next generation of SAMOA coastal circulation model applications. This research has focused on improving the SAMOA forecasts through three main objectives:

(i) Enhancing the ocean model set-up.
(ii) Updating the atmospheric forcings and the methodology to introduce their effect on the solution.
(iii) Testing a new nesting methodology by benefiting from a new Copernicus IBI-MFC regional 3D hourly forecast product.

A complete description of the sensitivity tests performed to improve the coastal model that support the SAMOA operational service, joint with the quality assessments via extensive model validation (using different local available in-situ and remote sense observations), is herein provided.

The paper is organised as follows: Section 2 outlines the PdE SAMOA initiative, describing its objectives, components and major benefits as well as the amelioration axes followed in its currently on-going second development phase (SAMOA-2). Section 3 describes the model sensitivity tests performed aimed to improve the SAMOA model solution. Section 4 presents a detailed discussion of the results with an analysis of the proposed SAMOA updates in the different port domains. Finally, Section 5 summarises main conclusions and gives a look ahead to on-going and future research topics that may enhance SAMOA operational forecast capabilities.

2. The Operational Context: The PdE SAMOA Coastal Forecast Service and Its Second Development Phase (SAMOA-2)

Since the mid-1990s, PdE has acted as an operational oceanographic centre, sustaining an integrated multiplatform ocean monitoring and forecasting system that provides service from regional to coastal scales. Met-ocean parameters, such as wind, waves, sea-level and currents have been extensively monitored and forecasted. Main applications range (i) the safety of port operations, (ii) impacts on ship draught allowance, (iii) water quality and (iv) navigation and piloting activities.

The SAMOA initiative, co-financed by PdE and the Spanish Port Authorities, was born in 2014 as an answer to the complex port needs in terms of coastal and local met-ocean information (Álvarez Fanjul et al., 2018 [6]). SAMOA aims to enhance the PdE traditional operational oceanographic service, implementing new operational systems for each Port Authority, with tailor-made operational met-ocean information, as well as some added value products, previously requested by each Port Authority.

The SAMOA service is organized as a modular infrastructure, with some of the SAMOA modules focused on enhancing near-real-time monitoring (i.e., by means of local upgrading of current PdE observational networks with specific instrumentation), whereas some others are focused on implementing new local high-resolution forecast applications for atmosphere, waves and ocean circulation. The enhancement of PdE model capabilities within SAMOA, and its follow-up in SAMOA-2, has been certainly significant, being a total of 17 new high-resolution atmospheric models (1 km resolution, based on HARMONIE model), 20 wave agitation models (5 m resolution mild slope model applications) and 31 circulation models (70 m ROMS model set-ups) developed and operationally implemented. The SAMOA modular architecture allows each Port Authority to decide what specific SAMOA services are needed to be implemented in their ports, mainly addressing their specific needs for met-ocean information. A total

47 ports from 28 Spanish Port Authorities benefited from these new SAMOA modelling and monitoring capabilities.

To properly exploit new SAMOA added-value products, a specific tool was developed for the Ports: The CMA (Cuadro de Mando Ambiental or Environmental Control Panel). This CMA dashboard is a tailor-made web service that provides easy and customized access to operational met-ocean information in the ports, as well as advanced viewing capabilities (see snapshot in Figure 1a) and other end-user services (such as an alert system (through email/SMS), or a tool to automatically generate daily bulletin reports with environmental conditions at selected locations). Managers in the ports grant access to this CMA tool and they can define different levels of user permissions. A growing community of 1913 port registered users benefits from the CMA tool.

With respect to ocean circulation forecast, and after running regional ocean circulation model applications for more than a decade (i.e., the Spanish ESEOO system; Sotillo et al., 2008 [20]), PdE does not run anymore a regional forecast component, relying since 2014 for these scales on the Copernicus Marine IBI-MFC regional forecast service (Aznar et al., 2016 [18]). This use of Copernicus regional products (currently delivering forecast products at 1/36° resolution that covers all the Spanish waters) has allowed PdE to focus their ocean modelling resources on running very high-resolution models at specific hot spots (p.eg. the Gibraltar Strait and nearby areas, operationally covered by the PdE SAMPA forecast service; Sanchez Garrido et al., 2013 and Sotillo et al., 2016 [21,22]) and on developing the new SAMOA operational downscaled systems for ports, here described.

The related PdE ocean circulation forecast products, updated on a daily basis, can be downloaded through the PdE catalogue (https://opendap.puertos.es/thredds/catalog.html accessed on 18 January 2022 [23]) and visualized through the mentioned CMA Dashboard and the PdE PORTUS web page https://portus.puertos.es/ accessed on 18 January 2022 [24] (see Figure 1b shows a snapshot of the modelled surface current field forecasted for coastal waters nearby the Bilbao harbour).

In SAMOA-2, it was decided to enhance the coastal/port circulation module by adding an extra oil spill forecast service linked to the implementation of a local high-resolution circulation model. This new option allows to have an on-demand use of the PdE oil spill forecast service (based on MedSlik model, De Dominicis et al., 2013 [25]) and to visualize their model outputs through the CMA dashboard (see snapshot of the output tool in Figure 1c). This added oil spill module played a relevant role on fostering the interest of Port Authorities in high-resolution coastal circulation products for their coastal fringes. This Ports' interest on currents, and mainly on its advective and dispersive effects, is partially explained by the legal commitment that Port Authorities need for ensuring good water quality within harbours (European Parliament, Directive 2019/883 [26]).

In the first SAMOA development phase, nine high-resolution coastal circulation systems were implemented for different Spanish ports located in the Mediterranean, the Iberian Atlantic and the Canary Islands (see green dots in Figure 2). This first SAMOA implementation of high-resolution circulation systems was performed thanks to the collaboration between PdE and the Maritime Engineering Laboratory of the Polytechnic University of Catalonia (LIM/UPC). The same team has continued along SAMOA-2 with the R&D works needed to face the challenge of (i) improving currently existing SAMOA port circulation systems and (ii) to extend this SAMOA forecast service to other requested ports (i.e., multiplying by three the number of Spanish ports where SAMOA coastal forecasts are implemented by 2021).

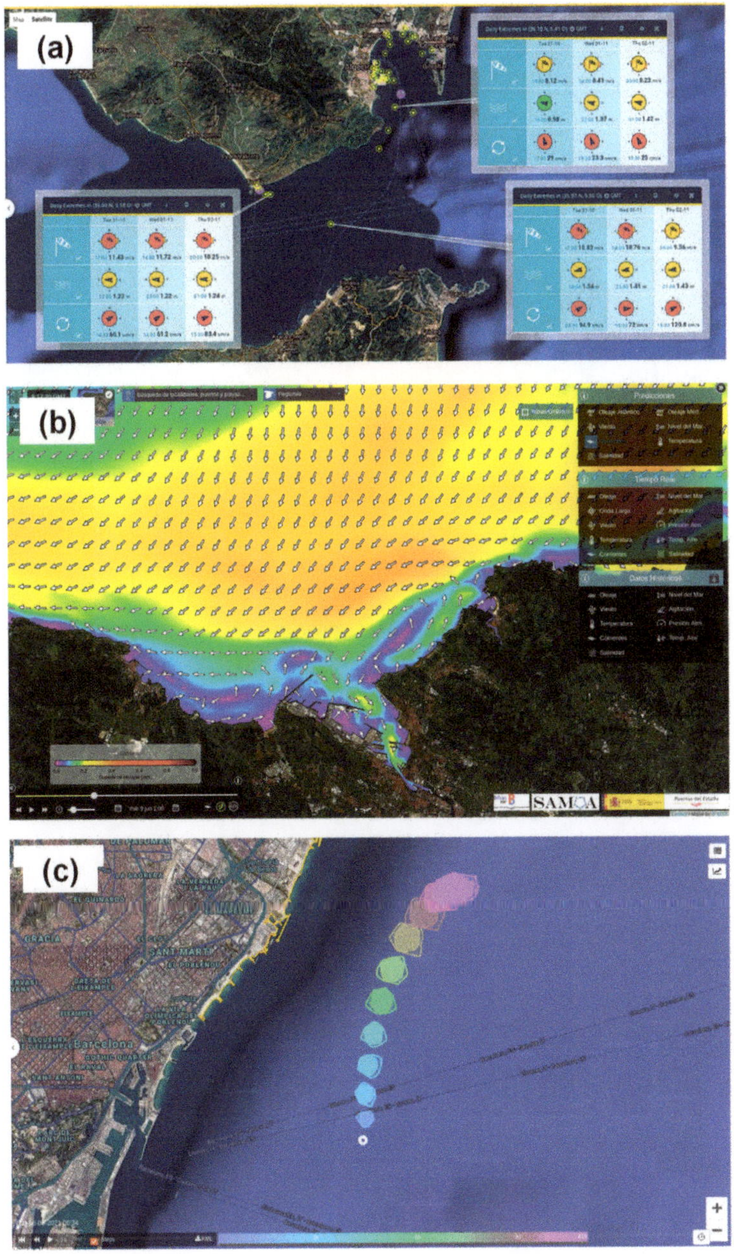

Figure 1. 3 Example of Port user-oriented viewing capability through the SAMOA CMA environmental dashboard: (**a**) 3-days forecast information for wind, waves and currents at Algeciras Bay; Depicted maximum forecasted values and coloured end-user alert symbols (green-yellow-red) following specific user-customized alert thresholds at location of interest for the Port operations; (**b**): Map of SAMOA high-resolution forecasted surface current field in nearby waters of the Bilbao Port. (**c**) Example of Oil spill forecast in front of Barcelona Port. Outputs from the PdE Oil Spill forecast service using SAMOA high resolution forecasted currents.

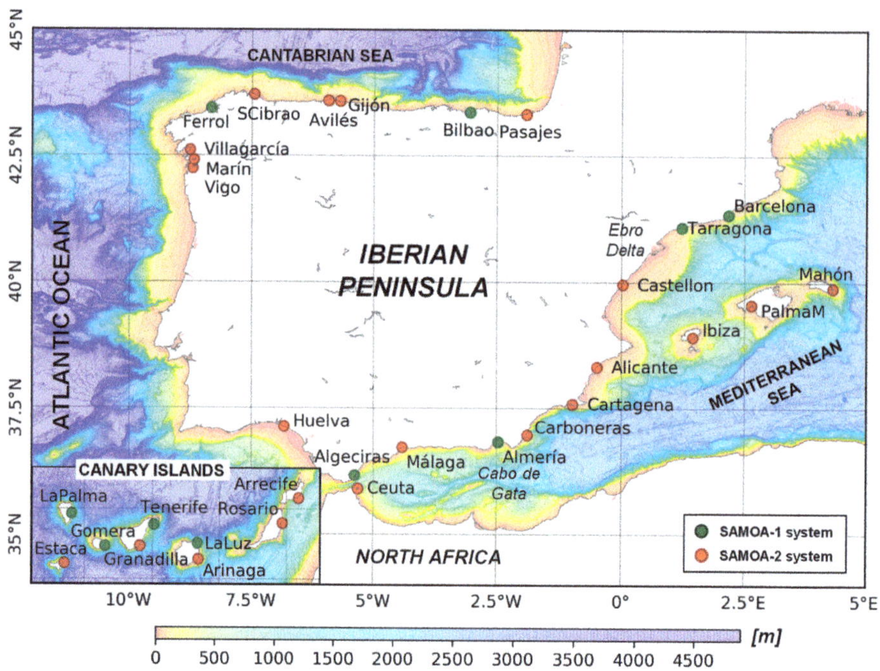

Figure 2. SAMOA Coastal/Port circulation forecast systems in operations. Green dots: Operational models systems developed within SAMOA1 (2014–2017). Red dots: Operational model systems developed within SAMOA2 (2018–2021). Bathymetry (in meters) is depicted. Note that systems in the Canary Islands are shown in the bottom left corner box (SW limit: 18.5° W, 27.5° N; NE limit: 13.2° W, 29.5° N).

This noticeable increase in the number of SAMOA Port systems (red dots in Figure 2 shows location of the Spanish ports where new circulation systems are deployed after the SAMOA second development phase) results in an almost complete coverage of the Spanish coast with high-resolution coastal circulation forecasts.

The operational SAMOA local circulation forecast systems daily produce short term (+3 Days) forecasts of 3-D currents and other oceanographic variables, such as temperature, salinity, together with sea level. Each SAMOA system uses the ROMS model (Shchepetkin and McWilliams (2005); present code source and documentation at the ROMS website [27,28]), and consists of two nested regular grids with spatial resolution of ~350 m and ~70 m for the coastal and harbour domains, respectively. The model is forced with very high-resolution atmospheric forcing (provided by the Spanish Meteorological Office AEMET) and it is nested into the core Copernicus IBI regional solution (IBI_PHY hereafter). Figure 3 shows a schematic view of the SAMOA system; for an end-to-end description of the SAMOA model set-up, see Sotillo et al., 2019 [7].

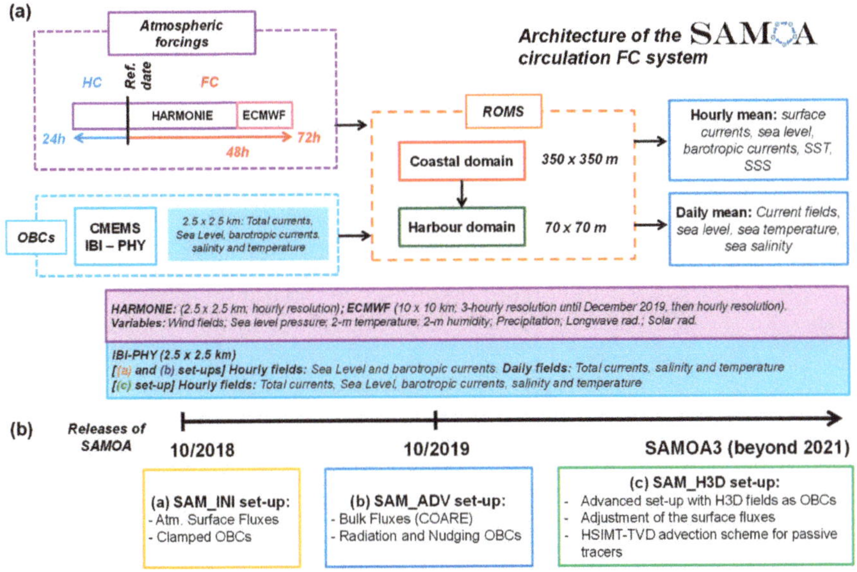

Figure 3. Architecture of the SAMOA circulation forecast system. (**a**) Operational scheme, detailing (i) input data sources, including atmospheric forcing and the ocean solution imposed at the Open Boundary Conditions (OBCs), (ii) the numerical core (ROMS) with domain organization, and (iii) delivered products. (**b**) SAMOA Operational Releases: The timeline shows main releases of this service since 2018 and their most prominent changes. Please note that the acronyms SAM_INI, SAM_ADV and SAM_H3D will be used throughout the text, to name the different model set-up experiments. Likewise with IBI_PHY (the SAMOA parent solution).

Product quality assessment is a key issue for any operational forecast service (Ryan et al., 2015 [29]). Nevertheless, one of the main bottlenecks identified in the recent review performed by EuroGOOS for European coastal operational ocean model services (Capet et al., 2019 [16]) is the lack of an adequate delivery in near-real-time (NRT) of operational observations. This lack of operational observations, especially on coastal areas, restricts the systematic exploitation of (i) the data assimilation capacities in operational marine modelling systems and (ii) the provision of NRT assessments of operational ocean model products. Note that only 20% of operational models currently available for European seas provide a dynamic uncertainty together with their forecast products.

In the case of SAMOA, an exhaustive NRT validation of the forecast products operationally generated is performed. This operational validation is based on a routine monitoring of the quality of the SAMOA forecast products on a daily and monthly basis. This SAMOA model forecast validation is using all operational ocean observation available in the port and nearby coastal waters. To this aim, an extension of the NARVAL tool (originally developed for the CMEMS IBI MFC regional model solutions (Lorente et al., 2019 [30])) was implemented to validate the SAMOA circulation model products. The comprehensive multi-parametric ocean model skill assessment is performed by using all available operational observational sources in the coastal domains. The list of observational data sources includes satellite L3 and L4 SST products, together with in-situ observations from moorings and tide-gauges, as well as HF-Radars. Further details about the complete list of observational products and platforms used by NARVAL to validate SAMOA can be seen in Sotillo et al., 2019 [7].

Another SAMOA module, consisting of observational campaigns at harbour waters was offered to Port Authorities. These campaigns include at the same time (and for a three month period) deployment of 1 ADCP current-meter, fixed (and repeated) CTD temperature and salinity stations, as well as additional meteorological stations. It is expected that this SAMOA extensive monitoring campaign will help the Ports to increase the knowledge on their waters and to validate the SAMOA met-ocean forecast applications. In that sense, it is remarkable that all Port Authorities (9) who have requested this SAMOA specific in-situ observational port campaign count with a SAMOA Port Circulation Forecast service. However, to assess the model sensitivity tests shown in the present work, only operational observational data sources were used (from PdE HF Radar, tide gauges and coastal or deep-water mooring stations), since data from the new SAMOA observational campaigns were not available when SAMOA model tests were performed.

3. Methodology: Model Sensitivity Tests for Improving the SAMOA Circulation Forecast Services

After more than 4-years running operationally the first 9 SAMOA port forecast services, joint with the continuous quality assessment of their model products (performed throughout the NARVAL validation tool described in the previous section), the following main concerns about the SAMOA solutions were identified:

(i) Existence of some systematic biases in passive tracers, such as the temperature and salinity fields.
(ii) Some inconsistencies between SAMOA coastal modelled circulation patterns, compared with those exhibited by the CMEMS IBI-MFC parent solution.

Specific research was conducted with the aim to minimize these limitations via improving the SAMOA model set-up for the forthcoming releases.

External forcing, and particularly the atmospheric one, remains as one of the major issues restricting the accuracy of operational coastal systems. In this sense, improving external atmospheric forcing has been identified as a major concern for enhancing the reliability of high-resolution forecast services (Capet et al., 2019 [16]). Furthermore, the improvement of atmospheric forcing (and a methodology for a better usage in coastal domains) is highlighted in the SAMOA roadmap (outlined in Sotillo et al., 2019 [7]), as one of its three main research lines.

On the other hand, Kourafalou et al., (2015) [19] point out that coastal ocean forecasting systems requires continuous scientific progress in a primary science topic as it is the downscaling from larger scale models, with a need of ad-hoc nesting procedures. This includes assessment of the boundary conditions from the larger scale parent systems to the coastal nested systems, and refinement of the model set-up (including grids, topographic details, forcing data and schemes used as open boundary condition). A challenge for these appropriate nesting procedures is to ensure consistency in the fluxes between the downscaled coastal model and its coarser parent solution. This consistency will avoid problems such as triggering unrealistic gravity transient currents that may unleash spurious dynamical coastal model features.

Consequently, the following three model sensitivity tests were performed:

(i) To update SAMOA atmospheric forcing data (experiment termed as SAM_INI, hereafter).
(ii) To improve the ROMS model set-up, with special attention to the open boundary condition treatment; joint with the computation of surface ocean-atmosphere fluxes through a new bulk formulation (SAM_ADV test experiment). Finally also,
(iii) To test a new nesting methodology taking advantage of the availability of a Copernicus regional 3D IBI hourly forecast product (SAM_H3D). This product allows enhancement of the temporal frequency (from daily to hourly resolution) of the data imposed at boundaries. Moreover, specific physics upgrade was implemented in this configuration experiment.

Table 1 summarises the main differences among the three different model sensitivity datasets. As denoted in Figure 3b, SAM_INI corresponds to the former operational version of SAMOA (active since October 2018 to October 2019), whereas SAM_ADV is the operational set-up currently in production. SAM_H3D will be the forthcoming SAMOA set-up version (to be operationally released in 2022).

Table 1. Summary of the main differences among the sensitivity tests conducted. The columns denote the three SAMOA model configurations (SAM_INI, SAM_ADV and SAM_H3D) and the rows the different upgrades. ✔ means included in each model set-up, and ✖ not included.

Feature	SAM_INI	SAM_ADV	SAM_H3D
SAMOA-1 model set-up (as in Sotillo et al. [7])	✔	✔	✔
Use of 2.5 km HARMONIE Atmospheric Forcing	✔	✔	✔
Use of the COARE bulk formulae to compute surface fluxes	✖	✔	✔
Changes in the treatment of Open Boundary Conditions	✖	✔	✔
Use of the new hourly-3D CMEMS IBI-MFC product	✖	✖	✔
Use of new advection scheme for passive tracers and improved surface fluxes physics	✖	✖	✔
Entry into Service (Operational Release)	October 2018	October 2019	To be in 2022

All the SAMOA model set-up novelties have been extensively tested. The proposed sensitivity model runs were validated using in-situ and remote sensing observations (i.e., from PdE coastal and deep-water mooring buoys for surface temperature, salinity and currents, tide gauges for sea level and HF-Radars for surface currents, available the latest only in some of the coastal model domains).

A complete description of the different sensitivity model tests performed to improve the operational coastal model applications that sustain the SAMOA forecast service for Spanish ports is provided below. Results from the different model tests are provided in the next section.

3.1. Upgrade of the Atmospheric Forcing: The Initial SAMOA-2 Model Configuration (SAM_INI)

SAM_INI is the first model configuration that is used in the sensitivity test. SAM_INI set-up is closely related with the SAMOA-1 set-up (Sotillo et al., 2019 [7]), except for a key issue: SAM_INI uses as atmospheric forcing the new AEMET HARMONIE data (Bengtsson et al., 2017 [31]), rather than the deprecated HIRLAM forcing data (also provided by the Spanish Met Office AEMET) that were used in the SAMOA-1 service. As shown in the operational scheme depicted in Figure 3, HARMONIE is used for the first 48 h of forecast, then completing the forecast range between 48 to 72 h with the ECMWF-IFS data (ECMWF 2019 [32]) provided by the European Centre for Medium-Range Weather Forecasts (ECMWF). The HARMONIE data have a spatial resolution of 2.5 km and an hourly frequency, whilst the ECMWF-IFS forcing have a resolution of 10 km and tri-hourly frequency. It should be noted that ECMWF-IFS, with a global coverage, is the parent solution for the AEMET HARMONIE. The latter is a limited area-system with two domains: (i) a first one that spans the whole Iberian Peninsula and Balearic Islands, and (ii) a second one that covers the Canary Islands.

For both atmospheric products, the following atmospheric variables are used: wind fields, sea-level pressure, 2-m air temperature, 2-m relative humidity, precipitation, latent heat fluxes and sensible heat fluxes. In this first SAM_INI model configuration, the atmospheric fluxes are computed with the same methodology presented in Sotillo et al., 2019 [7]).

As pointed out in Table 1, the SAM_INI model configuration was used in the SAMOA operations since October 2018, being the SAMOA operational set-up for one year until the last release (held in October 2019).

3.2. Upgrade of the Open Boundary Condition (OBC) Scheme and Bulk Fluxes: The Current Operational SAMOA Set-Up (SAM_ADV)

As mentioned above, the NARVAL validation system led to detect inconsistencies in the surface current fields forecasted by the SAMOA-1 systems and the parent solution (IBI-MFC). Additionally, the update of the atmospheric forcing in the SAM_INI configuration also intensified error metrics in the Sea Surface Temperature (SST).

With the aim of minimizing some of these identified operational shortcomings, a new SAMOA model set-up (termed SAM_ADV in Figure 3, and hereafter) was proposed. This new SAM-ADV configuration had two main objectives:

(i) To improve consistency between the parent solution (IBI-MFC) and the SAMOAs forecasts, via Open Boundary Conditions (OBCs) exchanges.
(ii) To improve of the SST solution through two synergistic approaches: first, improving the circulation fields; second, substituting with bulk formulas (COARE 3.0, Fairall et al., 2003 [33]) the SAMOA-1 treatment of the atmospheric forcings.

The SAM_INI set-up for the OBC treatments was featured by:

(i) Implicit Chapman for the free-surface.
(ii) Flather conditions for the barotropic currents (Flather, 1976 [34]).
(iii) Clamped conditions for the total velocities and passive tracers.

However, the SAM_ADV set-up proposed the following changes:

(i) Explicit Chapman for the free-surface (Chapman, 1985 [35]).
(ii) Shchepetkin (Mason et al., 2010 [36]) and Radiation with nudging (Marchesiello et al., 2001 [37]) for barotropic currents.
(iii) Radiation with nudging for total velocities and passive tracers.

This new OBC treatment was proposed to ensure that, both normal and tangential, components of the velocity and passive tracers were conserved from the IBI_PHY parent data to the SAMOA solution. Note that the OBC scheme used in the SAM_INI set-up (and in the SAMOA-1 one) provided systematic inconsistencies with the tangential component of the velocities. These boundary inconsistencies affected the SAMOA solutions most especially in mesotidal and macrotidal regimes (that means in the SAMOA systems implemented in the Canary Islands and in the Cantabrian Sea, respectively).

The performance of this new SAM_ADV configuration (and the difference with the SAM_INI one) was tested in all systems by a one-month simulation period (June 2019). The time interval, though limited, was enough to highlight the benefits of this proposed upgrade.

3.3. Use of a Higher Temporal Frequency Imposed Data at the Boundary and Upgrade of the Model Physics: The Forthcoming SAMOA Operational Set-Up (SAM_H3D)

Despite that the SAM_ADV set-up remarkably enhanced the consistency between the SAMOA and IBI_PHY at the boundaries; in some specific cases (i.e., for the Gran Canaria and Almería systems) the SAMOA performance was not satisfactory enough. A possible reason may come from the mismatch among the temporal frequency of the different variables in the OBCs. For instance, in previous set-ups (SAM_INI and SAM_ADV), sea-level and barotropic currents are updated on an hourly basis; but the total velocities, temperature and salinity have daily-mean frequency. Such treatment, usual in downstream services, may lead to non-conservation issues of the momentum and passive tracers.

The Copernicus service recently started delivering a new IBI forecast 3D hourly. This new product comprises hourly fields of total velocities, temperature and salinity along the whole water column for coastal and shelf areas of the IBI domain. The delivery of this new CMEMS IBI product means an opportunity for SAMOA to improve its downscaling. Hence, its use in the SAMOA nesting has been extensively tested in this SAM_H3D experiment, aimed to provide insights on the forthcoming SAMOA release.

This new nesting has been tested in five SAMOA systems: three microtidal Mediterranean ones (Barcelona, Tarragona and Almería), and two Atlantic ones (a mesoscale one, Gran Canaria, and a macrotidal one: Bilbao). The SAM_H3D testing period ranges one year (from October 2019 to October 2020), in order to assess how the SAMOA systems would behave throughout different season conditions. The OBC treatment in SAM_H3D is analogous to the one used in the SAM_ADV (previously described), but adapted to the higher temporal frequency of the IBI_PHY imposed data.

Additionally, specific changes in the physics have been implemented in the operational ROMS version (3.7) that currently uses the SAMOA system, namely:

(i) A new scheme for the advection term for the passive tracers (HSIMT-TVD, Wu and Zhu, 2010 [38]), based in the latest versions of the ROMS code.
(ii) A reduction of the wind-stresses due to the effect of surface currents.
(iii) An update of the COARE 3.0 bulk-formula to the COARE 3.5 (Edson et al., 2013 [39]).

These changes aim to provide consistent surface fields (be it heat fluxes or surface stresses), by including last advances in air-sea interaction processes.

3.4. Evaluation Criteria and Error Metrics

From each of the three SAMOA model test experiments (in the different port domains selected in each case), it was produced a daily set of daily-averaged and hourly-averaged data. In this contribution, the model validation was focused on the hourly-averaged products, because they are more suitable to assess two main issues:

(i) How each system handles the intra-daily cycles related to the atmospheric conditions (with special focus on surface heat fluxes or wind stresses)?
(ii) How each system simulates tidal-induced currents?

The hindcasted hourly-averaged products (the ones produced in the first 24 h of each forecast cycle; see Figure 3a) have been compared with hourly-averaged observations from the PdE Monitoring Network. This network comprises deep water and coastal buoys (REDEXT and REDCOS, respectively [40,41]), tidal stations (from the REDMAR network [42]) and HF Radar [43]. All these measurements have passed state-of-the-art Quality Control processes and can be accessed in NRT basis from the PORTUS [24] website.

Performance of the different SAMOA model tests have been assessed by comparing the different models with the available observational datasets at each coastal and port domain (see in Table 2 the observational coverage available at each SAMOA domain).

The following hourly-averaged error metrics has been computed for each dataset in each domain (when available observations): mean bias, correlation, Root-Mean Square Error (RMS) and Coefficient of Efficiency (COE). The metrics that will be shown in Tables 3–5 are representative of the whole analysed period in each sensitivity test. Metrics in Section 4.1 were computed from time-series enclosing June 2019; whereas those for the Section 4.2 range from October 2019 to October 2020.

The Coefficient of Efficiency (COE) (Legates and McCabe, 1999, 2013 [44,45]) is obtained as (Equation (1)):

$$COE = 1 - \frac{\sum_{i=1}^{N} |O_i - P_i|}{\sum_{i=1}^{N} |O_i - \overline{O}|}, \quad (1)$$

where Pi and Oi refer to the computed and observed signals respectively, N is the number of time records and ($\overline{\ }$) is the mean operator. A *COE* value of 1 means a perfect model. Despite *COE* having no lower bound, a *COE* value of 0 implies that the model is not more able to predict the measured values than the measured mean would do. Consequently, for negative *COE* values, the computed model signal would perform worse than the measured mean in predicting variations in the observed signal. For the sake of interpretation, please note that the *COE* and the skill score of Murphy (1988) [46] are both the same, in case the observed mean is used as the reference model in the skill score.

Table 2. Summary of the observational coverage available at each SAMOA domain. Rows denote the SAMOA domain, and the columns, the different elements of the PdE monitoring network (deep water buoys (EXT), coastal water buoys (CST), tidal gauge stations (TGS) and Radar HF) used in the model validation. The ✖ symbol means unavailability of a certain type of sensor in the area. If there exists a monitoring device within the SAMOA domain, its main features are summarized in brackets including: location (longitude, latitude), mooring depth (in m), spatial resolution (in km) for the HF Radars and measured variables (Sea Surface Temperature, SST; Sea Surface Salinity, SSS; Surface current speed, SC_S; Surface current direction, SC_D; Sea Level, SLev).

SAMOA Domain (Figure 2)	Deep Water Buoy (EXT)	Coastal Water Buoy (CST)	Tidal Gauge Station (TGS)	Radar HF
Barcelona (BCN)	✖	(2.20° E, 41.32° N; 68 m; SST)	(2.17° E, 41.34° N; SLev)	✖
Tarragona (TAR)	(1.47° E, 40.69° N; 688 m; SST, SSS, SC_S, SC_D)	(1.19° E, 41.07° N; 15 m; SST)	(1.21° E, 41.08° N; SLev)	(3 × 3 km; SC_S, SC_D)
Bilbao (BIL)	(3.04° W, 43.64° N; 580 m; SST, SSS, SC_S, SC_D)	(3.13° W, 43.40° N; 53 m; SST)	(3.05° W, 43.35° N; SLev)	✖
Almería (ALM)	(2.34° W, 36.57° N; 536 m; SST, SSS, SC_S, SC_D)	✖	(2.48° W, 36.83° N; SLev)	✖
Ferrol (FER)	✖	(8.56° W, 43.35° N; 60 m; SST)	(8.33° W, 43.46° N; SLev)	✖
Vigo (VIG)	✖	✖	(8.73° W, 42.24° N; SLev)	✖
La Gomera (GOM)	✖	✖	(17.11° W, 28.09° N; SLev)	✖
Gran Canaria (GCA)	✖	(15.39° W, 28.05° N; 30 m; SST)	(15.41° W, 28.14° N; SLev)	(0.75 × 0.75 km; SC_S, SC_D)
La Palma (PAL)	✖	✖	(17.77° W, 28.68° N; SLev)	✖
Tenerife (TEN)	✖	(16.23° W, 28.46° N; 56 m; SST)	(16.24° W, 28.48° N; SLev)	✖

Despite error metrics being computed with hourly-data, monthly average metrics were also calculated for Figures 6 and 7. These monthly metrics were computed specially in the case of surface currents versus HF Radar data (some figures in the Result section show monthly observational and model outputs, together with the bias field). The aim of these specific cases was to demonstrate that SAMOA can reproduce the spatial signature of surface-currents with a certain skill.

The model validation assessment has been held with the SAMOA coastal domains (350 m horizontal resolution, see Figure 3a) for surface currents, temperature and salinity, measured by met-ocean sensors on coastal and deep-water buoys (moored outside the ports) and by HF radars (covering nearby coastal areas but not inside the harbours). However, sea level has been addressed with the SAMOA local Port model solutions (70 m horizontal resolution), because coastal tidal stations are deployed within the harbours. Another remark for the sea level assessment, is that mean value has been removed from all the time series (be it models or observations), because each model has a different vertical system of reference. Consequently, estimated biases may be misleading and they are not comparable.

Table 3. Error metrics for the SAMOA SAM_ADV and SAM_INI test runs, plus the IBI_PHY parent solution (computed with hourly observations). The metrics comprise the June 2019 period. Variables: SST (Sea Surface Temperature in °C), SSS (Sea Surface Salinity, in PSU), SC_S (Surface current speed, in m/s), SC_D (Surface current direction, in °), SLev (sea level, in m). The domain column identifies the SAMOA system. The Type denotes the sensor type used: TGS (Tide Gauge Station), CST (Coastal Buoy), EXT (Deep-water buoy). N counts the size of the sample (conditioned by the availability of hourly observation). Each error metric (Bias, Correlation, Root-Mean-Square Error (RMS) and Coefficient of Efficiency (COE)) provided for each model solution (SAM_INI (INI), SAM_ADV (ADV) and IBI_PHY(IBI)). The last row of each variable represents the average error metric of all the sensors given a specific test run. Bold numbers highlight the best performing dataset.

Variable	Domain	Type	N	Bias (INI/ADV/IBI)	Correlation (INI/ADV/IBI)	RMS (INI/ADV/IBI)	COE (INI/ADV/IBI)
SST (°C)	BCN	CST	811	−0.21/−0.36/**0.02**	0.90/**0.93**/**0.93**	0.76/0.68/**0.62**	0.61/0.59/**0.67**
	TAR	EXT	1289	−0.11/−0.19/**0.03**	0.63/0.7/**0.72**	0.41/0.39/**0.29**	−0.27/−0.27/**0.14**
		CST	847	**−0.48**/−0.6/−0.67	0.82/**0.84**/0.79	0.28/**0.26**/0.3	**−0.25**/−0.54/−0.73
	BIL	EXT	864	−0.09/−0.14/**0.09**	**0.96**/0.95/**0.96**	**0.34**/**0.34**/0.35	**0.69**/0.68/0.67
		CST	1343	0.24/**−0.09**/0.25	0.91/0.92/**0.93**	0.6/**0.43**/0.47	0.45/**0.66**/0.54
	ALM	EXT	1493	0.7/**0.62**/0.72	0.73/0.77/**0.81**	0.88/0.79/**0.73**	0.02/**0.18**/0.13
	FER	CST	722	0.42/**0.14**/0.38	0.64/0.71/**0.84**	1.19/0.84/**0.54**	−0.39/−0.08/**0.17**
	GCA	CST	717	**0.15**/−0.23/−0.26	0.91/0.9/**0.92**	0.31/0.31/**0.29**	**0.57**/0.55/0.51
	TEN	CST	1224	1.17/0.54/**0.11**	0.54/0.83/**0.88**	0.69/0.44/**0.38**	−0.77/0.11/**0.53**
		AVERAGE		0.20/**−0.03**/0.07	0.78/0.84/**0.86**	0.61/0.50/**0.44**	0.07/0.21/**0.29**
SSS (PSU)	TAR	EXT	1289	−0.08/**−0.06**/−0.07	**0.09**/0.01/0.05	0.12/0.13/**0.1**	−1.91/−2.16/**−1.6**
	BIL	EXT	864	0.16/**0.14**/0.16	0.14/0.16/**0.17**	**0.54**/**0.54**/**0.54**	−0.47/**−0.36**/−0.51
	ALM	EXT	1493	**0.06**/0.1/**0.06**	**−0.09**/−0.12/−0.1	**0.3**/**0.3**/**0.3**	**−0.27**/−0.34/**−0.27**
		AVERAGE		**0.05**/0.06/**0.05**	**0.05**/0.02/0.04	0.32/0.32/**0.31**	−0.88/−0.95/**−0.79**
SC_S (m/s)	TAR	EXT	1289	0.16/0.2/**0.15**	**0.13**/0.11/0.08	0.13/0.14/**0.12**	−2.14/−2.79/**−1.96**
	BIL	EXT	864	0.06/0.07/**0.04**	0.13/**0.15**/0.08	0.13/0.13/**0.12**	−0.3/−0.35/**−0.18**
	ALM	EXT	1493	**−0.05**/−0.08/−0.12	0.05/**0.08**/−0.05	0.17/**0.16**/0.17	**−0.29**/−0.33/−0.5
		AVERAGE		0.06/0.06/**0.02**	0.10/**0.11**/0.04	**0.14**/**0.14**/**0.14**	−0.91/−1.16/**−0.88**
SC_D (°)	TAR	EXT	1289	**3**/8/**3**	**−0.01**/−0.06/**−0.01**	**129**/131/**129**	**−0.06**/−0.09/**−0.06**
	BIL	EXT	864	**81**/84/88	0.21/0.2/**0.23**	136/136/**134**	**−0.08**/**−0.08**/**−0.08**
	ALM	EXT	1493	86/93/**83**	0.12/**0.19**/0.04	124/**116**/122	**−0.91**/−0.92/−0.94
		AVERAGE		**57**/62/58	**0.11**/**0.11**/0.09	130/**128**/**128**	**−0.35**/−0.36/−0.36
SLev (m)	BCN	TGS	865	Not applicable	0.93/0.92/**0.96**	0.03/0.03/**0.02**	0.51/0.5/**0.61**
	TAR	TGS	837		0.89/**0.91**/0.9	0.05/**0.04**/0.05	0.43/**0.51**/0.44
	BIL	TGS	955		**0.99**/**0.99**/0.97	**0.09**/**0.09**/0.26	**0.92**/**0.92**/0.74
	ALM	TGS	744		**0.87**/0.86/0.84	**0.05**/**0.05**/**0.05**	**0.51**/0.48/0.46
	FER	TGS	1272		**0.99**/**0.99**/0.96	**0.07**/**0.07**/0.26	**0.92**/**0.92**/0.72
	VIG	TGS	812		**0.99**/**0.99**/0.96	**0.07**/**0.07**/0.23	0.93/**0.94**/0.73
	GOM	TGS	1268		**0.99**/**0.99**/0.96	0.05/**0.04**/0.13	0.91/**0.93**/0.72
	GCA	TGS	720		**0.99**/**0.99**/0.97	**0.05**/0.06/0.13	**0.91**/**0.91**/0.77
	PAL	TGS	560		**0.99**/**0.99**/0.97	**0.04**/**0.04**/0.13	**0.93**/**0.93**/0.76
		AVERAGE			**0.96**/**0.96**/0.94	0.06/**0.05**/0.14	0.77/**0.78**/0.66

Table 4. Error metrics for the SAMOA SAM_H3D and SAM_ADV test runs, plus the IBI_PHY parent solution (computed with hourly observations). The metrics comprise the time range from October 2019 to October 2020. Variables: SST (Sea Surface Temperature in °C), SSS (Sea Surface Salinity, in PSU), SC_S (Surface current speed, in m/s), SC_D (Surface current direction, in °), SLev (sea level, in m). The domain column identifies the SAMOA system. The Type denotes the sensor type used: TGS (Tide Gauge Station), CST (Coastal Buoy), EXT (Deep-water buoy). N counts the size of the sample (conditioned by the availability of hourly observation). Each error metric (Bias, Correlation, Root-Mean-Square Error (RMS) and Coefficient of Efficiency (COE)) provided for each model solution (SAM_ADV (ADV), SAM_H3D (H3D) and IBI_PHY(IBI)). The last row of each variable represents the average error metric of all the sensors given a specific test run. Bold numbers highlight the best performing dataset.

Variable	Domain	Type	N	Bias (ADV/H3D/IBI)	Correlation (ADV/H3D/IBI)	RMS (ADV/H3D/IBI)	COE (ADV/H3D/IBI)
SST (°C)	BCN	CST	7709	−0.26/−0.14/**0.1**	0.97/**0.99**/**0.99**	1.07/0.74/**0.65**	0.81/0.86/**0.88**
	TAR	EXT	8336	−0.24/**−0.06**/0.19	**0.99**/**0.99**/**0.99**	0.73/0.75/**0.57**	0.88/**0.89**/**0.89**
		CST	7763	−0.6/−0.28/**0.2**	0.94/**0.98**/**0.98**	1.49/0.89/**0.81**	0.73/0.83/**0.86**
	BIL	EXT	6696	−0.38/−0.32/**0.08**	0.98/**0.99**/**0.99**	0.78/0.53/**0.41**	0.83/0.88/**0.91**
		CST	6835	−0.61/−0.31/**−0.22**	0.96/**0.99**/**0.99**	1.06/0.63/**0.59**	0.7/0.83/**0.84**
	ALM	EXT	7683	−0.22/**−0.02**/0.05	**0.96**/0.95/0.95	**1.05**/1.12/1.15	**0.77**/0.76/0.76
	GCA	CST	8403	−0.31/**−0.02**/−0.15	0.95/**0.97**/**0.97**	0.59/0.4/**0.38**	0.6/**0.76**/**0.76**
		AVERAGE		−0.37/−0.16/**0.04**	0.96/**0.98**/**0.98**	0.97/0.72/**0.65**	0.76/0.83/**0.84**
SSS (PSU)	TAR	EXT	8333	**−0.13**/−0.14/−0.18	**0.54**/0.49/0.43	**0.23**/0.24/0.28	**−0.18**/−0.26/−0.49
	BIL	EXT	5229	0.16/0.21/**0.14**	**0.6**/0.58/0.58	**0.3**/**0.3**/0.31	**0.16**/0.09/0.15
	ALM	EXT	7670	**−0.16**/−0.17/**−0.16**	0.2/0.16/**0.21**	**0.51**/0.52/**0.51**	−0.09/−0.12/**−0.08**
		AVERAGE		−0.04/**−0.03**/−0.07	**0.45**/0.41/0.41	**0.35**/**0.35**/0.37	**−0.04**/−0.10/−0.14
SC_S (m/s)	TAR	EXT	8336	0.06/0.08/**0.05**	**0.36**/0.34/0.35	**0.16**/0.18/0.17	**−0.29**/−0.45/**−0.29**
	BIL	EXT	5858	−0.06/**−0.03**/−0.07	**0.19**/0.15/0.15	**0.14**/0.15/**0.14**	−0.2/−0.22/**−0.19**
	ALM	EXT	7619	0.14/0.17/**0.11**	**0.06**/−0.02/0.02	0.25/0.27/**0.24**	−0.67/−0.91/**−0.56**
		AVERAGE		0.05/0.07/**0.03**	**0.20**/0.16/0.17	**0.18**/0.20/**0.18**	−0.39/−0.53/**−0.35**
SC_D (°)	TAR	EXT	8336	6/**−0.1**/6	0.2/0.15/**0.22**	97/99/**93**	−0.1/−0.12/**−0.05**
	BIL	EXT	5858	9/−3/6	0.28/**0.31**/**0.31**	125/123/**122**	0.08/**0.11**/**0.11**
	ALM	EXT	7619	−35/**−29**/−32	0.06/**0.07**/0.06	138/139/**134**	−0.19/−0.18/**−0.15**
		AVERAGE		−7/−11/−7	0.18/0.18/**0.20**	120/120/**116**	−0.07/−0.06/**−0.03**
SLev (m)	BCN	TGS	8487	Not applicable	0.85/0.84/**0.92**	0.06/0.06/**0.04**	0.44/0.43/**0.59**
	TAR	TGS	8423		0.89/0.88/**0.92**	0.05/0.05/**0.04**	0.52/0.5/**0.6**
	BIL	TGS	8394		**0.99**/**0.99**/0.97	**0.13**/0.16/0.27	**0.9**/0.89/0.74
	ALM	TGS	8475		**0.88**/0.87/**0.88**	**0.06**/**0.06**/**0.06**	0.52/0.52/**0.53**
	GCA	TGS	8481		**0.99**/**0.99**/0.98	**0.08**/0.09/0.12	**0.87**/0.86/0.8
		AVERAGE			0.92/0.91/**0.93**	**0.08**/**0.08**/0.11	**0.65**/0.64/**0.65**

Table 5. Surface current error metrics at Tarragona and Gran Canaria domains vs. HF Radar, at a subset of representative points. The metrics comprise the time range from October 2019 to October 2020. Variables: SC_S (Surface current speed from HF Radar, in m/s), SC_D (Surface current direction from HF Radar, in °). The domain column identifies each SAMOA system. The Pnt denotes the point ID shown in subpanel c in Figures 6 and 7. The N column counts the available observations (hourly resolution) that have been used for computing the metrics. Error metrics (Bias, Correlation and Root-Mean-Square Error (RMS)) provided for each model solution (SAM_ADV (ADV), SAM_H3D (H3D) and IBI_PHY (IBI)). The last row of each variable represents the average error metric of all the sampling points given a specific SAMOA and test run. Bold numbers highlight the best performing dataset.

Variable	Dom	Pnt	N	Bias (ADV/H3D/IBI)	Correlation (ADV/H3D/IBI)	RMS (ADV/H3D/IBI)
SC_S (m/s)	TAR	P1	7457	0.02/0.05/−**0.01**	0.27/**0.30**/0.20	0.12/0.12/**0.11**
		P2	7696	**0.01**/0.02/−**0.01**	**0.42**/0.41/0.28	**0.11**/0.12/**0.11**
		P3	7561	0.02/0.03/**0.01**	0.32/**0.37**/0.25	0.12/0.12/**0.11**
		P4	6466	0.02/0.05/**0.01**	**0.40**/**0.40**/0.27	0.12/0.12/**0.11**
		P5	7158	**0.06**/0.08/**0.06**	**0.33**/0.32/0.26	**0.14**/**0.14**/0.15
		AVERAGE		0.03/0.05/**0.01**	0.35/**0.36**/0.25	**0.12/0.12/0.12**
	GCA	P1	7639	0.12/0.06/**0.04**	0.03/0.19/**0.28**	0.18/0.14/**0.11**
		P2	7669	0.06/0.09/**0.05**	0.19/0.2/**0.26**	0.15/0.14/**0.12**
		P3	7681	0.08/0.05/**0.04**	0.02/0.24/**0.30**	0.18/0.13/**0.11**
		P4	7660	0.11/0.1/**0.06**	0.16/0.24/**0.28**	0.17/0.15/**0.12**
		P5	7914	0.06/0.06/**0.01**	0.29/**0.41**/**0.41**	0.17/0.15/**0.13**
		AVERAGE		0.09/0.07/**0.04**	0.14/0.26/**0.31**	0.17/0.14/**0.12**
SC_D (°)	TAR	P1	7457	−4/−4/−**3**	0.21/**0.28**/0.25	105/**100**/102
		P2	7696	**2**/6/3	0.38/0.40/**0.47**	89/89/**84**
		P3	7561	17/**16**/24	0.22/0.24/**0.31**	108/108/**104**
		P4	6466	30/22/**20**	**0.29**/0.28/0.26	108/105/**104**
		P5	7158	21/19/**18**	0.16/**0.22**/0.20	107/**101**/**101**
		AVERAGE		13/**12**/**12**	0.25/0.28/**0.30**	103/101/**99**
	GCA	P1	7639	−30/−**12**/−16	0.07/**0.20**/0.18	97/**91**/97
		P2	7669	−6/4/**2**	0.16/0.34/**0.37**	81/**68**/73
		P3	7681	−26/−**7**/−13	0.09/**0.22**/0.20	94/**89**/96
		P4	7660	−2/3/**1**	0.14/**0.36**/0.32	74/**67**/77
		P5	7914	7/5/**3**	0.12/**0.37**/0.28	62/**56**/69
		AVERAGE		−11/−**1**/−5	0.12/**0.30**/0.27	82/**74**/82

4. Results

This Section shows the main results from the comparison of the different SAMOA model tests (SAM_INI, SAM_ADV and SAM_H3D; please refer to Table 1 and Figure 3b) together with their parent solution: IBI_PHY. Results are split in two Subsections: Section 4.1 gives information on the SAM_INI, SAM_ADV (and IBI_PHY) performances for the June 2019 period, at the 9 original SAMOA-1 domains (green dots at Figure 2): Barcelona (BCN), Tarragona (TAR), Bilbao (BIL), Almería (ALM), Ferrol (FER), La Gomera (GOM), Gran Canaria (GCA), Santa Cruz de la Palma (PAL), Tenerife (TEN); whereas Section 4.2. assesses SAM_ADV, SAM_H3D and IBI_PHY from October 2019 to October 2020 at 5 SAMOA domains (i.e., BCN, TAR, BIL, ALM and GCA).

4.1. Impacts in SAMOA Solutions Related to Changes in the OBC Treatment and the Use of a Bulk Formula to Deal with Atmospheric Forcing (SAM_ADV vs. SAM_INI)

This first SAM_INI configuration was carried out based on the SAMOA-1 model set-up after the update of the HARMONIE atmospheric forcing (see Section 3.1). The upgraded SAMOA set-up (SAM_ADV test) related to changes in the OBC treatment and the use of a Bulk formula is further described in Section 3.2.

The error metrics from both SAMOA tests and the IBI_PHY solution are summarized in Table 3. In general, it is seen that SAM_ADV improves considerably the SAM_INI Sea Surface Temperature (SST):

(i) Bias drop from 0.20 °C (SAM_INI) to −0.03 °C (SAM_ADV), even moderately better than IBI_PHY (0.07 °C).
(ii) Correlations were close to 0.8 for all three models.
(iii) IBI_PHY had lower RMS (0.44 °C) than SAM_INI (0.6 °C) and SAM_ADV (0.5 °C)
(iv) COE for IBI_PHY (0.29) and SAM_ADV (0.21) were similar, substantially higher than SAM_INI (0.07).

At a more detailed level, Figure 4 presents the SST at Gran Canaria coastal buoy. Despite that the three models are able to capture the intradaily cycles, SAM_ADV (green line) and IBI_PHY (black line) show similar trends. Additionally, for the first part of the month, SAM_ADV shows more agreement with the observations than SAM_INI (blue line). However, also note that at the second half of the month, there was a sudden rise in temperatures that only SAM_INI captured (though overestimating it).

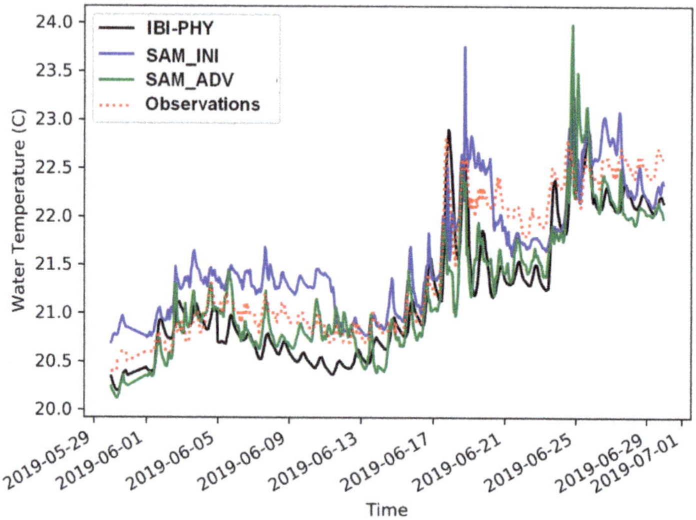

Figure 4. Observed and modelled Sea Surface Temperature (SST) at Gran Canaria PdE EXT mooring buoy (see location in Table 2) in June 2019. Units in Celsius degrees. The dotted red line represents the observations; the solid black line is the parent solution (CMEMS IBI_PHY); the solid blue line is the initial SAMOA set-up (SAM_INI); and the solid green line represents the SAMOA solution obtained with the advanced set-up (SAM_ADV).

Regarding mean sea-level, SAM_INI and SAM_ADV show similar performance, improving considerably IBI_PHY, especially at meso-tidal (GOM, GCA and PAL domains) and macro-tidal environments (BIL, FER, VIG). The three models show very high correlations (close to 0.96); but the RMS is lower (close to 0.05 m) for the SAMOA set-ups, whereas IBI_PHY reaches 0.14 m. The same pattern is also found for the COE: both SAMOA have values close to 0.77, but IBI_PHY does not surpass 0.66.

Sea Surface Salinity (SSS) remains with similar metrics between SAMOAs, not improving the IBI_PHY performance. RMS are identical and the COE is negative at all three. Hence, SAMOA does not add new information on this variable.

Surface currents also exhibit similar metrics amongst the different models. Note though, that at a qualitative assessment level, SAM_ADV has shown more consistency with the IBI_PHY patterns than SAM_INI. These phenomena especially happened at the SAMOA boundaries, where the presence of spurious rim-currents was partially alleviated. The consistency of the SAM_ADV OBC scheme was confirmed with the SAM_H3D test (next Subsection).

4.2. Impact in the SAMOA Solutions Related to Changes in the Physics and in the Temporal Frequency of the Imposed Data along Boundaries (SAM_H3D vs. SAM_ADV)

As pointed in the previous section, SAM_ADV seemed to enhance coherency between the SAMOA systems and their parent solutions (IBI_PHY). However, differences were still high in some cases, affecting the whole domain and specially along the boundaries. The availability of new hourly-3D CMEMS IBI-MFC products, joint with the upgrade of some physical parametrizations, allowed building the SAM_H3D set-up at 5 systems that were compared with the SAM_ADV and the IBI_PHY parent solution. Their results are summarized in Tables 4 and 5 and Figures 5–7.

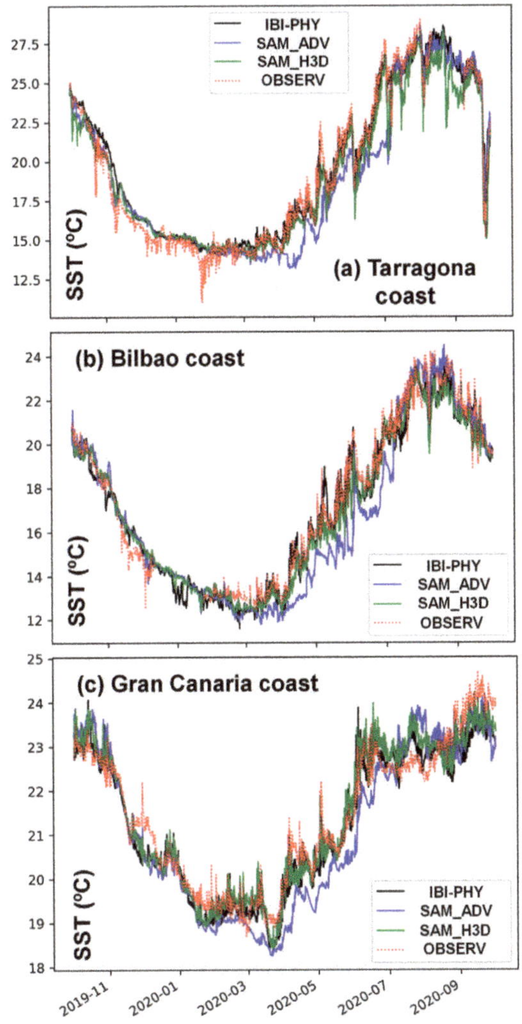

Figure 5. Observed and modelled Sea Surface Temperature (SST) at 3 coastal buoys. (**a**): Tarragona coast. (**b**): Bilbao coast. (**c**): Gran Canaria coast. Period shown: October 2019–October 2020. Units in Celsius degrees. The dotted red line represents the observations; the solid black line is the CMEMS IBI-MFC; the solid blue line is the current operational SAMOA set-up (SAM_ADV); the solid green line is the proposed set-up nested into hourly 3D IBI_PHY data (SAM_H3D). Further information about the buoy stations (location and mooring depth) can be found in Table 2.

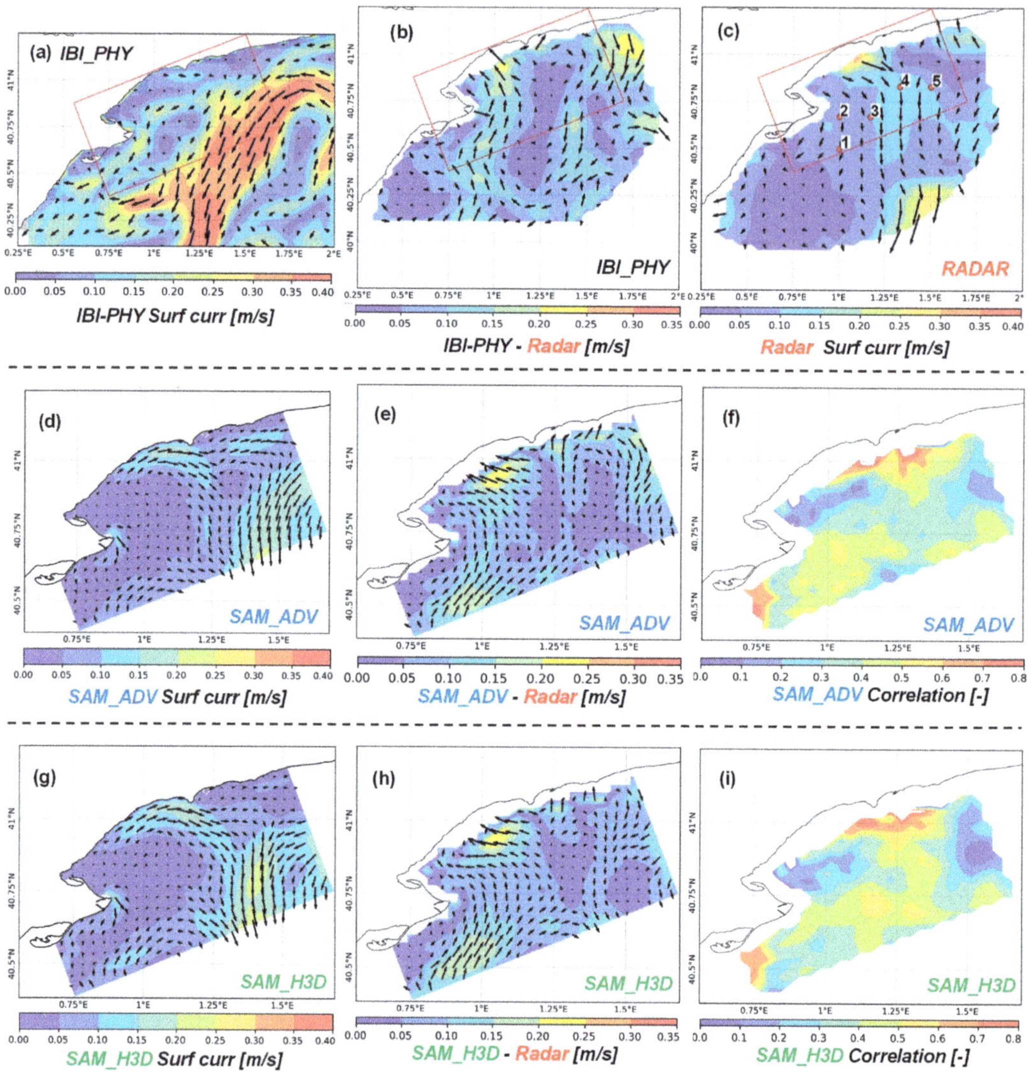

Figure 6. HF Radar vs. model solutions (SAM_H3D and SAM_ADV experiments, together with the Copernicus IBI parent solution) at Tarragona during March 2020. Monthly-averaged surface current field modelled and observed, depicted at subpanels (**a**) (IBI_PHY)-(**d**) (SAM_ADV)-(**g**) (SAM_H3D), joint with (**c**) (HF Radar), respectively. However, the error metrics subpanels (**b**,**e**,**f**,**h**,**i**) are computed from hourly-averaged fields. Model-observation monthly biases for IBI_PHY, SAM_ADV and SAM_H3D are depicted in the central column (subpanels **b**,**e**,**h**); SAM_ADV and SAM_H3D model correlation fields with HF Radar observations are shown in subpanels (**f**,**i**). Units in m/s. The red rectangle in subpanels (**a**–**c**) the SAMOA coastal domain. Red dots at subpanel (**c**) represent locations where time series metrics are computed (summarized in Table 5).

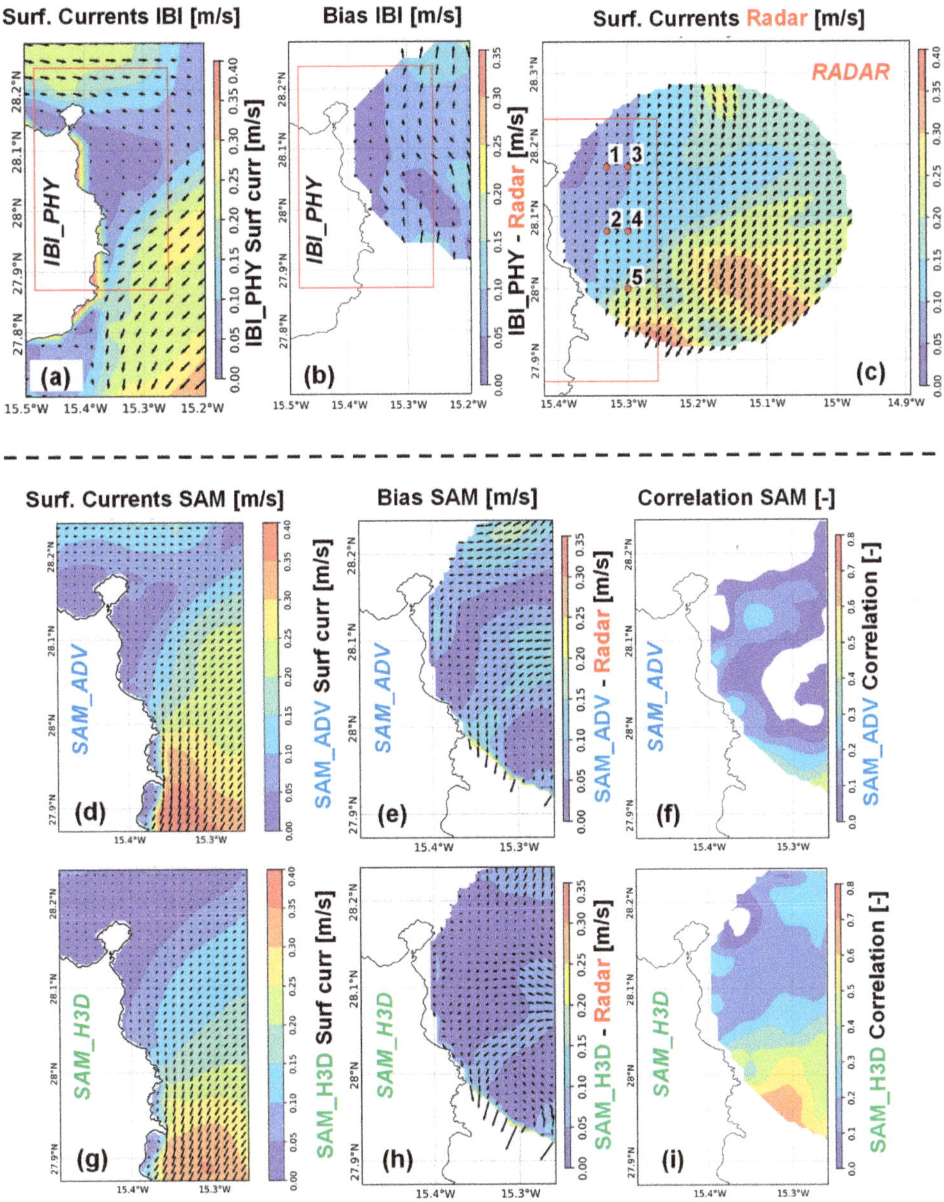

Figure 7. HF Radar vs. model solutions (SAM_H3D and SAM_ADV experiments, together with the Copernicus IBI_PHY parent solution) at Gran Canaria during September 2020. Monthly-averaged surface current field modelled and observed, depicted at subpanels (**a**) (IBI_PHY)-(**d**) (SAM_ADV)-(**g**) (SAM_H3D), joint with (**c**) (HF Radar), respectively. However, the error metrics subpanels (**b**,**e**,**f**,**h**,**i**) are computed from hourly-averaged fields. Model-observation monthly biases for IBI_PHY, SAM_ADV and SAM_H3D are depicted in the central column (subpanels **b**,**e**,**h**); SAM_ADV and SAM_H3D model correlation fields with HF Radar observations are shown in subpanels (**f**,**i**). Units in m/s. The red rectangle in subpanels (**a**–**c**) the SAMOA coastal domain. Red dots at subpanel (**c**) represent locations where time series metrics are computed (summarized in Table 5).

Regarding SST (Table 4), lower biases are found in the SAM_H3D (deep water TAR buoy, ALM and GCA) and IBI_PHY (BCN, BIL and coastal TAR buoy). Correlation is close to 0.98 at IBI_PHY and SAM_H3D, slightly higher than for the SAM_ADV case (0.96); pointing the correct capture of the intraday variability. IBI_PHY presents lower RMS (0.65 °C) than SAM_H3D (0.72 °C) and SAM_ADV (0.97 °C). The COE shows good agreement between SAM_H3D (0.83) and IBI_PHY (0.84), whilst SAM_ADV is significantly lower (0.76).

Gran Canaria (GCA) is the system in which SAM_H3D clearly outperforms IBI_PHY and SAM_ADV: (i) the bias is lower (−0.02 °C) than in IBI_PHY (−0.15 °C) and SAM_ADV (−0.31 °C) cases; (ii) SAM_H3D RMS (0.4 °C) is in the same line than IBI_PHY (0.38 °C), and lower than SAM_ADV (0.59 °C).

Error metrics are consistent with the time-series plots (Figure 5), as the model performance between IBI_PHY (black line) and SAM_H3D (green line) are fairly similar at the coastal buoys. Both systems capture the main trends and seasonal changes. However, SAM_ADV (blue line) presented a consistent negative/positive bias in the Winter-Spring/Summer, most probably due to heat-fluxes mismatches at the model domain. Note also that the error metrics between SAM_ADV and SAM_H3D present more agreement at deep waters (closer to the SAMOA domain boundary) than at the coastal locations (far away from the boundaries, heavily influenced by coastal circulation processes and a proper advection scheme).

Surface salinity is improved in Tarragona with the SAMOA solution, though SAM_ADV shows slightly better metrics (−0.13 PSU bias, 0.54 correlation) than SAM_H3D (−0.14 PSU, 0.49) and IBI_PHY (−0.18 PSU, 0.43). At Bilbao and Almería, the error metrics are quite similar for the three models, although SAM_ADV and IBI_PHY moderately outperform SAM_H3D.

As in Section 4.1, SAMOA Sea Level solution outperforms notably IBI_PHY at meso-tidal and macro-tidal environments: the RMS at these two places is significantly lower with SAMOA (0.13 m Bilbao, 0.08 m Gran Canaria) than IBI_PHY (0.27 m Bilbao, 0.12 m Gran Canaria). COE presents important differences (close to 0.9 in SAMOA, whilst IBI_PHY has 0.74 in Bilbao and 0.8 at Gran Canaria).

Sea Level at Barcelona and Tarragona exhibit similar behaviour on the three models. However, at these two domains, IBI_PHY solution exhibits a higher COE (around 0.6) and the long-term metrics underperform both at SAM_ADV and SAM_H3D.

Apart from SST, the most prominent differences at the SAM_H3D are found in the sea surface currents, both at current speed and direction (Table 4). Current direction metrics from SAMOA outperform at deep-water buoys: (i) SAM_H3D (−1.4°) bias is lower than IBI_PHY (−4.6°); (ii) correlation and RMS are also better at SAM_H3D (0.3, 74.2°) than at IBI_PHY (0.27, 84.2°).

Regarding surface current speed at Tarragona deep-water buoy, significant improvement in the correlation can be found (0.35 (SAM_ADV) vs. 0.36 (SAM_H3D) vs. 0.25 (IBI_PHY)). RMS is similar for all three datasets (0.16 vs. 0.18 vs. 0.17, respectively); but the bias is lower in IBI_PHY (0.01 m/s), and higher in SAM_H3D (0.05 m/s) than in SAM_ADV (0.03 m/s).

Tarragona and Gran Canaria are the two SAMOA that have coverage from Radar HF. Hence, special emphasis would be given. Error metrics have been computed at five specific points (Table 5), representative of the area (see their position at subpanels (c) in Figures 6 and 7).

Figure 6 shows the monthly-averaged HF-radar measurements, model results and error metrics in Tarragona during March 2020. In that month, there were four clustered NW inland wind-jets in the first week of the month, reaching hourly wind-speeds up to 20 m/s. From 10th March until the end of the month, the atmospheric conditions were fairly stable, though: wind speed had a mean average close to 3–4 m/s. Note however, that at Figure 6c, even at a monthly-averaged scale, these inland wind-jets are the most remarkable signature.

Both SAMOA systems reproduce the origin and the propagation of wind-jet. But SAM_H3D is the solution that shows more resemblance with the surface HF radar (Figure 6g–i), exhibiting correlations close to 0.5 at the wind-jet influence area. IBI_PHY does not reproduce the wind-jet and it tends to overestimate the Northern Current and General circulation, remaining some influence from circulation features associated to extreme storm events occurred in previous months (i.e., Mediterranean Storm Gloria).

At the Gran Canaria case (Figure 7) there were persistent, but moderate, NE wind speed (lower than 8 m/s, with monthly-averaged values close to 3 m/s). The monthly-averaged HF radar exhibits consistency (Figure 7c) with these wind field conditions. Surface circulation is mainly SW, exhibiting higher values at the South (close to 0.30 m/s) than in the North (around 0.1 m/s). IBI_PHY (Figure 7a) exhibits a clockwise gyre near the Gran Canaria harbour. This eddy is also reproduced in SAM_ADV set-up (Figure 7d–f), but it is not in SAM_H3D (Figure 7g–i).

SAM_ADV overestimates currents at the Southern part of the domain (Figure 7d). Biases at the SAM_H3D boundary are lower than SAM_ADV and even IBI_PHY. Additionally, SAM_H3D exhibit higher correlation at the Southern part of the domain (close to 0.7). Note that in SAM_ADV, the correlation is negative at the central part of the Eastern boundary. These findings are consistent regardless of the measurement devices, as the analysis performed with RadarHF and deep-water buoy observational data denote similar patterns in terms of error metrics.

5. Discussion

In the previous section, the main results were shown from the different SAMOA model set-up tests (SAM_INI, SAM_ADV, SAM_H3D) performed at different coastal and port domains. All the SAMOA simulations were validated with observational data sources for sea level, surface currents, SST and SSS, and compared with its parent solution (IBI_PHY).

The main aim of these SAMOA sensitivity tests is to assess the impacts that potential upgrades in the SAMOA set-up have on its solution at different domains. Once the quality control is performed (i.e., by comparing SAMOA solutions with IBI_PHY solutions, joint with the available observations at each coast/port domain), it can be verified whether the proposed set-up may result in a new operational SAMOA release.

Prior to deciding a new operational SAMOA release, it would be needed to verify if the new solution improves/degrades the control one (and at what level). Furthermore, it is not enough to test the model set-up upgrade in a single SAMOA domain; it is needed to be assessed how the positive impacts extend across the different SAMOA port systems.

Table 6 summarizes the results of the SAM_INI vs. SAM_ADV and the IBI_PHY parent solution (see SAMOA model set-up descriptions in Section 3.2. and the main validation results in the Section 4.1). In general terms, SAM_ADV improves sea level significantly in respect to IBI_PHY, but remains similar to the SAMOA control run (SAM_INI). The SST simulated with the SAM_ADV set-up improves the SAM_INI one, but the solution is worse than the one from the Copernicus IBI_PHY product. Surface salinity remains similar in all the model solutions, with SAM_ADV outperforming in the Bilbao domain. With respect to surface current speed, a slightly better performance of IBI_PHY is seen in the three cases where observations are available, but both SAMOA experiments outperform the surface current direction; the three solutions may be considered as similar for currents.

The proposed SAMOA SAM_ADV set-up improves notably the Sea Level respect the IBI-MFC solution, especially at meso and macro-tidal environments. The higher spatial resolution of SAMOA systems allows to better capture of the harbour geometry and coastal surroundings, being a plausible explanation of such improvement in terms of sea level. The use of the new COARE bulk formula introduced in the SAM_ADV test improves (significantly) the sea surface temperature metrics with respect to the SAM_INI control one in 5 (4) out of the 7 cases analysed. This significant improvement led SAM_ADV SAMOA set up to be transitioned into operations in October 2019, thus becoming the current (end of 2021) operational set-up.

Table 6. Summary of the impacts of the SAM_ADV set-up. Performance of SAM_ADV versus the control (SAM_INI) one and the parent solution (IBI_PHY), comparing in those domains where observations are available (for sea level, SST, SSS and surface currents). Legend: (SAM_ADV relative performance with respect to SAM_INI set-up/SAM_ADV relative performance with respect to IBI_PHY). Acronyms to define the relative performance of the proposed set-up: (I *: relevant improvement/I: improvement/S: similar performance/W: mild worsening).

SAMOA System	Sea Level	Surface Currents	SST	SSS
Barcelona (BCN)	(S/W)	-	(W/S)	-
Tarragona (TAR)	(I/I)	(S/S)	(W/W)	(S/S)
Bilbao (BIL)	(S/I *)	(S/S)	(I/S)	(I/I)
Ferrol (FER)	(S/I *)	-	(I */W)	-
Vigo (VIG)	(S/I)	-	-	-
Almería (ALM)	(S/I)	(S/W)	(I */I)	(S/S)
Gran Canaria (GCA)	(S/I *)	-	(I */W)	-
La Gomera (GOM)	(I/I)	-	-	-
St. Cruz Palma (PAL)	(S/I *)	-	-	-
Tenerife (TEN)	-	-	(I */W)	-

Improvements in operational systems tend to be incremental. The proposed SAM_H3D configuration enhances the SAM_ADV one in Surface Currents and SST; whereas Sea Level metrics remain similar (see summary of relative performances in Table 7). SAM_H3D also outperforms IBI_PHY at the following issues: (i) Sea Level in meso-tidal and macro-tidal environments; (ii) Surface current speed correlation and surface current direction; (iii) SST at intermediate and shallow waters (i.e., coastal buoys); (iv) Surface Salinity at Tarragona. The main reasons for this behaviour are discussed below.

Table 7. Summary of the impacts of the SAM_H3D proposed set-up. Performance of SAM_H3D versus the control (SAM_ADV) one and the parent solution (IBI_PHY), comparing in those domains where observations are available (for sea level, SST, SSS and surface currents). Legend: (SAM_H3D relative performance with respect to SAM_ADV set-up/SAM_H3D relative performance with respect to IBI_PHY). Acronyms to define the relative performance of the proposed set-up: (I *: relevant improvement/I: improvement/S: similar performance/W: mild worsening).

SAMOA System	Sea Level	Surface Currents	SST	SSS
Barcelona (BCN)	(S/W)	-	(I */S)	-
Tarragona (TAR)	(S/W)	(I/I)	(I */I)	(S/I *)
Bilbao (BIL)	(S/I *)	(S/S)	(I */S)	(W/W)
Almería (ALM)	(S/S)	(S/W)	(I/S)	(W/W)
Gran Canaria (GCA)	(S/I *)	(I */I)	(I */I)	-

First, better surface current correlations at SAM_H3D (respect to SAM_ADV) are related to three factors: (i) tidal-induced currents are correctly nested with the parent solution; (ii) wind-induced currents are better modelled with the COARE 3.6 parametrization; and (iii) SAMOA solutions exhibit more spatial heterogeneity than IBI_PHY. This last issue (heterogeneity) can be partly explained due to SAMOA higher spatial resolution. For instance, the model grid is able to reproduce sharp bathymetric pressure gradients; joint with the atmospheric forcing data that features local orographic constraints.

In this sense, atmospheric forcings are an important cause of the mismatch. SAMOA uses HARMONIE that has a systematically higher wind speed bias than ECMWF: around 1.5 m/s yearly-averaged difference along the coastal and deep waters of Iberian Peninsula, Balearic and Canary Islands. At the inland stations, biases between the two forcings are similar, mostly due to more stations to be assimilated. Moreover, HARMONIE overestimation can increase up to 7 m/s during extreme events, such as the recent Storm Gloria (January 2020) (Sotillo et al., 2021a [47]); whilst ECMWF tend to lower overestimations (close to 3 m/s). Such HARMONIE overestimation is one of the main reasons for the higher bias and RMS at some stations in the SAMOA surface currents, with respect to IBI_PHY.

Wind direction, though, provides a trade-off: whereas ECMWF biases can reach close to 30° at coastal locations, HARMONIE remains more accurate (biases around 10–20°). The higher horizontal resolution of HARMONIE (2.5 km vs. 10 km) is able to solve the orographic constraints, island boundaries and intraday processes (such as sea-breezes) more accurately. For instance, the Tarragona system is prone to wind jets (Grifoll et al., 2016 [48]), mainly constrained by the mountain-ranges close to the coastal fringe. As can be seen in the Tarragona case (Figure 6a), IBI_PHY forced with ECMWF-IFS cannot reproduce the wind jet. ECMWF-IFS coarse resolution (10 km) is not enough for modelling the narrow sharp topographic gradients between the mountain ranges near the Ebro Delta, but HARMONIE does it (see the SAMOA solutions in Figure 6d,g).

Hence, proper current direction is an important remark, especially when tidal-currents act concomitant to wind-induced currents. Errors in the wind direction can lead to wind-induced stresses; that albeit accurate in modulus, incorrectly reinforce/weakens surface currents driven by other physical processes.

IBI_PHY uses CORE bulk formula (Large and Yaeger, 2004 [49]), SAM_ADV the COARE 3.0 (Fairall et al., 2003 [33]), and SAM_H3D the COARE 3.5 (Edson et al., 2013 [39]). Charnock coefficient in COARE 3.5 (and its parametrized surface-drag coefficient) is lower for moderate winds than in COARE 3.0 (Brodeau et al., 2017 [50]). In COARE 3.0, Charnock coefficient is set to constant (0.011) at wind speeds from 0 to 10 m/s; whereas in COARE 3.5, Charnock is set to 0.006 until 6 m/s. Due to the systematic overestimation of the HARMONIE wind fields, the reduction of the wind-stresses can be a factor for the enhancement of SAM_H3D surface currents metrics, especially in correlation and general bias, when moderate wind conditions prevail.

Another difference between SAM_ADV and SAM_H3D is that the gustiness parameter is lower in COARE 3.5 than in the former release; and that may boost surface stresses at extreme wind regimes. Gustiness effects are not included in CORE, that may be a secondary factor for better IBI_PHY RMS metrics (ECMWF atmospheric forcings influence, being a primary one). Nonetheless, CORE drag coefficients tend to be higher than COARE at moderate wind speeds; and vice versa, from moderate to high winds (Pelletier et al., 2018 [51]). Further research in air-sea interaction would shed some light on this key issue.

SAM_H3D is the solution that shows more resemblance with the spatial signature from surface HF radar (Figures 6 and 7). Note however, that both SAMOA solutions are heavily constrained by the IBI_PHY solution. For instance, the spurious SW fluxes at the Eastern corner at the Tarragona model (Figure 6) are mainly due to the IBI parent solution. Eddies in the boundaries and corners, such as the clockwise one close to the Southern corner can be also problematic, especially when IBI patterns are unrealistic. The SAM_H3D nesting ensures mass and momentum conservation from the parent to child grids, but errors in the former can be easily propagated.

This latter phenomenon may partly explain the low predictive skill in the Almería system, both with IBI_PHY and SAMOA. A possible reason may be transient spurious mesoscale structures in IBI_PHY at Cabo de Gata (Figure 2). High surface current direction bias is found at the deep-water EXT buoy (almost 30°, in comparison with the 3° bias found when compared at the Bilbao or Tarragona EXT buoys). Note that these errors in direction are inherited from the parent solution (IBI_PHY), suggesting that the transport direction is incorrect. This phenomenon also affects both heat and salinity fluxes. Hence, a plausible solution to minimize undesired influences related to spurious dynamic features identified in the IBI_PHY parent solution may be to expand the SAMOA coastal domain, fixing the SAMOA boundaries outside of the areas where such spurious structures are favoured in the IBI_PHY solution; then generating a SAMOA intrinsic solution, less constrained by the limited-area domain and the IBI_PHY data imposed as boundary condition.

Regarding SST, the enhancement of SAM_H3D solution may be due to three factors: (i) better characterization of the heat fluxes with COARE 3.5; (ii) consistent circulation fields and (iii) the use of HSIMT-TVD advection scheme for passive tracers. Despite that SAMOA does not have data assimilation, the one-year metrics are close, or even better (at some cases) than IBI_PHY. Note that the Copernicus Regional System counts with a data assimilation scheme. Then, satellite SST fields and in-situ profiles data are assimilated in the IBI_PHY analyses.

In this sense, Gran Canaria highlights the role of coastal circulation in shallow-depths heat fluxes, as the enhancements of the circulation fields are consistent with SST ones. The RMS is lower in IBI_PHY than SAMOA, though; most probably due to the radiation forcings (be it short or long-wave), that have higher variability in HARMONIE than ECMWF.

SAMOA significantly improves the sea level, especially in those areas with meso and macro-tides. There is no important difference between SAM_ADV and SAM_H3D, suggesting that the main improvements come from the increase of horizontal resolution, that lead to better capture of the harbour geometry and coastal surroundings. Water mass fluxes and piling-up effects are better reproduced.

However, the performance is substantially lower at the Barcelona and Tarragona cases (as shown in the Table 3 metrics), in which all the three models require further assessment on two aspects:

(i) Inclusion of low frequency astronomical constituents, such as the solar semi-annual (Ssa) and solar annual (Sa) constituents. For these two specific systems, low-frequency constituents have an equivalent order of magnitude than higher frequency harmonics: for instance, in Tarragona, the sum of the amplitudes of the M2 (3.97 cm) and K1 (3.71 cm) are close to the SA value (7.21 cm) (Pérez-Gómez, 2014 [42]). Low-frequency constituents require long-term observations for its correct estimation; and this remains a main shortcoming for their inclusion in operational models.

(ii) The non-tidal residual term is not properly modelled in IBI_PHY at specific synoptic conditions, due to no inclusion of the whole Mediterranean basin subinertial transport. The IBI_PHY storm-surge solution for the recent storm Gloria was analysed in Pérez-Gómez et al., 2021 [52], then identifying a persistent underestimation of the positive surge. Such underestimation was mainly related to the fact that IBI system does not include barotropic mass-transport from the Eastern Mediterranean subbasin towards the Western Mediterranean one. Correction for this misbalance in the MSLP-driven transports between the Mediterranean sub-basins is in the IBI-MFC roadmap, and it is expected to be overcome in forthcoming IBI_PHY operational releases.

Wave-driven currents are another point in which SAMOA will devote further efforts. At present (late 2021), the Copernicus IBI_PHY solution is one-way coupled with the IBI-WAV system (based on the WAM (WAMDI Group, 1988 [53]) spectral wave model) by adding (i) the Stokes drift, (ii) wave-induced mixing and (iii) wind drag coefficients formulas based on the sea-state. This addition may be particularly important during wave storms, as suggested in Sotillo et al., 2021a [47] and Lorente et al., (2021) [54]. Current developments of the SAMOA system, developed within the EuroSea Project, are going in the same direction, with strong focus on the local processes beneath intermediate and shallow waters.

Same focus on local coastal processes, that has become the leitmotif of the SAMOA forecasting system, will also be taken on improving the land-sea connection. Proper modelling of salinity fluxes require operational run-off discharge products, such as the one analysed in this issue (Sotillo et al., 2021b [55]), rather than climatological values. All these improvements, albeit ambitious, represent a clear SAMOA roadmap for the coming years. Moreover, as suggested in the case of Gran Canaria (i.e., SST improves when surface currents do it), the synergistic gain from addressing all of them, would enhance the predictive skill of SAMOA operational products at an integrated level.

6. Conclusions

6.1. Assessment of the SAMOA Coastal Forecasting Service Upgrades

The Puertos del Estado SAMOA coastal and port ocean forecast service has delivered operational ocean forecast products to Spanish Port Authorities since 01/2017. The service provides daily forecasts of sea-level, circulation, temperature and salinity fields at horizontal resolution that range from 350 m (Coastal domains) to 70 m (Port domains). The system was successfully implemented at nine ports in the SAMOA-1 project (2014–2017); and it was expanded, improving the model set-up, to 31 ports at the end of SAMOA-2 project (2018–2021).

This contribution has assessed the incremental upgrades that have been developed in the SAMOA system from 2018 to 2021. In chronological order, it has been tested (at different ports) the following three main set-ups:

(i) The SAM_INI configuration, analogous to the one used in SAMOA-1, but substituting the former AEMET HIRLAM atmospheric forcing by the new HARMONIE 2.5 km forcing data.
(ii) The SAM_ADV set-up, focused on upgrading the open boundary condition scheme treatment and the computation of surface atmosphere-ocean fluxes through the COARE 3.0 bulk formula.
(iii) The SAM_H3D set-up, in which it is proposed (a) the use of the hourly-3D IBI-MFC forecast dataset, (b) the update of the bulk formula for surface fluxes (COARE 3.5), (c) the use of a new advection scheme (HSIMT-TVD) for passive tracers and (d) the reduction of the wind-stresses due to the effect of surface currents.

The protocol proposed to assess these SAMOA model set-up upgrades focuses on evaluating in the Spanish ports the simulation of ocean variables (typically, the sea level, currents, surface temperature and salinity), validating them with available in-situ and remote-sensed observations (from tide gauges, mooring buoys and HF Radar stations). To check model sensitivities, as well as to evaluate the potential added value related to the proposed modifications, each SAMOA model test run is always compared with a control SAMOA run (i.e., the operational version of the SAMOA system at the time of the test), and with the CMEMS IBI_PHY parent solution.

Applying this protocol, it was demonstrated that:

(i) The three proposed SAMOA tests improve notably the simulation of sea level with respect to the IBI_PHY parent solution, especially at meso- and macro-tidal environments. The SAMOA higher spatial resolution allows to better capture of the harbour geometry and coastal surroundings.
(ii) The use of the COARE bulk formula (introduced in the SAM_ADV set-up) improved sea surface temperature (SST) metrics respect to SAM_INI; but SAM_H3D outperforms SAM_ADV and the skill is close to the IBI_PHY solution. This SAM_H3D improvement in SST can be explained by (a) improvements in the surface fluxes physics, (b) the use of HSIMT-TVD for passive tracers and especially (c) the overall improvement of the surface circulation fields.
(iii) This improvement of the circulation fields arises from two factors: (a) the SAM_H3D nesting ensures mass and momentum conservation from the IBI_PHY parent solution (hourly updated) to the SAMOA child grids along boundaries; and (b) SAM_H3D benefits from the high-resolution HARMONIE forcing. These two factors allow proper modelling of joint tidal and wind-induced currents. Consequently, local effects such as in-land wind-jets (case of Tarragona HF radar), and those influenced by topographic gradients exhibit good metrics in SAM_H3D, even slightly better than in IBI_PHY. At the Gran Canaria case, SAM_H3D is the model solution that shows more resemblance with the spatial signature from surface HF radar. However, the Almería case also highlights how the SAM_H3D nesting scheme can propagate errors or spurious circulation features from IBI_PHY into the SAMOA solution.

(iv) Finally, this contribution summarizes the impact of the incremental model upgrades performed in the SAMOA operational circulation system since 2018. Indeed, two of the SAMOA model set-ups here shown (SAM_INI and SAM_ADV) have been the base for the last two operational SAMOA releases (launched in October 2018 and October 2019, respectively). The SAM_ADV model configuration is the current SAMOA operational set-up, but it is expected to be upgraded in 2022 by the also here tested SAM_H3D one. This contribution demonstrates that the proposed SAM_H3D configuration enhances the current SAMOA solution, especially in terms of surface currents and SST.

6.2. Recommendations for Coastal Forecasting Services and Future SAMOA Perspectives

This contribution has presented the SAMOA circulation system as a downstream service that benefits from Copernicus regional products. The authors recommend the use of products that ensure proper nesting between the parent solution and the local application. In this sense, the use of the hourly-3D IBI-MFC forecast dataset has contributed to improve the circulation and sea temperature fields (as shown in the SAM_H3D set-up). It has been shown, that the error metrics decrease when the boundary conditions have hourly-averaged resolution (be it sea-level, currents, sea temperature and salinity). This product ensures more consistency with the parent solution than the previous SAMOA releases (hourly-averaged sea-level and barotropic currents; but daily-averaged total currents, temperature and salinity).

High resolution atmospheric forcings and the proper air-sea parameterizations have also proven to be relevant in modelling local processes, such as wind-jets. The COARE 3.5 bulk-formula (implemented in SAM_H3D) has shown better skill than COARE 3.0. The HSIMT-TVD scheme contributed to improve sharp gradients in the sea surface temperature solution.

This research has also identified the following future perspectives for the development of SAMOA:

(i) To improve the modelling of air-sea interaction processes, with focus on surface stresses and heat fluxes.
(ii) To expand the coastal domains, with the aim of generating intrinsic solutions that are less dependent on limitations from the parent solution (such as its inability to reproduce accurately certain topographic gradients, due to its coarser horizontal resolution).
(iii) To include wave-driven effects such as (a) the Stokes drift interaction with the currents, (b) wave-induced mixing parameterizations and (c) wind drag coefficients formulas based on the sea-state.
(iv) To include low-frequency astronomical tide constituents. These low-frequency harmonics are especially relevant in those SAMOA systems located at the Western Mediterranean sub-basin.
(v) To verify and to correct (if needed) the subinertial transport at specific SAMOA systems, then enhancing the forecast skill under certain synoptic conditions.
(vi) To couple the SAMOA systems with operational coastal run-off discharge systems.

The joint action of these synergistic developments will reinforce SAMOA leitmotif: a downstream service, fed by Copernicus regional products, that enhances predictive capabilities on local coastal processes at Spanish Port zones.

Author Contributions: Conceptualization, M.G.S., E.Á.F. and M.G.-L.; methodology, M.G.-L., M.G.S.; software, M.G.S., M.G.-L. and M.M.; validation, M.G.-L.; formal analysis, M.G.S., M.G.-L., E.Á.F.; investigation, M.G.S., M.G.-L., M.M.; resources, E.Á.F. and M.G.S.; data curation, M.G.-L. and M.M.; writing—original draft preparation, M.G.-L., M.G.S.; visualization, M.G.-L., M.G.S.; supervision, E.Á.F., M.G.S. and M.E.; project administration, E.Á.F., M.G.S. and M.E.; funding acquisition, E.Á.F. All authors have read and agreed to the published version of the manuscript.

Funding: This research received no external funding.

Institutional Review Board Statement: Not applicable.

Informed Consent Statement: Not applicable.

Data Availability Statement: The SAMOA circulation forecast data and the observations used in this contribution can be downloaded from Puertos del Estado OpenDAP catalog: (https://opendap.puertos.es/thredds/catalog.html, accessed on 18 January 2022). The Copernicus Regional IBI-MFC solution can be downloaded at Copernicus Marine Environmental Monitoring Service (CMEMS) (http://marine.copernicus.eu/, accessed on 18 January 2022).

Acknowledgments: The authors acknowledge support from the SAMOA-2 initiative (2018–2021), co-financed by Puertos del Estado (Spain) and the Spanish Port Authorities. This contribution has been conducted using E.U. Copernicus Marine Service Information. Specifically, from its NRT forecast products at the IBI area. Likewise, ocean in-situ and HF-radar observations from the Puertos del Estado monitoring network are also duly acknowledged.

Conflicts of Interest: The authors declare no conflict of interest.

Abbreviations

The following abbreviations are used in this manuscript:

ADCP	Acoustic Doppler Current Profiler
AEMET	Agencia Estatal de Meteorología de España (Spanish Met Office)
CMA	Cuadro de Mando Ambiental (Environmental Control Panel, see Section 2)
CMEMS	Copernicus Marine Service
COARE	Coupled Ocean-Atmosphere Response Experiment
COE	Coefficient of Efficiency (see Section 3.4, Equation (1))
CORE	Coordinated Ocean-Ice Reference Experiment
CST	Coastal buoy from the REDCOS network (Tables 3–5)
CTD	Conductivity, Temperature and Depth
ECMWF	European Centre for Medium-Range Weather Forecasts
ECMWF-IFS	European Centre for Medium-Range Weather Forecasts—Integrated Forecast System
ESEOO	Establecimiento de un Sistema Español de Oceanografía Operacional (Project)
EXT	Deep-water buoy from the REDEXT network (Tables 3–5)
EuroGOOS	European Global Ocean Observing System
EU	European Union
HARMONIE	HIRLAM-ALADIN Research on Mesoscale Operational NWP in Euromed
HF Radar	High Frequency Radar
HIRLAM	High Resolution Limited Area Model
HSIMT-TVD	High-Order Spatial Interpolation at the Middle Temporal level coupled with a Total Variation Diminishing limiter
IBI-MFC	Iberian-Biscay-Ireland—Monitoring and Forecasting Centre
IBI_PHY	IBI-MFC Circulation Forecasting System
IBI-WAV	IBI-MFC Wave Forecasting System
NARVAL	North Atlantic Regional VALidation
NRT	Near-Real Time
OBC	Open Boundary Condition
PdE	Puertos del Estado (Spanish State Ports, public enterprise for port management)
REDEXT	Deep-water Buoy Monitoring Network from Puertos del Estado
REDCOS	Coastal Buoy Monitoring Network from Puertos del Estado
REDMAR	Tidal Station Monitoring Network from Puertos del Estado
RMS	Root-Mean Square Error
ROMS	Regional Ocean Modeling System
SAM_INI	SAMOA initial set-up (see Section 3.1 and Table 1)
SAM_ADV	SAMOA advanced set-up (current release, see Section 3.2 and Table 1)
SAM_H3D	SAMOA H3D set-up (forthcoming release, see Section 3.3 and Table 1)

SAMOA	Sistema de Apoyo Meteorológico y Oceanográfico a las Autoridades Portuarias (System of Meteorological and Oceanographic Support for Port Authorities)
SAMPA	Sistema de Apoyo Meteorológico y Oceanográfico al Puerto de Algeciras (System of Meteorological and Oceanographic Support for Algeciras Harbour)
SC_S	Surface current speed (Tables 3–5)
SC_D	Surface current direction (Tables 3–5)
SLev	Sea Level (Tables 3–5)
SSS	Sea Surface Salinity
SST	Sea Surface Temperature
TGS	Tide Gauge Station from the REDMAR network (Tables 3–5)
WAM	WAve Model

References

1. European Global Observing Systems (EuroGOOS). Available online: http://eurogoos.eu/ (accessed on 18 January 2022).
2. Schiller, A.; Mourre, B.; Drillet, Y.; Brassington, G. An overview of operational oceanography. In *New Frontiers in Operational Oceanography*; Chassignet, E., Pascual, A., Tintoré, J., Verron, J., Eds.; GODAE OceanView: Madrid, Spain, 2018; pp. 1–26. [CrossRef]
3. Davidson, F.; Alvera-Azcárate, A.; Barth, A.; Brassington, G.B.; Chassignet, E.P.; Clementi, E.; De Mey-Frémaux, P.; Divakaran, P.; Harris, C.; Hernandez, F.; et al. Synergies in Operational Oceanography: The Intrinsic Need for Sustained Ocean Observations. *Front. Mar. Sci.* **2019**, *6*, 450. [CrossRef]
4. De Mey-Frémaux, P.; Ayoub, N.; Barth, A.; Brewin, R.; Charria, G.; Campuzano, F.; Ciavatta, S.; Cirano, M.; Edwards, C.A.; Federico, I.; et al. Model-Observations Synergy in the Coastal Ocean. *Front. Mar. Sci.* **2019**, *6*, 436. [CrossRef]
5. EU EuroStat. EuroStat Explaining Statistics: Maritime Ports Freight and Passenger Statistics. 2017. Available online: http://ec.europa.eu/eurostat/statistics-explained/index.php/Maritime_ports_freight_and_passenger_statistics (accessed on 18 January 2022).
6. Álvarez Fanjul, E.; Sotillo, M.G.; Pérez, B.; García-Valdecasas, J.M.; Pérez, S.; Lorente, P.; Rodríguez-Dapena, A.; Martínez, I.; Luna, Y.; Padorno, E.; et al. Operational oceanography at the service of the ports. In *New Frontiers in Operational Oceanography*; Chassignet, E., Pascual, A., Tintoré, J., Verron, J., Eds.; GODAE OceanView: Madrid, Spain, 2018; pp. 729–736. [CrossRef]
7. Sotillo, M.G.; Cerralbo, P.; Lorente, P.; Grifoll, M.; Espino, M.; Sanchez-Arcilla, A.; Álvarez-Fanjul, E. Coastal ocean forecasting in Spanish ports: The SAMOA operational service. *J. Oper. Oceanogr.* **2019**, *13*, 37–54. [CrossRef]
8. Lellouche, J.-M.; Greiner, E.; Le Galloudec, O.; Garric, G.; Regnier, C. Recent updates to the Copernicus Marine Service global ocean monitoring and forecasting real-time 1/12° high-resolution system. *Ocean Sci.* **2018**, *14*, 1093–1126. [CrossRef]
9. Lellouche, J.-M.; Le Galloudec, O.; Drévillon, M.; Régnier, C.; Greiner, E.; Garric, G.; Ferry, N.; Desportes, C.; Testut, C.-E.; Bricaud, C.; et al. Evaluation of global monitoring and forecasting systems at Mercator Océan. *Ocean Sci.* **2013**, *9*, 57–81. [CrossRef]
10. Liu, H.; Lin, P.; Zheng, W.; Luan, Y.; Ma, J.; Ding, M.; Mo, H.; Wan, L.; Ling, T. A global eddy-resolving ocean forecast system in China—LICOM Forecast System (LFS). *J. Oper. Oceanogr.* **2021**, *14*, 2781–2799. [CrossRef]
11. Bell Michael, J.; Lefebvre, M.; Le Traon, P.Y.; Smith, N.; Wilmer-Becker, K. GODAE The Global Ocean Data Assimilation Experiment. *Oceanography* **2009**, *22*, 14–21. [CrossRef]
12. Bell, M.J.; Schiller, A.; Le Traon, P.-Y.; Smith, N.R.; Dombrowsky, E.; Wilmer-Becker, K. An introduction to GODAE OceanView. *J. Oper. Oceanogr.* **2015**, *8* (Suppl. 1), s2–s11. [CrossRef]
13. Tonani, M.; Balmaseda, M.; Bertino, L.; Blockley, E.; Brassington, G.; Davidson, F.; Drillet, Y.; Hogan, P.; Kuragano, T.; Lee, T.; et al. Status and future of global and regional ocean prediction systems. *J. Oper. Oceanogr.* **2015**, *8* (Suppl. 2), s201–s220. [CrossRef]
14. Copernicus Marine Environmental Monitoring Service (CMEMS). Available online: http://marine.copernicus.eu/ (accessed on 18 January 2022).
15. Le Traon, P.Y.; Reppucci, A.; Alvarez Fanjul, E.; Aouf, L.; Behrens, A.; Belmonte, M.; Bentamy, A.; Bertino, L.; Brando, V.E.; Kreiner, M.B.; et al. From observation to information and users: The Copernicus Marine service perspective. *Front. Mar. Sci.* **2019**, *6*, 234. [CrossRef]
16. Capet, A.; Fernández, V.; She, J.; Dabrowski, T.; Umgiesser, G.; Staneva, J.; Mészáros, L.; Campuzano, F.; Ursella, L.; Nolan, G.; et al. Operational Modeling Capacity in European Seas—An EuroGOOS Perspective and Recommendations for Improvement. *Front. Mar. Sci.* **2020**, *7*, 129. [CrossRef]
17. Sotillo, M.G.; Cailleau, S.; Lorente, P.; LeVier, B.; Aznar, R.; Reffray, G.; Amo-Baladrón, A.; Chanut, J.; Benkiran, M.; Alvarez-Fanjul, E. The MyOcean IBI Ocean Forecast and Reanalysis Systems: Operational products and roadmap to the future Copernicus Service. *J. Oper. Oceanogr.* **2015**, *8*, 63–79. [CrossRef]
18. Aznar, R.; Sotillo, M.G.; Cailleau, S.; Lorente, P.; Levier, B.; Amo-Baladrón, A.; Reffray, G.; Álvarez-Fanjul, E. Strengths and weaknesses of the CMEMS forecasted and reanalyzed solutions for the Iberia–Biscay–Ireland (IBI) waters. *J. Mar. Syst.* **2016**, *159*, 1–14. [CrossRef]
19. Kourafalou, V.H.; De Mey, P.; Staneva, J.; Ayoub, N.; Barth, A.; Chao, Y.; Cirano, M.; Fiechter, J.; Herzfeld, M.; Kurapov, A.; et al. Coastal Ocean Forecasting: Science foundation and user benefits. *J. Oper. Oceanogr.* **2015**, *8* (Suppl. 1), s147–s167. [CrossRef]

20. Sotillo, M.G.; Álvarez-Fanjul, E.; Castanedo, S.; Abascal, A.J.; Menendez, J.; Emelianov, M.; Olivella, R.; García-Ladona, E.; Ruiz-Villarreal, M.; Conde, J.; et al. Towards an operational system for oil-spill forecast over Spanish waters: Initial developments and implementation test. *Mar. Pollut. Bull.* **2008**, *56*, 686–703. [CrossRef]
21. Sánchez-Garrido, J.C.; García-Lafuente, J.; Álvarez-Fanjul, E.; Sotillo, M.G.; de los Santos, F.J. What does cause the collapse of the Western Alboran Gyre? Results of an operational ocean model. *Prog. Oceanogr.* **2013**, *116*, 142–153. [CrossRef]
22. Sotillo, M.G.; Amo-Baladrón, A.; Padorno, E.; Garcia-Ladona, E.; Orfila, A.; Rodríguez-Rubio, P.; Conti, D.; Jiménez Madrid, J.A.; de los Santos, F.J.; Álvarez Fanjul, E. How is the Surface Atlantic water inflow through the Gibraltar Strait forecasted? A lagrangian validation of operational oceanographic services in the Alboran Sea and the Western Mediterranean. *Deep. Sea Res. Part II Top. Stud. Oceanogr.* **2016**, *133*, 100–117. [CrossRef]
23. PdE Opendap Catalogue. Available online: https://opendap.puertos.es/thredds/catalog.html (accessed on 18 January 2022).
24. PdE PORTUS Ocean Monitoring & Forecasting System. Available online: https://portus.puertos.es/ (accessed on 18 January 2022).
25. De Dominicis, M.; Pinardi, N.; Zodiatis, G.; Lardner, R. MEDSLIK-II, a Lagrangian marine surface oil spill model for short-term forecasting–Part 1: Theory. *Geosci. Model Dev.* **2013**, *6*, 1851–1869. [CrossRef]
26. European Union. Directive (EU) 2019/883 of the European Parliament and of the Council of 17 April 2019 on port reception facilities for the delivery of waste from ships, amending Directive 2010/65/EU and repealing Directive 2000/59/EC. *Rom. J. Intell. Prop. L.* **2019**, *151*, 116–142.
27. Shchepetkin, A.; McWilliams, J.C. The regional oceanic modeling system (ROMS): A split-explicit, free-surface, topography-following-coordinate oceanic model. *Ocean Model.* **2005**, *9*, 347–404. [CrossRef]
28. ROMS Model Website. Available online: http://www.myroms.org/ (accessed on 18 January 2022).
29. Ryan, A.G.; Regnier, C.; Divakaran, P.; Spindler, T.; Mehra, A.; Smith, G.C.; Davidson, F.; Hernandez, F.; Maksymczuk, J.; Liu, Y. GODAE OceanView Class 4 forecast verification framework: Global ocean inter-comparison. *J. Oper. Oceanogr.* **2015**, *8* (Suppl. 1), s98–s111. [CrossRef]
30. Lorente, P.; Sotillo, M.G.; Amo-Baladrón, A.; Aznar, R.; Levier, B.; Aouf, L.; Dabrowski, T.; Pascual, A.; Reffray, G.; Dalphinet, A.; et al. The NARVAL Software Toolbox in Support of Ocean Models Skill Assessment at Regional and Coastal Scales. In *Computational Science—ICCS 2019, Proceedings of the 19th International Conference, Faro, Portugal, 12–14 June 2019*; Springer: Cham, Switzerland, 2019; Part IV. [CrossRef]
31. Bengtsson, L.; Andrae, U.; Aspelien, T.; Batrak, Y.; Calvo, J.; De Rooy, W.; Gleeson, E.; Hansen-Sass, B.; Homleid, M.; Hortal, M.; et al. The HARMONIE-AROME model configuration in the ALADIN-HIRLAM NWP system. *Mon. Weather Rev.* **2017**, *145*, 1919–1935. [CrossRef]
32. ECMWF, Evolution of the IFS Cycle. 2019. Available online: https://www.ecmwf.int/en/forecasts/documentation-and-support/changes-ecmwf-model (accessed on 9 June 2021).
33. Fairall, C.W.; Bradley, E.F.; Hare, J.E.; Grachev, A.A.; Edson, J.B. Bulk parameterization of air–sea fluxes: Updates and verification for the COARE algorithm. *J. Clim.* **2003**, *16*, 571–591. [CrossRef]
34. Flather, R.A. A tidal model of the northwest European continental shelf. *Mem. Soc. R. Sci. Liege* **1976**, *6*, 141–164.
35. Chapman, D.C. Numerical Treatment of Cross-Shelf Open Boundaries in a Barotropic Coastal Ocean Model. *J. Phys. Oceanogr.* **1985**, *15*, 1060–1075. [CrossRef]
36. Mason, E.; Molemaker, J.; Shchepetkin, A.; Colas, F.; McWilliams, J.C.; Sangrà, P. Procedures for offline grid nesting in regional ocean models. *Ocean Model.* **2010**, *35*, 1–15. [CrossRef]
37. Marchesiello, P.; McWilliams, J.C.; Shchepetkin, A.F. Open boundary conditions for long-term integration of regional oceanic models. *Ocean Model.* **2001**, *3*, 1–20. [CrossRef]
38. Wu, H.; Zhu, J. Advection scheme with 3rd high-order spatial interpolation at the middle temporal level and its application to saltwater intrusion in the Changjiang Estuary. *Ocean Model.* **2010**, *33*, 33–51. [CrossRef]
39. Edson, J.B.; Jampana, V.; Weller, R.A.; Bigorre, S.P.; Plueddemann, A.J.; Fairall, C.; Miller, S.D.; Mahrt, L.; Vickers, D.; Hersbach, H. On the Exchange of Momentum over the Open Ocean. *J. Phys. Oceanogr.* **2013**, *43*, 1589–1610. [CrossRef]
40. Álvarez Fanjul, E.; de Alfonso, M.; Ruiz, M.I.; López, J.D.; Rodriguez, I. Real time monitoring of Spanish coastal waters: The deep water network. In *Elsevier Oceanography Series*; Elsevier: Amsterdam, The Netherlands, 2003; Volume 69, pp. 398–402.
41. Alfonso, M.; Lopez, J.D.; Alvarez, E.; Ruiz, M.I. Real time buoy data quality control and exploitation. In Proceedings of the The Fifth COPRI International Conference on Ocean Wave Measurement and Analysis (WAVES), Madrid, Spain, 3–7 July 2005; Coasts, Oceans, Ports, and Rivers Institute, American Society of Civil Engineers: Madrid, Spain, 2005.
42. Pérez Gómez, B. Design and Implementation of an Operational Sea Level Monitoring and Forecasting System for the Spanish Coast. Ph.D. Thesis, Cantabria University, Santander, Spain, 2014; 242p.
43. Roarty, H.; Cook, T.; Hazard, L.; Doug, G.; Harlan, J.; Cosoli, S.; Wyatt, L.; Álvarez-Fanjul, E.; Terrill, E.; Otero, M.; et al. The Global High Frequency Radar Network. *Front. Mar. Sci.* **2019**, *6*, 164. [CrossRef]
44. Legates, D.R.; McCabe, G.J. Evaluating the use of 'goodness-of-fit' measures in hydrologic and hydroclimatic model validation. *Water Resour. Res.* **1999**, *35*, 233–241. [CrossRef]
45. Legates, D.R.; McCabe, G.J. A refined index of model performance: A rejoinder. *Int. J. Climatol.* **2013**, *33*, 1053–1056. [CrossRef]
46. Murphy, A.H. Skill Scores Based on the Mean Square Error and Their Relationships to the Correlation Coefficient. *Mon. Weather Rev.* **1988**, *116*, 2417–2424. [CrossRef]

47. Sotillo, M.G.; Mourre, B.; Mestres, M.; Lorente, P.; Aznar, R.; García-León, M.; Liste, M.; Santana, A.; Espino, M.; Álvarez-Fanjul, E. Evaluation of the operational CMEMS and coastal downstream ocean forecasting services during the storm Gloria (January 2020). *Front. Mar. Sci.* **2021**, *8*, 300. [CrossRef]
48. Grifoll, M.; Navarro, J.; Pallares, E.; Ràfols, L.; Espino, M.; Palomares, A. Ocean–atmosphere–wave characterisation of a wind jet (Ebro shelf, NW Mediterranean Sea). *Nonlinear Processes Geophys.* **2016**, *23*, 143–158. [CrossRef]
49. Large, W.G.; Yeager, S.G. Diurnal to decadal global forcing for ocean and sea-ice models: The data sets and flux climatologies, NCAR Technical Note. *Natl. Cent. Atmos. Res.* **2004**, *11*, 324–336.
50. Brodeau, L.; Barnier, B.; Gulev, S.K.; Woods, C. Climatologically significant effects of some approximations in the bulk parameterizations of turbulent air–sea fluxes. *J. Phys. Oceanogr.* **2017**, *47*, 5–28. [CrossRef]
51. Pelletier, C.; Lemarié, F.; Blayo, E. Sensitivity analysis and metamodels for the bulk parametrization of turbulent air–sea fluxes. *Q. J. R. Meteorol. Soc.* **2018**, *144*, 658–669. [CrossRef]
52. Pérez-Gómez, B.; García-León, M.; García-Valdecasas, J.; Clementi, E.; Aranda, C.M.; Pérez-Rubio, S.; Masina, S.; Coppini, G.; Molina-Sánchez, R.; Muñoz-Cubillo, A.; et al. Understanding Sea Level Processes During Western Mediterranean Storm Gloria. *Front. Mar. Sci.* **2021**, *8*, 647437. [CrossRef]
53. WAMDI Group. The WAM model—A third generation ocean wave prediction model. *J. Phys. Oceanogr.* **1988**, *18*, 1775–1810. [CrossRef]
54. Lorente, P.; Lin-Ye, J.; García-León, M.; Reyes, E.; Fernandes, M.; Sotillo, M.G.; Espino, M.; Ruiz, M.I.; Gracia, V.; Perez, S.; et al. On the Performance of High Frequency Radar in the Western Mediterranean During the Record-Breaking Storm Gloria. *Front. Mar. Sci.* **2021**, *8*, 205. [CrossRef]
55. Sotillo, M.G.; Campuzano, F.; Guihou, K.; Lorente, P.; Olmedo, E.; Matulka, A.; Santos, F.; Amo-Baladrón, A.; Novellino, A. River Freshwater Contribution in Operational Ocean Models along the European Atlantic Façade: Impact of a New River Discharge Forcing Data on the CMEMS IBI Regional Model Solution. *J. Mar. Sci. Eng.* **2021**, *9*, 401. [CrossRef]

Journal of
Marine Science and Engineering

MDPI

Article

Modeling of the Turkish Strait System Using a High Resolution Unstructured Grid Ocean Circulation Model

Mehmet Ilicak [1,*], Ivan Federico [2], Ivano Barletta [2,3], Sabri Mutlu [4], Haldun Karan [4], Stefania Angela Ciliberti [2], Emanuela Clementi [5], Giovanni Coppini [2] and Nadia Pinardi [3]

1 Eurasia Institute of Earth Sciences, Istanbul Technical University, Istanbul 34469, Turkey
2 Fondazione Centro Euro-Mediterraneo sui Cambiamenti Climatici, Ocean Predictions and Applications Division, 73100 Lecce, Italy; ivan.federico@cmcc.it (I.F.); ivano.barletta@cmcc.it (I.B.); stefania.ciliberti@cmcc.it (S.A.C.); giovanni.coppini@cmcc.it (G.C.)
3 Department of Physics and Astronomy, Universita di Bologna Alma Mater Studiorum, 40126 Bologna, Italy; nadia.pinardi@unibo.it
4 TUBITAK MRC Environment and Cleaner Production Institute, Kocaeli 41470, Turkey; sabri.mutlu@tubitak.gov.tr (S.M.); haldun.karan@tubitak.gov.tr (H.K.)
5 Fondazione Centro Euro-Mediterraneo sui Cambiamenti Climatici, Ocean Modeling and Data Assimilation Division, 40126 Bologna, Italy; Emanuela.clementi@cmcc.it
* Correspondence: milicak@itu.edu.tr

Abstract: The Turkish Strait System, which is the only connection between the Black Sea and the Mediterranean Sea, is a challenging region for ocean circulation models due to topographic constraints and water mass structure. We present a newly developed high resolution unstructured finite element grid model to simulate the Turkish Strait System using realistic atmospheric forcing and lateral open boundary conditions. We find that the jet flowing from the Bosphorus Strait into the Marmara creates an anticyclonic circulation. The eddy kinetic energy field is high around the jets exiting from the Bosphorus Strait, Dardanelles Strait, and also the leeward side of the islands in the Marmara Sea. The model successfully captures the two-layer structure of the Sea of Marmara. The volume transport at the Bosphorus is around 120 km^3/year which is consistent with the recent observations. The largest bias in the model is at the interface depth due to the shallower mixed layer.

Keywords: sea of Marmara; Bosphorus strait; Dardanelles strait; Turkish strait system; SHYFEM

Citation: Ilicak, M.; Federico, I.; Barletta, I.; Mutlu, S.; Karan, H.; Ciliberti, S.A.; Clementi, E.; Coppini, G.; Pinardi, N. Modeling of the Turkish Strait System Using a High Resolution Unstructured Grid Ocean Circulation Model. *J. Mar. Sci. Eng.* **2021**, *9*, 769. https://doi.org/10.3390/jmse9070769

Academic Editor: Matthew Lewis

Received: 16 June 2021
Accepted: 7 July 2021
Published: 12 July 2021

1. Introduction

The Turkish Strait System (TSS) consisting of the Marmara Sea, the Bosphorus, and the Dardanelles Strait is the only connection between two fundamentally different marginal seas: the Black Sea and the Mediterranean Sea. While the Black Sea has the dynamics of a typical estuarine circulation i.e., precipitation plus river input surpass evaporation ($E - P - R < 0$) [1,2], the Mediterranean Sea is an example of an inverse–estuary ($E - P - R > 0$) [3,4]. The TSS has unique dynamics since the flow is both density-driven between the salty (≈38 psu) Mediterranean Sea and the brackish (≈17 psu) Black Sea, and also barotropic-driven because of the permanent sea level difference between these two marginal seas. The sea level height in the Black Sea is approximately 0.3 m higher than the Marmara Sea due to its lower density. Climatological northeasterly winds weakly influence the general circulation in the Sea of Marmara [5].

The Marmara Sea is a relatively small inland sea which covers an area of 11,500 km^2 with three deep basins (>1000 m) and an extended shelf in the south. The Bosphorus and the Dardanelles Straits share common physical and geographical properties. They are both relatively narrow (Bosphorus 0.7–3.5 km, the Dardanelles 1.2–7 km), long (31 km for the Bosphorus and 61 km for the Dardanelles), and shallow (30–100 m for the Bosphorus and 50–120 m for the Dardanelles) channels [6–8]. Both Straits enable a water exchange between two different marginal seas with significant density differences.

The fresh water inflow from the Marmara Sea into the Aegean is crucial for the northern part of the Aegean Sea. Convection in the northern part of the Aegean is regulated by the brackish waters of the Black Sea coming from the Dardanelles. In the Black Sea, the river input is balanced by the salty and warm flow coming from the Mediterranean Basin through the Bosphorus Strait. This salty dense water (with the Cold Intermediate Water) is also important for the ventilation process of the deep, oxygen depleted Black Sea waters. Salty Mediterranean Sea waters are equilibrated at the depth of the suboxic and anoxic layers of the Black Sea and play an important role in the redox potential of the chemistry of the Black Sea [8].

Most ocean circulation models use structured grids with finite difference/volume discretizations. These types of grids are particularly challenging in modeling the TSS system because of the narrow and long Bosphorus and Dardanelles Straits, and the deep basins of the Marmara Sea. The complex topography and two-layer density structure require multiple hydraulic controls in both Straits [9]. These narrow Straits thus require a high resolution (less than 100 m) to resolve their dynamics. In addition, the TSS salinity and heat balances are controlled by the Black Sea and the Mediterranean Sea. Hence, the entire TSS region needs to be part of the modeling effort.

Some studies have modeled the transports of individual straits using either 2D idealized reduced gravity or non-hydrostatic models or 3D regional idealized models [10–14]. Some of these studies have attempted to model the Marmara Sea with or without the Turkish Straits using structured grid ocean models. Demyshev et al. conducted a finite difference numerical simulation without atmospheric forcing and reproduced the S-shaped circulation of the jet current exiting the Bosphorus and crossing the Marmara Sea with a basin scale anti-cyclonic circulation [15]. However, the model does not contain strait dynamics. Similarly, the Regional Ocean Modeling System (ROMS [16]) is used to model the Marmara Sea using realistic atmospheric forcing but with open boundaries at the Straits nudged to observation fields [17]. ROMS is a terrain following a structured grid model which needs bathymetry smoothing due to pressure gradient error. Without including the Straits in the model, the authors assumed that the surface circulation depends solely on the strength and directional pattern of the wind force in the Sea of Marmara. Sannino et al. used the structured grid MITgcm model [18] with curvilinear coordinates for a high resolution around the Bosphorus Strait [19]. Their model also did not use any atmospheric forcing. They investigated the circulation of the TSS changing the barotropic flow between the Black Sea and the Aegean Sea.

Few studies have used unstructured grid models to simulate the three-dimensional baroclinic circulation of the TSS. Stanev et al. used the SCHISM model for interconnected basins including the Black Sea, the TSS, and the northern Aegean Sea [20]. They used realistic atmospheric forcing with lateral open boundaries in the south. Their aim was to accurately represent the transport at the straits and the resulting circulation dynamics of the Black Sea. An implicit advection scheme was used in SCHISM for larger time steps. This leads to a coarser model resolution, with only 53 vertical levels at the deepest point of the Black Sea.

Similarly, the Finite Element Sea-Ice Ocean Model (FESOM), another unstructured grid model, was used by Aydogdu et al. to study the circulation of the TSS [21]. They analyzed the combined response of the Sea of Marmara with atmospheric forcing and strait dynamics. Although the FESOM model has a high resolution of up to 65 m in the horizontal and 110 vertical levels, the setup has a closed boundary, and a volume correction is needed to maintain the sea level difference between the Black Sea and the Aegean Sea sides.

Our aim in this study is to simulate the entire TSS region using an unstructured grid model with high resolution in both horizontal and vertical directions. The model has been forced by realistic atmospheric reanalysis and open boundaries with ocean analysis data sets. The main goal is to develop a regional model with an adequate representation of the mean and variability of the TSS circulation. We plan to use the output from the new model as lateral boundary conditions for future Copernicus Marine Service (CMEMS) Mediterranean Sea and Black Sea models.

This paper is organized as follows: the numerical model and details of the experiment are introduced in Section 2. The main results including mean circulation properties, validation of water mass structure, and volume fluxes across the straits are presented in Section 3. Finally, we summarize and conclude in Section 4.

2. Model Setup

We used the System of HydrodYnamic Finite Element Modules (SHYFEM) as the numerical model. SHYFEM is a free-surface, hydrostatic, primitive equations ocean model which uses an unstructured grid finite element scheme in the horizontal and vertically layered in the vertical [22–24]. A key advantage of an unstructured grid is to obtain a varying resolution which allows for a finer mesh in the coastal areas, which is also sufficiently accurate in the open ocean and a comparable resolution with parent models. The model domain covers the area between longitudes 22.54° E and 31.41° E, latitudes 38.79° N and 43.41° N (see Figure 1). In this new model, there is a high resolution mesh around 50 m in the shallow areas to resolve the Turkish Straits and 2500 m in the deep areas. Given that the Rossby radius of deformation ($R_d = \sqrt{\frac{g\Delta\rho}{\rho}H/f}$) is around 18 km in the Sea of Marmara, our new model can be considered as an eddy resolving model. The model has 93 geopotential coordinate levels in the vertical in order to resolve the complex hydraulics of the rapidly changing bathymetry of the Marmara Sea. The vertical resolution is 1 m in the first 50 m of depth and increases to 100 m at the bottom boundary layer in the deepest part of the model domain.

We have used the high resolution bathymetry and initial conditions provided by Aydogdu et al. [21]. We used April temperature and salinity profiles since they had a minimum bias compared to observations. A Smagorinsky-type dynamical momentum closure scheme with a non-dimensional constant of 2.2 was used in the horizontal closure for momentum. Scale aware momentum closures, especially Smagorinsky, significantly reduce the numerical mixing while keeping the energetics of the mesoscale eddies [25]. For the vertical mixing scheme, a k-epsilon type second order turbulence closure with the Canuto-A stability function was chosen. This scheme performs significantly better in density driven flows (i.e., gravity currents) than the other closures such as k-omega and K-Profile Parameterization [26].

A nonlinear equation of state is used to compute the density [27]. A quadratic bottom drag formulation with a drag coefficient of $C_d = 2 \times 10^{-3}$ is used to compute the bottom friction. No slip boundary conditions are applied in the horizontal except open boundaries. A total variation diminishing (TVD) scheme is used for tracer advection to ensure conservation properties.

We conducted a four-year simulation between 1 January 2016 and 31 December 2019. Atmospheric surface boundary conditions were provided by the European Center for Medium-Range Weather Forecasts (ECMWF) with 1/8° resolution. The forcing data were applied using bulk formulae with a frequency of 6 h. Clamped type open boundary conditions were employed at the southern and northeast boundaries for sea surface height and inflow active tracers. Total velocities were nudged at the open boundaries and zero gradient boundary conditions were used for outflow active tracers. We used daily averaged values of sea surface height, u-velocity, v-velocity, temperature, and salinity fields for the open boundaries. These data were provided by a CMEMS (https://marine.copernicus.eu/, analysis, in particular the Black Sea Forecasting System [28] for the northern boundary and Mediterranean Sea Forecasting System for the southern boundary [29].

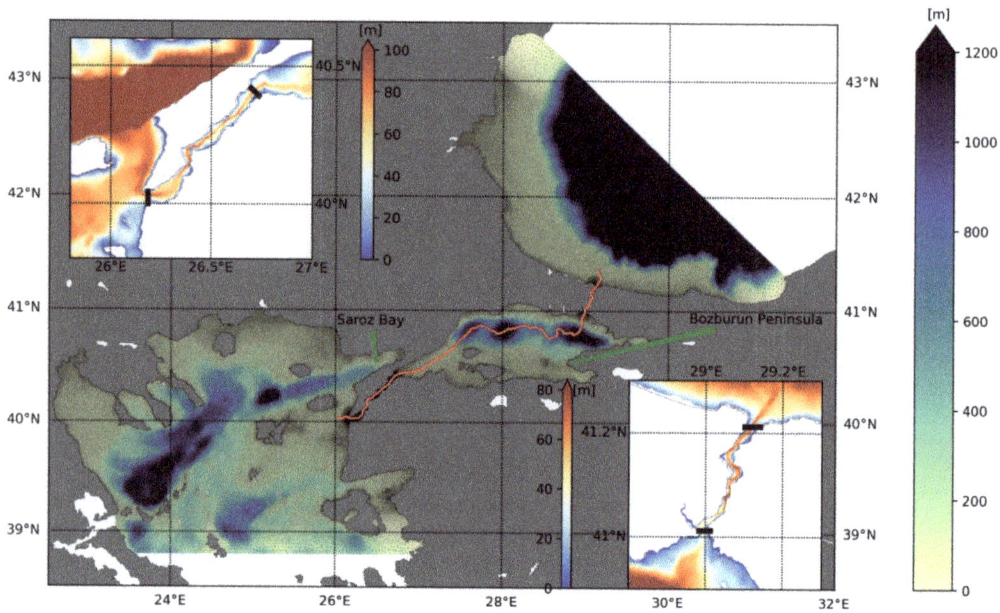

Figure 1. Mesh and bathymetry of the SHYFEM-TSS model domain. Bathymetry of the Dardanelles (top left) and Bosphorus (bottom right) Straits are shown in detail in the small panels. The red line shows the location of the cross sections used in Figure 6. Northern and southern sections to compute volume transports are shown in black lines at the top left and bottom right bathymetry figures.

3. Results

This section presents the results of the model from the SHYFEM-TSS setup between 2016 and 2019, including detailed analyses of the surface circulation, vertical water mass structure, and validation against the observations. All the time mean results used here are averaged over the whole period of the model simulation unless stated otherwise. Time series are presented of the volume transports at the two gateways of the Marmara Sea—northern and southern parts of the Bosphorus Strait, and northern and southern parts of the Dardanelles Strait.

3.1. Mean Surface Circulation and Water Mass Structure

Figure 2 shows the mean of the surface speed averaged over the simulation period. The Bosphorus Jet reaches the southern coast of the Marmara Sea (Bozburun Peninsula) and turns west towards the middle of the basin. The mean speed of the jet is around 0.6 m/s at the exit of the Bosphorus into the Marmara Sea. The jet separates into two branches due to the Imrali island at the west of the Bozburun Peninsula. The upper branch, which is also the main branch, also splits into two different branches. One part of the upper branch forms an anticyclonic gyre in the northern part of the Marmara Sea, whereas the second part crosses the Marmara Sea after turning south and exits the Marmara Sea through the Dardanelles Strait. The lower branch splits at the Imrali Island, circulates in between the Marmara islands and merges with the upper branch to exit the Marmara Sea.

Previous studies have shown similar circulation patterns [5,17,21]. One of the numerical studies which was lacking wind forcing demonstrated a cyclonic gyre in the central basin due to potential vorticity injection from the Bosphorus Jet [19]. This finding is similar to another previous study where a cyclonic circulation has been found in the Arctic Ocean using an idealized barotropic model [30]. However, the mean state in our model, with a more realistic forcing, is different from those idealized simulations. The mean circulation pattern in the central Marmara Sea was persistent with different strengths throughout the simulation.

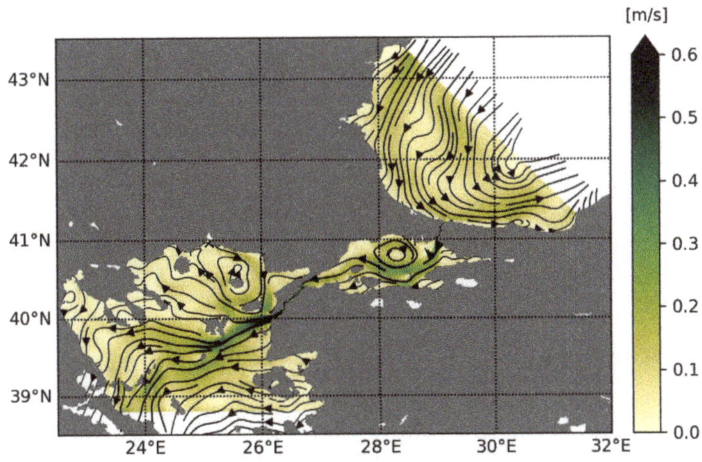

Figure 2. Averaged surface speed between 2016 and 2019. Stream lines are shown in black lines.

The mean kinetic energy (MKE) and eddy kinetic energy (EKE) fields are shown in Figure 3. We decompose the model velocity into low-pass mean (U) and fluctuation values (u') with respect to this low-pass mean (i.e., $u = U + u'$), and compute MKE as $0.5 \times (U^2 + V^2)$ and EKE as $0.5 \times (u'^2 + v'^2)$. We used a 60 day low pass filter, which is approximately the mean residence time at the surface layer [5], to compute U. The mean kinetic energy field presents intensities on the rim current in the Black Sea, the Bosphorus Jet, the current out of Dardanelles, and the cyclonic gyre in the Marmara Sea. These regions are also evident in the mean speed field in the previous figure. The EKE, which represents the mesoscale eddy kinetic energy field, differs from the MKE, especially in the Marmara Sea and the northwest of the Aegean Sea. The magnitude of the EKE field is comparable to that of the MKE field. In the Marmara Sea, the EKE field is high around the Bosphorus Jet, Izmit Bay, and southern shelf. MKE decreases and EKE increases downstream of the Bosphorus Jet where it starts to turn to the west in the Marmara Sea. This indication of energy for the EKE field is provided by the release of energy of the mean flow through baroclinic instability processes. There is another high EKE region in the northwestern part of the Aegean Sea exit of the Black Sea water from the Dardanelles. Once again, there is an energy conversion from the MKE field into the EKE field.

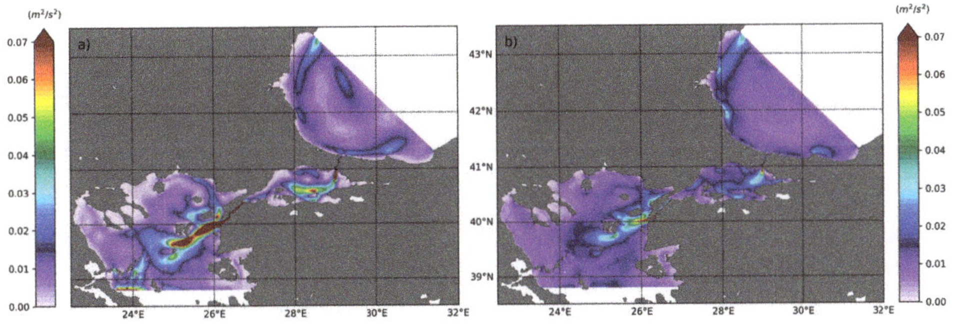

Figure 3. (**a**) Time-averaged mean kinetic energy (MKE); (**b**) time-averaged eddy kinetic energy (EKE) computed from four years of model velocities.

The sea surface temperature (SST) averaged over four years ranges from 15 °C at the northern part of the Black Sea to 21 °C at the eastern part of the Aegean Sea (Figure 4). The cold jet flow coming from the Bosphorus can be easily traced to the southern coast of the Marmara Sea. The jet is carrying the Black Sea water, approximately 16 °C, into the Marmara at the surface. This is consistent with the mean circulation mentioned above. One of the interesting features in the simulation is the eastern part of the Aegean Sea, outside of the Dardanelles Strait and Saroz Bay. The cold water upwelling process in the summer months dominates the averaged SST field in these regions.

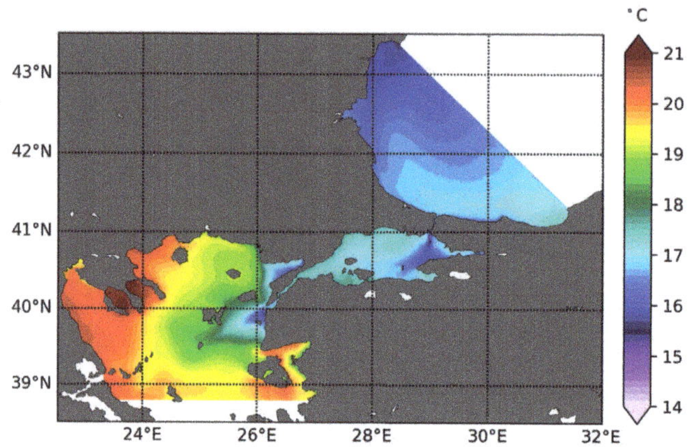

Figure 4. The mean sea surface temperature for 2016–2019.

Our model shows that the mean sea surface salinity (SSS) changes within 17 psu in the Black Sea and 38 psu in the Aegean Sea (Figure 5). SSS in the Marmara Sea (MS) ranges between 20 psu (Bosphorus exit) to 27 psu (west part of the MS). There is also a meridional gradient where SSS is around 22 psu in the northern MS and 25 psu in the south. Previous observations suggested that surface salinity is in a range of 23 ± 2 psu in the Marmara Sea, using the long-term measurements from 1986 to 1992 [5]. However, both our model simulation and recent observations show higher salinity ranges in the Marmara Sea (see the Water Mass Validation section).

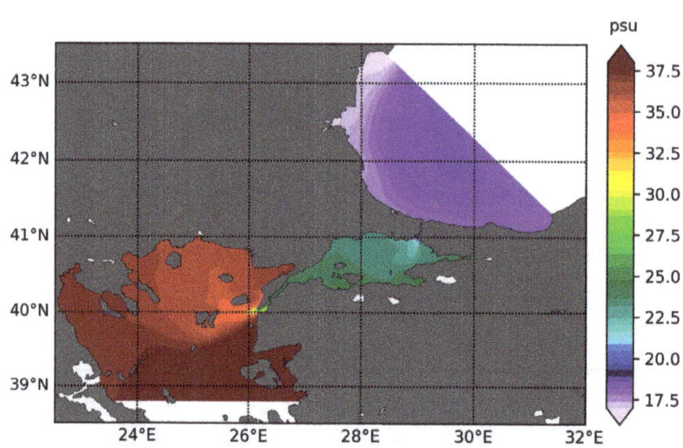

Figure 5. The mean sea surface salinity for 2016–2019. Note that the colorbar is not linear.

Figure 6 shows the vertical section of time averaged salinity and temperature fields across the section shown in Figure 1 (red line). This cross section follows the deepest part of the Bosphorus and Dardanelles Straits and also passes through three deep basins of the Marmara Sea. The interface depth between the fresh-warm surface layer and salty-cold lower layer is at around 25 m in the central basin of the Marmara Sea. The salinity below 50 m becomes uniform in the Marmara Sea due to the influence of the Mediterranean Sea. However, during the simulation period, deeper layers (i.e., below 200 m) were getting modified with the relatively cold Aegean Sea water coming from the Dardanelles (temperature field at 150 km in Figure 6). There is a distinct cold intermediate layer (CIL) extending from the Bosphorus Strait into the Marmara Sea at the halocline depth in the model. Previous observations also reported a cold temperature anomaly around 11–12 °C at the same depth in June–July between 1996 and 2000 [13]. We conclude that the model successfully reproduces CIL, which is an important feature for the Marmara Sea water mass structure.

The two-layered water mass structure can also be seen in the Dardanelles Strait (first 100 km of the section shown in Figure 6). The mean salinity and temperature vertical structure are similar to those of Sannino et al. [19] and Aydogdu et al. [21]. In the Bosphorus Strait, salinity in the bottom layer is mixed with the ambient water when the layer flows into the Black Sea at the northern exit (section between 320 km and 400 km shown in Figure 6). Previous observations and numerical studies reported a similar salinity distribution [9,14,19,21,31]. The flow in the Bosphorus Strait has two distinct hydraulic control locations which increase the amount of mixing between two layers. The cold upper layer of the Black Sea water mixes with the relatively warm bottom layer of the Marmara Sea water and merges into the CIL at the southern side of the Bosphorus Strait. Surface layer velocities are higher at the southern exits of the Dardanelles and Bosphorus Straits compared to their northern exits. This is due to increasing exit widths in the straits and also contraction regions. For more details of the hydraulic controls in the Bosphorus and Dardanelles Straits, see Sozer and Ozsoy [14], Sannino et al., [19] and references therein.

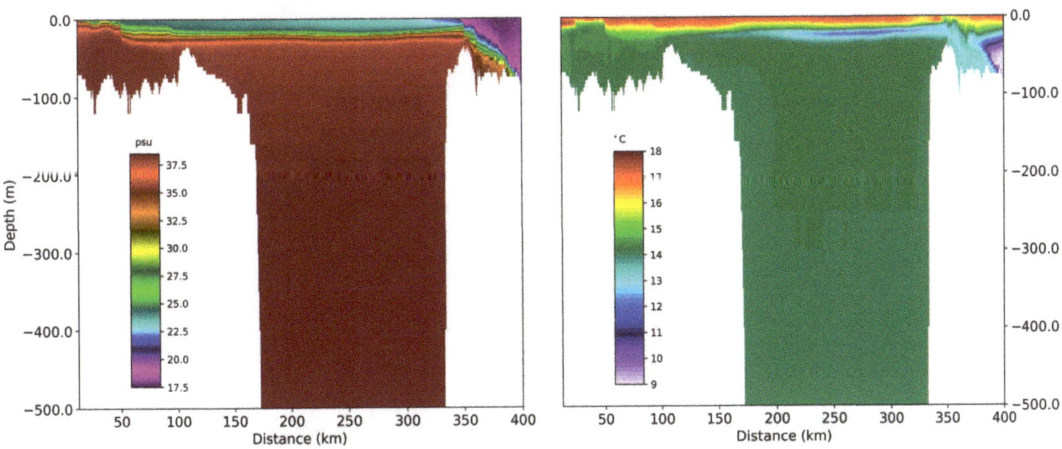

Figure 6. The mean salinity (**left**) and temperature (**right**) for 2016–2019 along the thalweg section shown as a red line in Figure 1.

3.2. Water Mass Validation

We used four data sets of in situ salinity and temperature observations which were collected by RV TUBITAK Marmara of the TUBITAK Marmara Research Center in order to validate the simulation. These observation datasets covered all over the Marmara Sea and were collected in different seasons; for the Summer 2017 on 8–16 August, for the Winter 2018 on 15–30 January, for the Spring 2018 on 24 April–3 May, and finally for the Summer 2018 10–28 August within the framework of the "Integrated Marine Pollution Monitoring 2017–2019 Program" by the Turkish Ministry of Environment and Urbanization and TUBITAK-MRC.

In each cruise, conductivity, temperature, and depth (CTD) measurements were done from the surface up to the sea bottom at the pre-defined same 90 stations at the speed of 16 Hz. The accuracy rates of the sensors were 0.001 °C and 0.0003 S/m for temperature and conductivity, respectively. After quality control, the high-resolution data were averaged to 1 m bin size during the post-processing for later use.

Figure 7a shows the spatial distribution of the interpolated sea surface temperature field in August 2017 (summer cruise). The black dots represent the station locations. The interpolation was performed using the Data-Interpolating Variational Analysis (DIVA) method. There is an approximately 3 °C temperature gradient between the east and west of the Marmara basin. Relatively cold Black Sea water (≈23 °C) has spread to the exit of the Bosphorus Strait (northeast side of the basin). However, the cold water at the coast of the Bozburun Peninsula (southeastern side of the basin) is probably due to an upwelling event, since the salinity at the same location is much higher than the Black Sea brackish water (Figure 8a). Observations show that the western part of the Marmara Sea is around 26–27 °C. Figure 7b shows the model SST field in August 2017. Cold jet water can be seen at the eastern side of the domain. One of the prominent features of the model is the cold upwelling at the Saroz Bay, north of the Dardanelles. The northern coast of the Marmara Sea is one degree warmer in the model than in the observation.

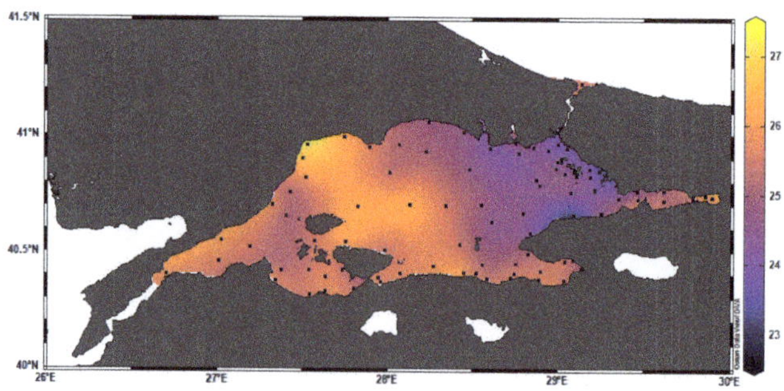

(a) Observed surface temperature field interpolated over the Marmara Sea.

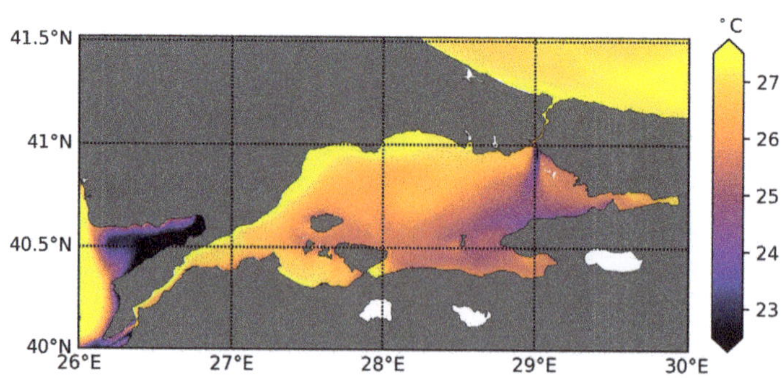

(b) Surface model temperature field.

Figure 7. Sea surface temperature field for the observations and model for the summer 2017 Marmara cruise.

Vertical salinity/temperature bias and root mean square (rms) error profiles compared to all observations are shown in Figures 9 and 10, respectively. The mean surface salinity bias improves over time. The largest bias in all seasons is around the mixed layer depth (20–30 m). This is probably due to the missing representation of mixed layer dynamics such as insufficient mixing. The surface temperature mean bias is around 0.1 °C in the top 10 m in August 2018. Similarly to salinity, temperature bias and rms error are highest around the seasonal thermocline depth. We believe that the bias below 40 m is due to the initial conditions.

The sea surface temperature at the interior of the Marmara Sea is close to the observed values. Figure 8a shows the sea surface salinity field from the observations. The surface salinity clearly shows a wide spread between 18 psu in the Black Sea entrance of the Bosphorus and 25 psu at the Dardanelles exit of the Marmara Sea. Once again, there is an east-western gradient in salinity in the basin. The exit of the Bosphorus and its surroundings have less than 21 psu of salinity, while the western side surface salinity is more than 24 psu. The low salinity jet flow can also be seen in the model simulation (Figure 8b). Interior salinity values of the model are similar to those in the observed field. The southern coast of the Marmara Sea is slightly saltier in the model due to the upwelling of the saltier deep waters to the surface. This is also likely due to relatively colder waters in the SST field.

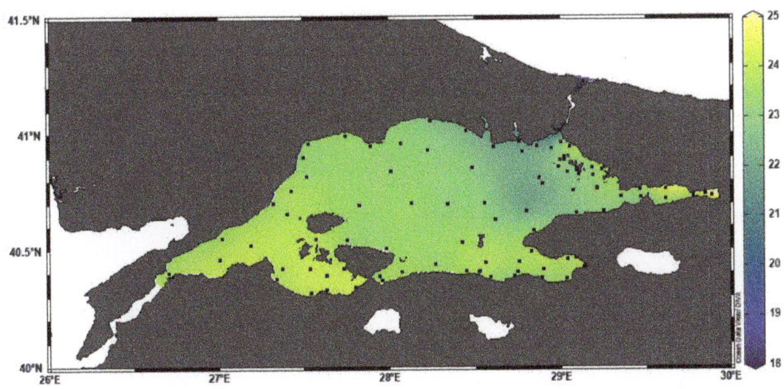

(**a**) Observed surface salinity field interpolated over the Marmara Sea.

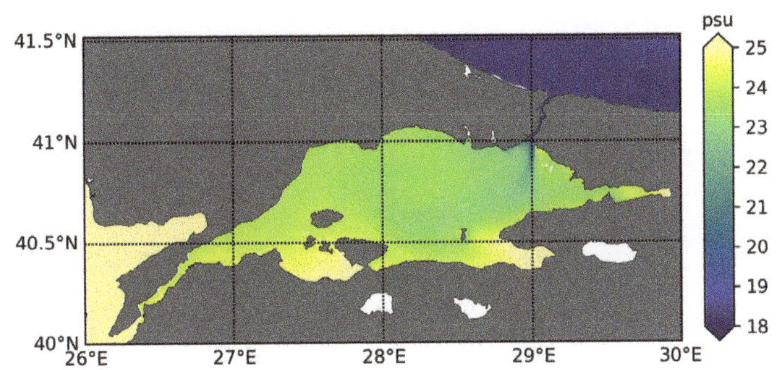

(**b**) Surface model salinity field.

Figure 8. Sea surface salinity field for the observations and model for the summer 2017 Marmara cruise

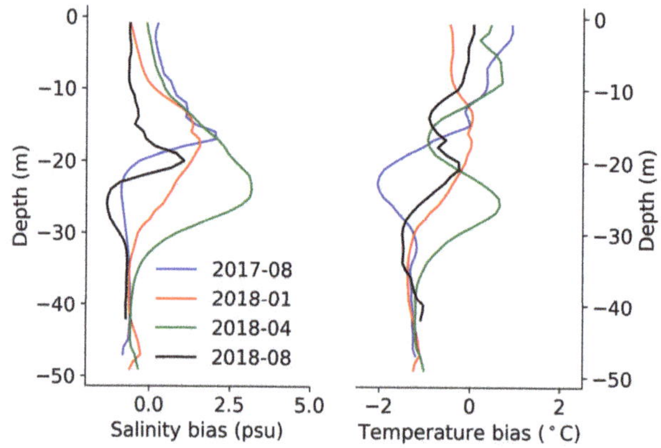

Figure 9. Vertical distributions of salinity (**left**) and temperature (**right**) bias in August 2017, January 2018, April 2018, and August 2018.

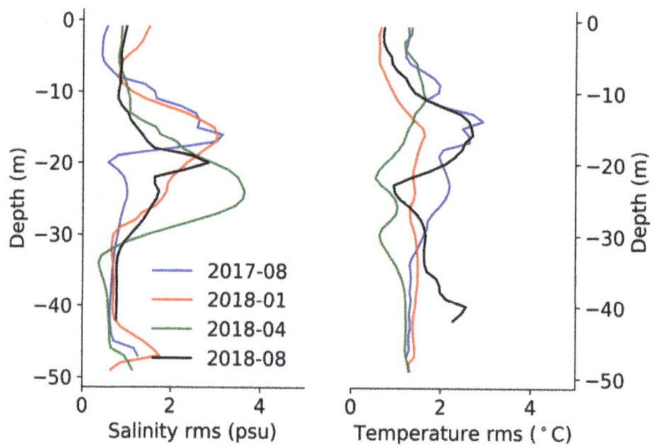

Figure 10. Vertical distributions of salinity (**left**) and temperature (**right**) RMS error in August 2017, January 2018, April 2018, and August 2018.

3.3. Volume Fluxes through the Straits

Figure 11 shows the daily and monthly averaged net, upper and lower layers' volume fluxes at different exits of the Bosphorus and Dardanelles Straits. The means of each time series are also shown in Table 1. Positive values indicate flows towards the Black/Marmara Sea for the Bosphorus/Dardanelles sections. Significant temporal variability can be seen in all sections. Daily fluctuations may be three times more than the mean value of the corresponding time series. The upper layer variability is higher than the lower layer along the Bosphorus Strait sections, while layer fluctuations are comparable in the Dardanelles. The net flow at the Bosphorus Strait reverses in the winter because of weakening and/or reversing of the upper layer. The lower layer at the Bosphorus Strait is at its minimum in the summer. Field observations show similar fluctuation differences and a reverse of the upper layer in the Bosphorus Strait [32]. The net volume transports in our model are around 140 km^3yr^{-1} and 170 km^3yr^{-1} at the Bosphorus Strait and the Dardanelles Strait, respectively. The difference may be due to the evaporation–precipitation (E–P) balance in the Sea of Marmara and numerical errors during the post-processing of the unstructured grid. Nevertheless, the net values are approximately half those of the historical estimation which is around 300 km^3yr^{-1} and which is based on the conservation of mass in the Black Sea. However, net transport using direct measurements obtained at the exits of the

Bosphorus Strait is around 110–120 km^3yr^{-1} [32]. We conclude that model transport values are within the uncertainty of the recent observations and other model studies.

The previous studies that obtained the net transport of 300 km^3yr^{-1} assumed that maximum salinity values of the deep layer can be used to compute a two-layer budget analysis with (E–P) balance in the Black Sea [5]. Here, we use mean salinity values of the bottom and top layers at the northern and southern sections of the Bosphorus Strait and show that the net transport cannot be that high. The net transport (Q_{net}) is the difference between the top layer transport (Q_1) and bottom layer transport (Q_2) at the northern section

$$Q_{net} = Q_1 - Q_2. \tag{1}$$

This is also true for the top layer transport (Q_3) and bottom layer transport (Q_4) at the southern section (i.e., $Q_{net} = Q_3 - Q_4$). We then follow the two-layer analytical model for the salt transport at each sections,

$$\begin{aligned} S_1Q_1 - S_2Q_2 &= 0 \\ S_3Q_3 - S_4Q_4 &= 0 \\ S_1Q_1 - S_2(Q_1 - Q_{net}) &= 0 \\ S_3Q_3 - S_4(Q_3 - Q_{net}) &= 0, \end{aligned}$$

where S_i is the average salinity of the respected layer. The last two equations above are equal to each other, thus we get

$$(S_1 - S_2)Q_1 + S_2Q_{net} = (S_3 - S_4)Q_3 + S_4Q_{net}. \tag{2}$$

Change in transport of upper layer between sections ($\Delta Q = Q_1 - Q_3$) is around 50 km^3yr^{-1} both in our model and in the analytical solution [5]. In addition, in the model, the averaged salinity differences between the upper and lower layers at a cross section are approximately equal to each other (i.e., $\Delta S = S_1 - S_2 \approx S_3 - S_4$). Hence,

$$\frac{\Delta S \Delta Q}{Q_{net}} = (S_4 - S_2). \tag{3}$$

If we choose the maximum salinity values in a layer as used in [5], $S_1 = 17$ psu, $S_2 = 35$ psu, $S_3 = 20$ psu and $S_4 = 37$ psu, then Equation (3) satisfies $Q_{net} = 300$ km^3yr^{-1}. However, the averaged values in a layer are significantly different to the maximum ones because of the mixing in the interfacial layer. The results of the model show that ΔS is around 8 psu rather than 17 psu. This ensures that Q_{net} should be approximately 150 km^3yr^{-1}.

The layer transports at the Dardanelles are higher than shown by the model results provided by [21] but still lower than the historical estimates which are around 850/550 km^3yr^{-1} for the upper/lower layer in the northern section and 1200/900 km^3yr^{-1} for the upper/lower layer in the southern section, respectively [1,5,6,33]. This disagreement in layer fluxes may be due to an inaccurate representation of turbulent mixing and/or forcing along the open boundaries.

Table 1. Annual mean of net, upper layer, and lower layer volume fluxes for the whole simulation period at different sections of the Turkish Straits.

Annual Mean (km^3yr^{-1})	Net Transport	Upper Layer	Lower Layer
Northern Bosphorus	−137.52	−457.96	317.64
Southern Bosphorus	−140.31	−498.37	360.85
Northern Dardanelles	−176.89	−589.97	413.08
Southern Dardanelles	−170.14	−850.92	680.78

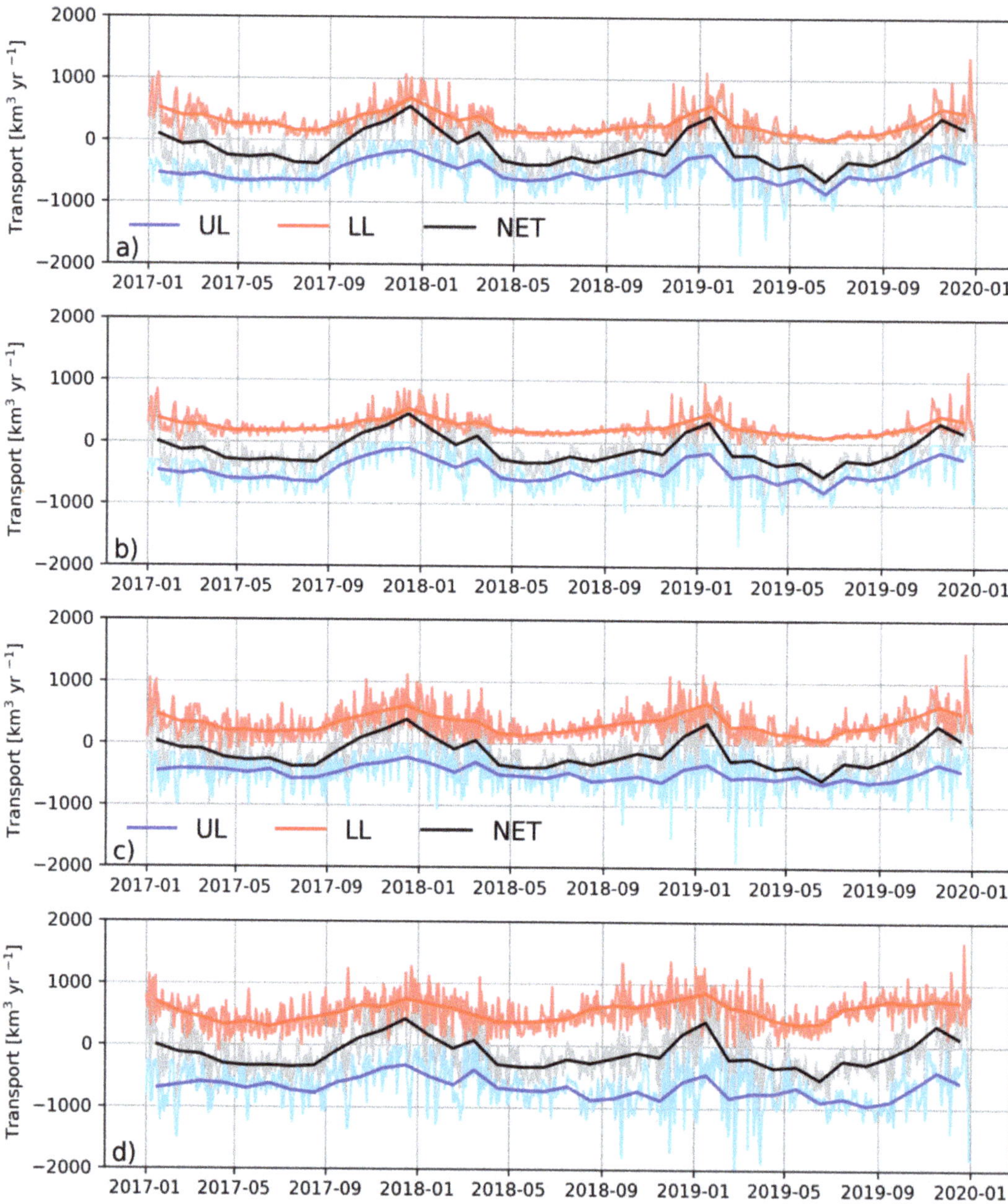

Figure 11. Daily upper layer (blue, UL), lower layer (red, LL), and net (grey, NET) volume transports through northern Bosphorus (**a**), southern Bosphorus (**b**), northern Dardanelles (**c**), and southern Dardanelles (**d**). Monthly averages are overlaid with a darker tone of the same color.

4. Summary and Conclusions

We have presented a new high resolution unstructured grid model for the Turkish Strait System using realistic atmospheric surface forcing and ocean lateral open boundary conditions. The SHYFEM-TSS model was run for four years and successfully reproduced vertical water mass structure, horizontal circulation, and volume fluxes through the Bosphorus and Dardanelles Straits.

The model captures the buoyant jet coming from the Bosphorus into the Marmara Sea on the surface. There is an anticyclonic gyre in the center of the Marmara Sea due to this jet and its potential vorticity injection into the semi-enclosed sea. There is also another fast current flowing out of the Dardanelles into the Aegean Sea carrying relatively fresh Black Sea water to the Eastern Mediterranean Sea. Previous models also observed similar circulation patterns [5,17,21]. Both mean and eddy kinetic energy fields are high around the jets. The mean kinetic energy also shows the main branch of the flow pattern in the Marmara Sea. The eddy kinetic energy field is high around the islands of the Marmara Sea, indicating the so-called island effect, eddies generated due to topographic features.

The temperature and salinity profiles were compared with the observations, and bias/rms fields were computed. Our model performs well compared to the observation, and reproduces the two-layer structure in the Marmara Sea. The SHYFEM-Marmara captures the realistic distribution of the water mass structure. The largest bias is around the pycnocline depth which indicates less mixing in the model. However, it is also possible that small changes at the interface depth show larger errors especially in a two-layer system.

The volume transports at the Straits were also computed. However, the net Bosphorus transport is smaller than the historical value of 300 km^3/year estimated based on estuary circulation equations using the maximum salinity of the layers [5], and agrees well with recent observation values 120 km^3/year [32] measured by multiple bottom-mounted acoustic Doppler current profilers. When we computed the estuary circulation equations using the averaged layer salinity values from the model results, we also get the 150 km^3/year which is in line with net transport in the model. The upper and lower layer transports in both Straits display significant variability at different time scales.

We have demonstrated that the new SHYFEM-Marmara model using a high resolution unstructured grid with realistic atmospheric forcing and lateral open boundary conditions is capable of simulating the main circulation and water characteristics of the Marmara Sea. The future of Marmara Sea modeling relies on the data-assimilation for forecasting capability and also coupling with a biological model to study the impact of man-made pollution coming from the river discharge into the Sea of Marmara.

Author Contributions: Conceptualization, M.I.; methodology, I.F., I.B., M.I.; validation, H.K., S.M., and M.I.; writing—original draft preparation, M.I., writing—review and editing, M.I., G.C., N.P.; upstream data: S.A.C., E.C. All authors have read and agreed to the published version of the manuscript.

Funding: M.I. is funded through a project named Development of the new Turkish Strait System Model for optimal interface between the Mediterranean and Black Sea models by ITUNOVA A.S., and is also supported by the Scientific and Technological Research Council of Turkey (TUBITAK) under project number 120C137. This research was funded by the Copernicus Marine Environment Monitoring Service for the Black Sea Monitoring and Forecasting Center (Contract No. 72-CMEMS-MFS-BS) and the Mediterranean Sea Monitoring and Forecasting Center (Contract No. 74-CMEMS-MFC-MED). Partial funding through the Strategic project #4 "A multihazard prediction and analysis testbed for the global coastal ocean" of CMCC is gratefully acknowledged.

Institutional Review Board Statement: Not applicable.

Informed Consent Statement: Not applicable.

Data Availability Statement: this work has been used CMEMS data from the Med-MFC and BS-MFC, in particular analysis and forecasting products for the physical component of the Mediterranean Sea and the Black Sea (https://marine.copernicus.eu/ (accessed on 9 July 2021)).

Acknowledgments: The Authors thank the "Integrated Marine Pollution Monitoring 2017–2019 Program" carried out by the Turkish Ministry of Environment and Urbanization/General Directorate of EIA, Permit and Inspection/ Department of Laboratory, Measurement and coordinated by TUBITAK-MRC ECPI for the provisioning of the data used in this study for model validation.

Conflicts of Interest: The authors declare no conflict of interest.

References

1. Özsoy, E.; Ünlüata, Ü. Oceanography of the Black Sea: A review of some recent results. *Earth-Sci. Rev.* **1997**. [CrossRef]
2. Kara, A.B.; Wallcraft, A.J.; Hurlburt, H.E.; Stanev, E.V. Air-sea fluxes and river discharges in the Black Sea with a focus on the Danube and Bosphorus. *J. Mar. Syst.* **2008**, *74*. [CrossRef]
3. Robinson, A.; Leslie, W.; Theocharis, A.; Lascaratos, A. Mediterranean Sea Circulation. In *Encyclopedia of Ocean Sciences*; Elsevier: Amsterdam, The Netherlands, 2001; doi:10.1006/rwos.2001.0376. [CrossRef]
4. Pinardi, N.; Zavatarelli, M.; Adani, M.; Coppini, G.; Fratianni, C.; Oddo, P.; Simoncelli, S.; Tonani, M.; Lyubartsev, V.; Dobricic, S.; et al. Mediterranean Sea large-scale low-frequency ocean variability and water mass formation rates from 1987 to 2007: A retrospective analysis. *Prog. Oceanogr.* **2015**, *132*. [CrossRef]
5. Besiktepe Sukru, T.; Sur, H.I.; Özsoy, E.; Latif, M.A.; Oğuz, T.; Ünlüata, Ü. The circulation and hydrography of the Marmara Sea. *Prog. Oceanogr.* **1994**. [CrossRef]
6. Ünlülata, Ü.; Oğuz, T.; Latif, M.A.; Özsoy, E. On the Physical Oceanography of the Turkish Straits. In *The Physical Oceanography of Sea Straits*; Springer: Dordrecht, The Netherlands, 1990; doi:10.1007/978-94-009-0677-8_2. [CrossRef]
7. Jarosz, E.; Teague, W.J.; Book, J.W.; Beiktepe, T. Observations on the characteristics of the exchange flow in the Dardanelles Strait. *J. Geophys. Res. Ocean* **2012**, *117*. [CrossRef]
8. Altiok, H.; Kayişoğlu, M. Seasonal and interannual variability of water exchange in the Strait of Istanbul. *Mediterr. Mar. Sci.* **2015**, *16*. [CrossRef]
9. Özsoy, E.; Di Iorio, D.; Gregg, M.C.; Backhaus, J.O. Mixing in the Bosphorus Strait and the Black Sea continental shelf: Observations and a model of the dense water outflow. *J. Mar. Syst.* **2001**, *31*. [CrossRef]
10. Johns, B.; Oguz, T. The modeling of the flow of water through the Bosphorus. *Dyn. Atmos. Ocean.* **1989**, *14*, 229–258. [CrossRef]
11. Oguz, T.; Özsoy, E.; Latif, M.A.; Sur, H.I.; Ünlüata, Ü. Modeling of hydraulically controlled exchange flow in the Bosphorus Strait. *J. Phys. Oceanogr.* **1990**, *20*.<0945:MOHCEF>2.0.CO;2. [CrossRef]
12. Ilicak, M.; Özgökmen, T.; Özsoy, E.; Fischer, P. Non-hydrostatic modeling of exchange flows across complex geometries. *Ocean Model.* **2009**, *29*. [CrossRef]
13. Altiok, H.; Sur, H.I.; Yüce, H. Variation of the cold intermediate water in the Black Sea exit of the Strait of Istanbul (Bosphorus) and its transfer through the strait. *Oceanologia* **2012**, *54*. [CrossRef]
14. Sozer, A.; Özsoy, E. Water exchange through canal Istanbul and Bosphorus strait. *Mediterr. Mar. Sci.* **2017**, *18*. [CrossRef]
15. Demyshev, S.G.; Dovgaya, S.V.; Ivanov, V.A. Numerical modeling of the influence of exchange through the Bosporus and Dardanelles Straits on the hydrophysical fields of the Marmara Sea. *Izv. Atmos. Ocean Phys.* **2012**, *48*. [CrossRef]
16. Haidvogel, D.B.; Arango, H.; Budgell, W.P.; Cornuelle, B.D.; Curchitser, E.; Di Lorenzo, E.; Fennel, K.; Geyer, W.R.; Hermann, A.J.; Lanerolle, L.; et al. Ocean forecasting in terrain-following coordinates: Formulation and skill assessment of the Regional Ocean Modeling System. *J. Comput. Phys.* **2008**, *227*. [CrossRef]
17. Chiggiato, J.; Jarosz, E.; Book, J.W.; Dykes, J.; Torrisi, L.; Poulain, P.M.; Gerin, R.; Horstmann, J.; Beşiktepe, Ş. Dynamics of the circulation in the Sea of Marmara: Numerical modeling experiments and observations from the Turkish straits system experiment. *Ocean Dyn.* **2012**, *62*, 139–159. [CrossRef]
18. Marshall, J.; Adcroft, A.; Hill, C.; Perelman, L.; Heisey, C. A finite-volume, incompressible navier stokes model for, studies of the ocean on parallel computers. *J. Geophys. Res. C Ocean* **1997**, *102*. [CrossRef]
19. Sannino, G.; Sözer, A.; Özsoy, E. A high-resolution modeling study of the Turkish Straits System. *Ocean Dyn.* **2017**, *67*. [CrossRef]
20. Stanev, E.V.; Grashorn, S.; Zhang, Y.J. Cascading ocean basins: Numerical simulations of the circulation and interbasin exchange in the Azov-Black-Marmara-Mediterranean Seas system. *Ocean Dyn.* **2017**, *67*. [CrossRef]
21. Aydoğdu, A.; Pinardi, N.; Özsoy, E.; Danabasoglu, G.; Gürses, Ö.; Karspeck, A. Circulation of the Turkish Straits System under interannual atmospheric forcing. *Ocean Sci.* **2018**, *14*, 999–1019. [CrossRef]
22. Umgiesser, G.; Canu, D.M.; Cucco, A.; Solidoro, C. A finite element model for the Venice Lagoon. Development, set up, calibration and validation. *J. Mar. Syst.* **2004**, *51*. [CrossRef]
23. Federico, I.; Pinardi, N.; Coppini, G.; Oddo, P.; Lecci, R.; Mossa, M. Coastal ocean forecasting with an unstructured grid model in the southern Adriatic and northern Ionian seas. *Nat. Hazards Earth Syst. Sci.* **2017**, *17*. [CrossRef]
24. Ferrarin, C.; Davolio, S.; Bellafiore, D.; Ghezzo, M.; Maicu, F.; Mc Kiver, W.; Drofa, O.; Umgiesser, G.; Bajo, M.; De Pascalis, F.; Malguzzi, P.; Zaggia, L.; Lorenzetti, G.; Manfè, G. Cross-scale operational oceanography in the Adriatic Sea. *J. Oper. Oceanogr.* **2019**, *12*. [CrossRef]
25. Ilicak, M.; Adcroft, A.; Griffies, S.; Hallberg, R. Spurious dianeutral mixing and the role of momentum closure. *Ocean Model.* **2012**, *45*. [CrossRef]
26. Ilicak, M.; Özgökmen, T.; Peters, H.; Baumert, H.; Iskandarani, M. Performance of two-equation turbulence closures in three-dimensional simulations of the Red Sea overflow. *Ocean Model.* **2008**, *24*. [CrossRef]
27. Jackett, D.R.; Mcdougall, T.J. Minimal Adjustment of Hydrographic Profiles to Achieve Static Stability. *J. Atmos. Ocean. Technol.* **1995**, *12*, 381–389.<0381:maohpt>2.0.co;2. [CrossRef]

28. Ciliberti, S.A.; Peneva, E.L.; Jansen, E.; Martins, D.; Cretí, S.; Stefanizzi, L.; Lecci, R.; Palermo, F.; Daryabor, F.; Lima, L.; et al. *Black Sea Analysis and Forecast (CMEMS BS-Currents, EAS3 system) (Version 1))*; Data Set; Copernicus Monitoring Environment Marine Service (CMEMS): 2020; doi:10.25423/CMCC/BLKSEA_ANALYSIS_FORECAST_PHYS_007_001_EAS3.
29. Clementi, E.; Pistoia, J.; Escudier, R.; Delrosso, D.; Drudi, M.; Grandi, A.; Lecci, R.; Cretí, S.; Ciliberti, S.; Coppini, G.; et al. *Mediterranean Sea Analysis and Forecast (CMEMS MED-Currents, EAS5 System)*; Data Set; Copernicus Monitoring Environment Marine Service (CMEMS): 2019. Available online: https://www.cmcc.it/wp-content/uploads/2021/02/CMEMS-MED-PUM-006-013.pdf (accessed on 9 July 2021).
30. Yang, J. The Arctic and Subarctic Ocean flux of Potential Vorticity and the Arctic Ocean circulation. *J. Phys. Oceanogr.* **2005**, *35*. [CrossRef]
31. Gregg, M.C. Flow, water mass changes, and hydraulics in the Bosphorus. *J. Geophys. Res.* **2002**, *107*. [CrossRef]
32. Jarosz, E.; Teague, W.J.; Book, J.W.; Beşiktepe, Ş. Observed volume fluxes in the Bosphorus Strait. *Geophys. Res. Lett.* **2011**, *38*. [CrossRef]
33. Tuğrul, S.; Beşiktepe, .T.; Salihoğlu, I. Nutrient exchange fluxes between the Aegean and Black Seas through the marmara sea. *Mediterr. Mar. Sci.* **2002**, *3*. [CrossRef]

Journal of
Marine Science and Engineering

MDPI

Article

Denoising Effect of Jason-1 Altimeter Waveforms with Singular Spectrum Analysis: A Case Study of Modelling Mean Sea Surface Height over South China Sea

Jiajia Yuan, Jinyun Guo *, Yupeng Niu, Chengcheng Zhu, Zhen Li and Xin Liu *

College of Geodesy and Geomatics, Shandong University of Science and Technology, Qingdao 266590, China; yuanjiajia2017@sdust.edu.cn (J.Y.); yupengniu1@163.com (Y.N.); cczhu2018@sdust.edu.cn (C.Z.); lizsk1994@163.com (Z.L.)

* Correspondence: guojy@sdust.edu.cn (J.G.); skd994268@sdust.edu.cn (X.L.)

Received: 11 May 2020; Accepted: 8 June 2020; Published: 10 June 2020

Abstract: Altimeter waveforms are usually contaminated due to nonmarine surfaces or inhomogeneous sea state conditions. The present work aimed to present how the singular spectrum analysis (SSA) can be used to reduce the noise level in Jason-1 altimeter waveforms to obtain SSA-denoised waveforms, improving the accuracy of a mean sea surface height (MSSH) model. Comparing the retracked sea surface heights (SSHs) by a 50% threshold retracker for the SSA-denoised waveforms with those for the raw waveforms, the results indicated that SSA allowed a noise reduction on Jason-1 waveforms, improving the accuracy of retracked SSHs. The MSSH model (called Model 1) over the South China Sea with a grid of $2' \times 2'$ was established from the retracked SSHs of Jason-1 by the 50% threshold retracker for the SSA-denoised waveforms. Comparing Model 1 and Model 2 (established from the retracked SSHs by the 50% threshold retracker for the raw waveforms) with the CLS15 and DTU18 models in the South China Sea, it was found that the accuracy of Model 1 was higher than that of Model 2, which indicates that using SSA to reduce noise level in Jason-1 waveforms can effectively improve the accuracy of the MSSH model.

Keywords: singular spectrum analysis; altimeter waveform; Json-1; mean sea surface height; threshold retracker; waveform retracking

1. Introduction

Satellite radar altimeters provide information on the Earth's surface by transmitting a series of radio-frequency pulses and recording their echo waveforms [1]. Satellite altimetry has been widely used in geodesy, geophysics, and oceanography [2]. Some satellite altimetry products have been obtained using satellite altimeter data, e.g., ocean tide models, gravity field models, and mean sea surface height (MSSH) models. Among them, the MSSH model is the time-averaged physical height of the ocean's surface [3] and is an essential and important parameter to support oceanographic and geophysical studies [4]. The accuracy of the MSSH model is affected by the quality of satellite altimeter data, as well as the theory and methods of data processing. With the improvement of satellite altimetry error correction theory and data processing methods [5], using waveform retracking to improve the quality of altimeter data has gradually become the key to improving the accuracy and application of altimeter data, especially in coastal regions [6].

Altimeter waveforms are usually contaminated due to land, island, sea reef, sea ice, seabed terrain, etc. If the extracted ranges from these corrupted waveforms are used, sea levels calculated from these ranges will be incorrect as well. Waveform retracking is a method to find the right tracking gate, cutting the midpoint of the leading edge from these corrupted waveforms to extract the actual ranges [7]. So far, the waveform retracking technique has developed a variety of retracking methods, which are mainly divided into two categories: One based on the empirical statistical properties of the waveform data, and the other based on fitting functional model [2,8]. These retracking methods can improve the quality of altimeter data, but the accuracy of the data still cannot meet the actual needs [9–11]. Besides, the noise information contained in altimeter waveforms is not considered during the process of waveform retracking.

Several altimeter waveforms are connected end-to-end to form a waveform series, which oscillates periodically with the number of the waveform samples (e.g., the waveform samples of Jason-1 are 104). The altimeter waveform can be regarded as composed of two parts: The main waveform information and the noise information. The main waveform information includes the thermal noise area, leading edge, and trailing edge of the altimeter waveform, while the noise information is caused by reflective surfaces such as land, sea, glaciers, etc. By performing singular spectrum analysis (SSA) on the waveform series, the noise information contained in altimeter waveforms can be reduced, so that useful waveform information is extracted.

SSA is a nonparametric method of time series analysis [12,13]. It can reduce noise information from time series containing noise and extract as much reliable information as possible [14,15]. This method has been widely used in meteorology, climatology, geophysics, and other fields [13]. Therefore, SSA can be used to decompose and reconstruct the waveform series, reduce the noise level in altimeter waveforms, and obtain SSA-denoised waveforms. Then, these SSA-denoised waveforms are reprocessed by the waveform retracking technique to improve the accuracy and application of altimeter data.

The goal of this work was to reduce the noise level in Jason-1 altimeter waveforms with SSA to improve the accuracy of Jason-1 altimeter data, and to validate whether SSA can effectively improve the accuracy of MSSH model over the South China Sea established from SSA-denoised waveforms retracking of Jason-1 data. The structure of the thesis is as follows: Section 2 mainly introduces the study area, the data used, and data processing methods; Section 3 involves the results; Section 4 presents MSSH model over the South China Sea established from SSA-denoised waveforms retracking of Jason-1 data and validates whether SSA can effectively improve the accuracy of MSSH model; and the main conclusions and perspectives are given in Section 5.

2. Study Area, Data, and Data Processing Methods

2.1. Study Area and Data

The area around the South China Sea, covering 0°–25° N, 105°–125° E, was selected as the study area. The South China Sea belongs to the western Pacific Ocean and is one of the three marginal seas in Asia. It is located at the intersection of the Eurasian plate, Indo-Australian plate, and Pacific plate. Covering an area of about 3.5 million square kilometers, it is the third largest sea in the world, following the Coral Sea in the South Pacific and the Arabian Sea of the Indian Ocean. With an average depth of 1212 m and the deepest point of 5377 m, the South China Sea is virtually surrounded by land, peninsulas, and islands. The South China Sea is linked in the northeast to the East China Sea and the Pacific Ocean through the Taiwan Strait; in the south to the Java Sea, the Andaman Sea, and the Indian Ocean through the Malacca Strait; and in the east to the Sulu Sea through the Bashi Strait. Located in the low-latitude region, the South China Sea has the warmest climate of all tropical deep seas of China, with a high surface water temperature (25 °C–28 °C), small annual temperature variation (3 °C–4 °C), year-round high temperature and humidity, and long summer without winter. It has a maximal salinity of 35‰ and tidal difference below 2 m.

Jason-1 was launched on 7 December 2001 and decommissioned on 1 July 2013, obtaining about 11.5 years of altimeter waveforms data [16]. These data were used in our study as listed in Table 1. The data, spanning from 2002 to 2013, are version E of sensor geophysical data records (SGDR) products (including the so-called measured 20-Hz waveforms) provided by Archiving, Validation, and Interpretation of Satellite Oceanographic Data (AVISO). The orbit of Jason-1 had three different phrases. From its launch, its orbit had the same ground tracks as TOPEX/Poseidon (T/P), which was the first phrase. In mid-February, 2009 (cycle 262), Jason-1 assumed a new orbit midway between its original ground tracks, which was the second phrase. At the end of February and in early March 2012, it began a series of maneuvers to reduce the orbit on a geodetic orbit, which was the third phrase. In the former two phrases, Jason-1 performed exactly repeated mission (ERM) (the mission of the first phrase hereafter called ERM1 and the second phrase ERM2) with a cycle of 9.9 days. In the third phrase, Jason-1 performed the geodetic mission (GM), and its orbit was a drifting orbit with a cycle of 406 days and some sub-cycles of 3.9–10.9–47.5–179.5 days.

Table 1. Altimeter waveforms data of Jason-1 missions.

Missions	Data Duration	Cycle Number	Orbit Altitude (km)	Mean Track Separation at the Equator (km)
ERM1	2002/01/15–2009/01/26	001-259	1336	315
ERM2	2009/02/10–2012/03/03	262-374	1336	315
GM	2012/05/07–2013/06/21	500-537	1324	7

Figure 1 is the ground tracks map of Jason-1 in South China Sea. Figure 1a is the ground tracks map of ERM1 and ERM2, and Figure 1b is that of GM.

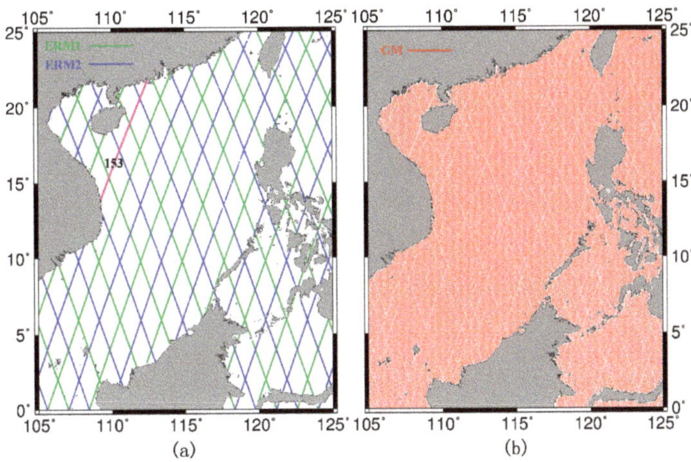

Figure 1. The ground tracks map of Jason-1 in South China Sea. (**a**) The ground tracks map of ERM1 and ERM2. The number 153 is the pass number, the green lines are the ERM1 ground tracks, and the blue lines are the ERM2 ground tracks; (**b**) the ground tracks map of the geodetic mission (GM).

2.2. Data Processing Methods

Jason-1 waveforms were connected end-to-end to form the waveform series (one pass corresponded to one waveform series), and the waveform series was denoised with SSA to obtain the SSA-denoised waveforms. Then, these SSA-denoised waveforms were retracked by a 50% threshold retracker to obtain the retracked sea surface heights (SSHs). The process of Jason-1 waveforms processing is shown in Figure 2.

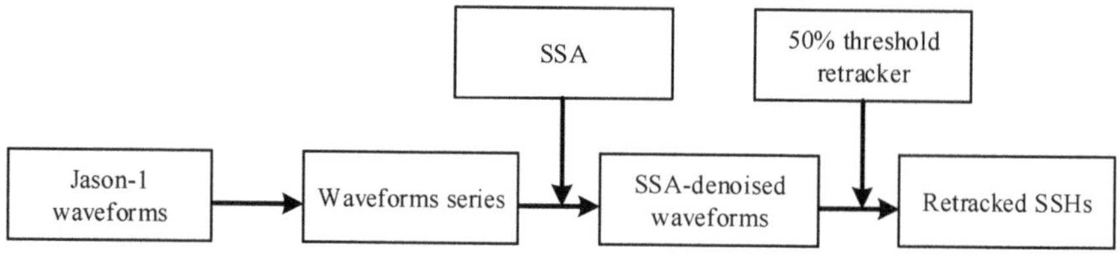

Figure 2. Processing flowchart of Jason-1 waveforms.

2.2.1. Singular Spectrum Analysis (SSA) Applied to Altimeter Waveforms

SSA constructs a multidimensional trajectory matrix for a waveform series and decomposes and reconstructs the trajectory matrix to extract signals that represent the main waveform information and noise information.

Assume that a waveform series consisting of i waveforms is $X(t)\{x_t : 1 < t < N\}$, $N = i \times 104$ (104 is the number of waveform samples of Jason-1). SSA was used to denoise this waveform series, and the process was mainly divided into four steps [14,15]:

(1) Embedding

Because the oscillation period of the waveform series is 104, the SSA method with a window of 104 was adapted to reduce the noise information of the waveform series. Defining the delay vector $\mathbf{X}_k$ as $\mathbf{X}_k = \{x_k, \cdots, x_{k+M-1}\}^T$, $1 \le k \le N - M + 1$, the trajectory matrix $\mathbf{S}$ of the waveform series can be expressed as:

$$\mathbf{S} = [\ \mathbf{X}_1\ \ \mathbf{X}_2\ \ \cdots\ \ \mathbf{X}_{N-M+1}\ ] = \begin{bmatrix} x_1 & \cdots & x_{N-M+1} \\ \vdots & \ddots & \vdots \\ x_M & \cdots & x_N \end{bmatrix} \quad (1)$$

where N is the waveform series length and M is the window ($M = 104$).

(2) Singular Value Decomposition (SVD)

SVD was performed on the covariance matrix $\mathbf{R}=\mathbf{S}\mathbf{S}^{\mathrm{T}}$ of the trajectory matrix $\mathbf{S}$ to obtain M eigenvalues in descending order $\lambda_1 \ge \lambda_2 \ge \cdots \ge \lambda_M$ and corresponding eigenvectors $\mathbf{U}_1, \mathbf{U}_2, \cdots, \mathbf{U}_M$. Then, the principal components (PC) $\mathbf{V}_i$ can be expressed as $\mathbf{V}_i = \mathbf{S}^T\mathbf{U}_i/\sqrt{\lambda_i}, i = \{1, \cdots, M\}$, and the trajectory matrix $\mathbf{S}$ can be expressed as the sum of the elementary matrices $\mathbf{S}_i$ ($\mathbf{S}_i = \sqrt{\lambda_i}\mathbf{U}_i\mathbf{V}_i^T, i = \{1, \cdots, M\}$) obtained by SVD, i.e.,

$$\mathbf{S} = \mathbf{S}_1 + \mathbf{S}_2 + \cdots + \mathbf{S}_M \quad (2)$$

(3) Grouping

The waveform series can be seen as being composed of two parts: The main waveform information and the noise information. Therefore, the subscript $D = \{1, \cdots, M\}$ of the elementary matrix $\mathbf{S}_i$ was divided into two disjoint subsets, namely $D_1 = \{1, \cdots, l\}$, $1 < l < M$ containing the first few leading components that describe the main waveform information, and $D_2 = D - D_1$ containing the residual components that describe the noise information. Then, the trajectory matrix S can be decomposed into two parts, $\mathbf{S}^{D_1} = \mathbf{S}_1 + \mathbf{S}_2 + \cdots + \mathbf{S}_l$ and $\mathbf{S}^{D_2} = \mathbf{S}_{l+1} + \mathbf{S}_{l+2} + \cdots + \mathbf{S}_M$, namely:

$$\mathbf{S} = \mathbf{S}^{D_1} + \mathbf{S}^{D_2} \quad (3)$$

The parameter l denotes the leading components, which provides a good description of the main waveform information, and the lower M-l components represent the noise information. If l is too small, a part of the main waveform information will miss. Alternatively, if l is too large, a part of the main waveform information is approximated with the noise information. Currently, there is no clear standard for determining this parameter [13]. To properly choose the parameter l, the ratio calculated in Equation (4) was used to estimate the contribution of the i-th candidate principal component corresponding to the raw waveform series.

$$\text{Ratio}_i = \lambda_i / \sum_{i=1}^{M} \lambda_i \tag{4}$$

where λ_i is the i-th eigenvalue.

(4) Reconstruction

Reconstruction was performed to restore $\mathbf{S}^{D_1}$ and $\mathbf{S}^{D_2}$ in Equation (3) to new series $X^{D_1}(t)$ and $X^{D_2}(t)$ of length N, respectively. Taking reconstruction $\mathbf{S}^{D_1}$ as an exa $M^* = \min(M, N-M+1)$ mple to describe its specific process and assuming that the element in $\mathbf{S}^{D_1}$ is $y_{i,j}$, and $K^* = \max(M, N-M+1)$, the reconstruction of $\mathbf{S}^{D_1}$ into the new series $X^{D_1}(t)\{g_t : 1 < t < N\}$ can be obtained as follows:

$$g_t = \begin{cases} \frac{1}{t}\sum\limits_{i=1}^{t} y_{i,t-i+1}, & 1 \leq t < M^* \\ \frac{1}{M^*}\sum\limits_{i=1}^{M^*} y_{i,t-i+1}\,, & M^* \leq t \leq K^* \\ \frac{1}{N-t+1}\sum\limits_{i=t-K^*+1}^{N-K^*+1} y_{i,t-i+1}, & K^* < t \leq N \end{cases} \tag{5}$$

2.2.2. Waveform Retracking Method

Accurate range estimates were obtained using methods of waveform retracking (e.g., Ocean, Ice-2, Belta5, Threshold) [17]. These retracking methods and their applications to the coastal altimeter waveform have been reviewed in Gommenginger et al. [8]. Some studies have shown that the standard Ocean retracking method, Ice-2 retracking algorithm, and Belta5 retracker are not appropriate for complex coastal waveforms [6,8,10,17]. However, the threshold retracker is easy to implement and successful in producing valid SSHs from coastal altimeter waveform retracking [6,17–20]. Therefore, the threshold retracker was used in our study.

The threshold retracker was developed by Davis [21,22]. It is based upon the dimensions of the rectangle computed using the offset center of gravity (OCOG) retracking algorithm [6,8]. The threshold level is referenced to the OCOG amplitude as 25%, 50%, and 75% of the amplitude. The retracking gate estimate is determined by linearly interpolating between adjacent samples of a threshold crossing the steep part of the leading-edge slope of the waveform [8]. The selection of the appropriate threshold level (such as 25%, 50%, or 75% of the OCOG amplitude) is very important. However, a reasonable choice is very difficult and difficult to grasp [6,19]. For the coastal altimeter waveform, a 50% threshold level retracking can usually obtain better results [19]. The formula and explanations in the calculation of the 50% threshold retracker have been presented by Deng [6] and Gommenginger et al. [8].

2.2.3. Estimating Sea Surface Height

The retracked gate was estimated from 50% threshold retracker to determine retracked range correction by [5]:

$$\Delta R_{\text{Ret}} = \frac{c}{2} \times \tau \times (G_{\text{Ret}} - G_{\text{Ref}}) \tag{6}$$

where ΔR_{Ret} is the range correction derived from the waveform retracking methods (m), c is the light velocity in vacuum, τ is the time interval for one gate Jason-1 satellite (3.125 ns), G_{Ret} is the retracked gate estimated from the 50% threshold retracker, and G_{Ref} is the reference tracking gate on-board observation of Jason-1 satellite (32, in unit of gates) [23]. Then, the precise SSH value of retracking waveform was calculated by Equation (7) [5,16]:

$$SSH_{\text{corr}} = R_{\text{Alt}} - (R + \Delta R_{\text{ret}}) - corr \tag{7}$$

where SSH_{corr} is the retracked SSH (m); R_{Alt} is the satellite altitude (m); R is the range of satellite measurement (m); and $corr$ is the error correction (m), which includes wet troposphere correction, dry troposphere correction, ionosphere correction, sea-state correction, dynamic atmospheric correction, ocean tide correction, solid Earth-tide correction, loading tide correction, and pole tide correction. All these error corrections can be found in Jason-1 SGDR products.

3. Results

3.1. SSA-Denoised Waveform Series

We experimented with the 20-Hz waveforms data of Jason-1 cycle 340 pass 153 from 13.80° N to 21.75° N (pass with purple color in Figure 1, hereafter called c340-p153 track). This track had a total of 3118 waveforms and contained the altimeter data moving from both land to ocean and ocean to land. The deepest part of the waters that this track moved was about 1730 m, and the shallowest part was about 7 m. The c340-p153 track had sufficient data volume and a certain representativeness. The waveforms of the c340-p153 track were connected end-to-end to construct a waveform series with length of $N = 3118 \times 104$. Then, this waveform series was denoised with SSA.

The main purpose of the waveform retracker is to find the position of the leading-edge component with respect to the fixed nominal tracking point [6,8]. Therefore, it is critical to retain the slope information of the leading-edge component when performing noise reduction of the waveform series with SSA. It was assumed that a component with a ratio lower than 0.01% was considered as the noise information. That is, the ratio of 0.01% was used as a boundary value to distinguish the main waveform information and the noise information. The components with a ratio higher than 0.01% were classified as the main waveform information, and those lower than 0.01% were classified as the noise information. This means we can reconstruct the main waveform information from the components with a ratio higher than 0.01%.

Figure 3 presents the contribution ratio of the eigenvalues $\{\lambda_i\}_i^M$ corresponding to the raw waveform series. As can be seen from Figure 3, except for the ratio of the first eigenvalue reaching about 70.5%, the others were less than 10%, and the ratio of each eigenvalue starting from the 49th eigenvalue was within 0.01%. The ratio of the first 48 eigenvalues was more than 99%. Therefore, it was preferable to choose $l = 48$ in reconstruction to obtain the SSA-denoised waveform series. Thus, the main waveform information can be well preserved.

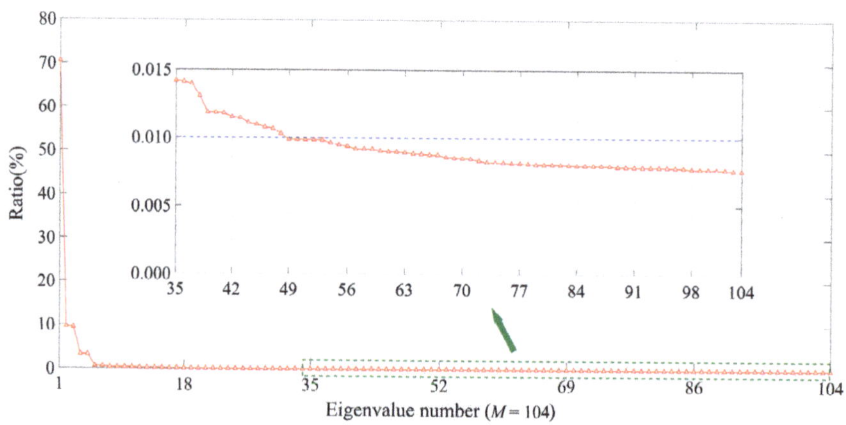

Figure 3. Ratio of the eigenvalues.

Figure 4a is the beginning part of the raw waveform series and the SSA-denoised waveform series, the latter corresponding to the main waveform information. Figure 4b is the beginning part of the residual series (i.e., noise information of the waveform series), defined as the difference between the raw and SSA-denoised waveform series. As can be seen from Figure 4, by comparison, the raw waveforms and the SSA-denoised waveforms mainly differed in the thermal noise component and amplitude, while the slope information of leading-edge component was well retained. This indicates that SSA allowed a noise reduction on Jason-1 waveforms. The larger amplitude of the residual series in Figure 4b was mainly concentrated on the junction of different waveforms. The reason is that the trailing edge component of the waveform was affected by the low-frequency signal of the thermal noise component of the following waveform. However, the leading edge was less affected.

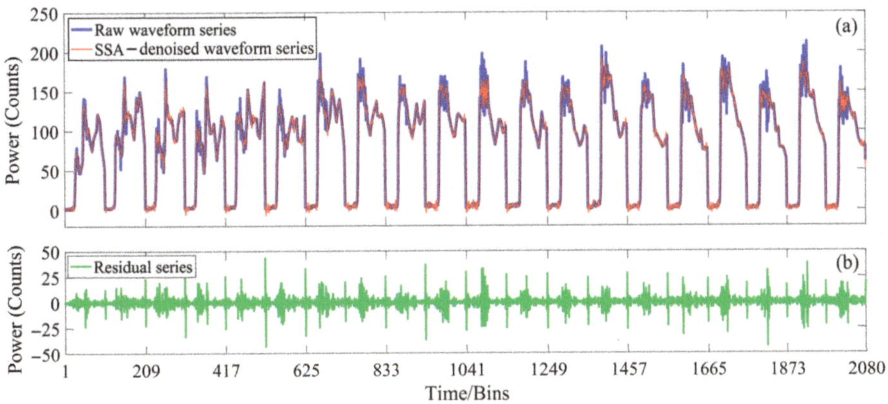

Figure 4. Beginning part of the raw waveform series, the SSA-denoised waveform series, and corresponding residual series defined as the difference between the raw and SSA-denoised waveform series. (**a**) Raw waveform series and SSA-denoised waveform series; (**b**) residual series.

3.2. Comparison of Retracked SSHs

The 20-Hz waveforms data of the c340-p153 track were denoised using the SSA algorithm described earlier to obtain the SSA-denoised waveforms. Both the raw and SSA-denoised waveforms were retracked by the 50% threshold retracker, respectively. Figure 5 compares the raw SSH, the retracked SSHs from the 50% threshold, and SSA + 50% threshold retracker, respectively, with referenced geoidal heights calculated by the EGM2008 model. Here, the SSA + 50% threshold retracker refers to using the 50% threshold retracker for the SSA-denoised waveforms. As can be seen from Figure 5, the deviation of SSH from geoidal height increased as the tracks approached the land. In the coastal region, the retracked SSH profile was smoother than the raw SSH profile, and the former was more similar to the geoidal height than the latter.

The success of the retracker in producing a better SSH estimate value was identified by computing the standard deviations of the difference between SSHs and geoid heights, and the improvement percentage (IMP). The calculation formula of IMP is as follows [18].

$$IMP = \frac{\delta_{raw} - \delta_{retracked}}{\delta_{raw}} \times 100\% \tag{8}$$

where, δ_{raw} and $\delta_{retracked}$ are the standard deviations of the differences between raw SSHs and geoidal heights, and retracked SSHs and geoidal heights, respectively. The geoidal heights are calculated by EGM2008 model [24] in the present study.

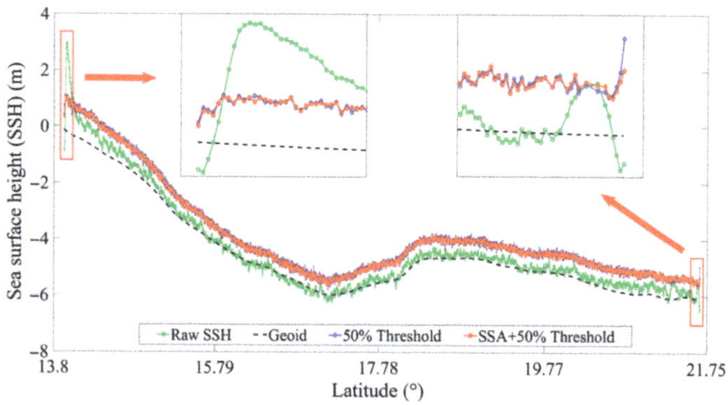

Figure 5. Comparison of raw SSH, retracked SSH, and referenced geoid height (calculated by EGM2008 model) along the track of Jason-1 cycle340 pass153.

The IMP was compared in three cases: (1) The entire c340-p153 track; (2) part of the c340-p153 track from land to ocean within 10 km from the coastline; and (3) part of the c340-p153 track from ocean to land within 10 km from the coastline. Table 2 shows the standard deviations (STDs) of the differences between raw SSH, retracked SSH, and geoidal heights and the IMP in these three cases. In Table 2, the STDs of differences between raw SSH and geoidal heights were smaller than that between the retracked SSH and geoidal heights. This means that waveform retracking iswas successful in improving the quality of altimeter data, especially in the coastal region. Whether in the open ocean or coastal region, the IMP values from the SSA + 50% threshold retracker were larger than those from the 50% threshold retracker. This indicates that SSA successfully improved the retracked SSHs estimate both in the open ocean and coastal region.

Table 2. Standard deviations of differences between raw SSH, retracked SSH, and geoidal heights and improvement percentage (IMP).

The c340-p153 Track	Retracker	δ_{raw} (m)	$\delta_{retracked}$ (m)	IMP (%)
Entire track	50% threshold	0.2954	0.1587	46.27
	SSA + 50% threshold	0.2954	0.1519	48.57
Land to Ocean (Distance < 10 km)	50% threshold	0.8813	0.1170	86.73
	SSA + 50% threshold	0.8813	0.1164	86.80
Ocean to Land (Distance < 10 km)	50% threshold	0.1681	0.0900	46.47
	SSA + 50% threshold	0.1681	0.0690	58.99

3.3. Comparison of Retracked SSHs Discrepancies at Crossover Points

SSA was used to reduce the noise information contained in the 20-Hz waveforms data of Jason-1 GM from cycle 500 to cycle 537 in the South China Sea to obtain the SSA-denoised waveforms, which was the same process as using SSA to denoise the waveforms of the c340-p153 track. Then, the 50% threshold retracker was used to retrack the raw and SSA-denoised waveforms.

There was an SSH difference at the crossover point between the ascending and descending tracks, e.g., Jason-1 GM ground tracks in Figure 1b. The crossover differences can be used to evaluate the quality of the retracked SSHs [10]. In order to compare the retracked SSHs from the 50% threshold retracker and that from the SSA + 50% threshold retracker quantitatively, the mean, STD, and root mean square (RMS) of the crossover differences were calculated (Table 3). The statistical results were categorized into two classes, according to the nearest distances from the crossovers to land: Distances less than 10 km and greater than 10 km. The number in the brackets in column one indicates the number of crossovers in these two classes.

Table 3. Statistical results of crossover differences between the retracked SSHs from the 50% threshold and SSA + 50% threshold retracker.

Distance	Retracker	Mean (m)	STD (m)	RMS (m)
d ≤ 10 km (2526)	50% threshold	0.0125	0.5052	0.5053
	SSA + 50% threshold	0.0059	0.4467	0.4466
d > 10 km (48,407)	50%threshold	−0.0007	0.2664	0.2664
	SSA + 50%threshold	−0.0002	0.2529	0.2529

As can be seen from Table 3, the accuracy of the retracked SSHs in the coastal region (distances less than 10 km from the land) was lower than that in the open ocean (distances more than 10 km from the land). Regardless of whether the distance was more than 10 km or less than 10 km, both the STDs and RMSs from the SSA + 50% threshold retracker were smaller than those from the 50% threshold retracker, which shows that the retracked result of the SSA + 50% threshold retracker was better than that of the 50% threshold retracker. This indicates SSA can effectively improve the precision of the retracked SSHs, whether in the open ocean or coastal region. This conclusion is consistent with Section 3.2.

4. MSSH Model and Validation

4.1. MSSH Model from SSA-Denoised Waveform Retracked SSHs

All the 20-Hz waveforms data from Jason-1 SGDR products, including ERM1, ERM2, and GM in Table 1, were denoised with the SSA described earlier to obtain SSA-denoised waveforms, and the 50% threshold retracker was performed for these SSA-denoised waveforms to obtain the 20-Hz retracked SSHs. These 20-Hz retracked SSHs were compressed by linear regression to obtain 1-Hz retracked SSHs. In this linear regression, the SSHs over three-times larger than the STD were eliminated by an iterative outlier detection, and data with less than 10 points were also not considered. Then, these 1-Hz SSHs were used to establish an MSSH model over the South China Sea with grid of $2' \times 2'$ (shows in Figure 6). The process of establishing MSSH model mainly includes data preprocessing, the removal of the temporal oceanic variability, crossover adjustment, and gridding, which has been detailed by Yuan et al. [25].

4.2. Validations

The MSSH model (shown in Figure 6) was called Model 1. The retracked SSHs by the 50% threshold retracker for the raw waveforms was also used to establish an MSSH model called Model 2. The fundament for both Model 1 and Model 2 was a seven-year (from 2002 to 2009) mean profile of Jason-1. The main difference between Model 1 and Model 2 was whether the SSA algorithm was used for noise reduction during data processing.

In order to validate the accuracy and reliability of the MSSH model established in the present study and whether SSA can effectively improve the accuracy of the MSSH model, Model 1 and Model 2 were compared in terms of SSHs with the CLS15 [26] and DTU18 [27] models in the South China Sea, respectively. CLS15 was published by the Collecte Localisation Satellites (CLS) and the French Centre National d'Etudes Spatiales (CNES), and DTU18 was released by the Technical University of Denmark (DTU). The fundament for CLS15 and DTU18 with a grid of $1' \times 1'$ was a 20-year (from 1993 to 2012) mean profiles of T/P, Jason-1, and Jason-2. The difference between the MSS models depends on the dataset used for calculation and the data processing method [25].

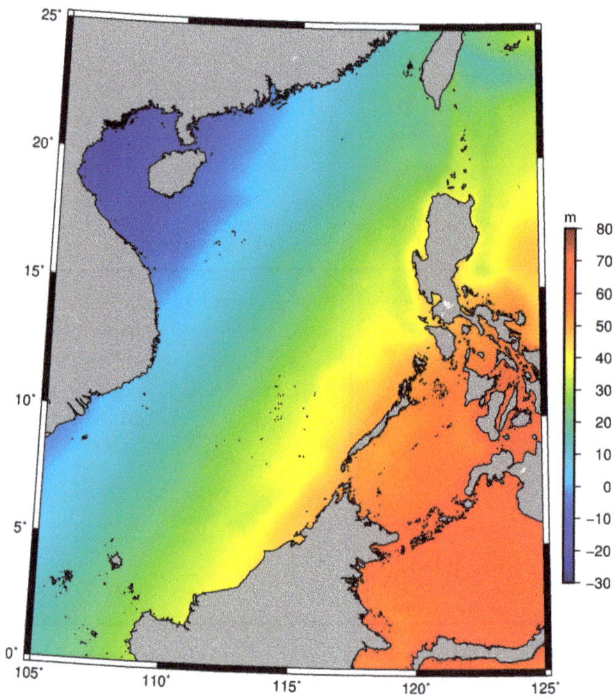

Figure 6. Mean sea surface height (MSSH) model over the South China Sea from the SSHs of the SSA-denoised waveform retracking of Jason-1.

Model 1 (represented by M1), Model 2 (represented by M2), CLS15 (represented by C), and DTU18 (represented by D) were compared in terms of SSHs with each other, as listed in Table 4. The table shows that means of Model 1 and Model 2 compared with CLS15 and DTU18 were obviously systematic biases. These systematic biases were mainly caused by two reasons: The differences between the SSHs measured by different altimetry satellites, and the impact of the oceanic variability. Both CLS15 and DTU18 were established from multi-satellite altimeter data, and these data were adjusted to have the same reference ellipsoid and frame as T/P. However, there was a systematic bias between the SSHs measured by T/P and those measured by Jason-1 of about 10.86 cm in the South China Sea [28]. These four models have different reference time periods: The reference periods of Model 1 and Model 2 span from 2002 to 2009, while the reference periods of CLS15 and DTU18 span from 1993 to 2012. Moreover, the sea level in the South China Sea showed an upward trend, with an increase rate of about 4.25 mm/yr [28,29].

Table 4. Statistics on the differences between different MSS models (Model 1 and Model 2 were established using the retracked sea surface heights (SSHs) from the SSA + 50% threshold and 50% threshold retracker, respectively; M1 for Model 1; M2 for Model 2; C for CLS15; D for DTU18) (in m).

Model Discrepancy	M1-C	M2-C	M1-D	M2-D	M1-M2	C-D
Max	2.2095	2.1881	4.3607	4.3481	0.3909	4.5890
Min	−1.2853	−1.3391	−0.6792	−0.7294	−0.3536	−1.5820
Mean	0.1414	0.1281	0.1607	0.1474	0.0133	0.0193
STD	0.0698	0.0709	0.1179	0.1188	0.0141	0.1090
RMS	0.1577	0.1464	0.1993	0.1893	0.0194	0.1107
Number of points	325,951	325,951	325,951	325,951	325,951	325,951

The STDs of Model 1 and Model 2 compared with CLS15 were much smaller than those compared with DTU18, and the differences of STDs among CLS15, Model 1, and Model 2 were within 1 cm of DTU18. This indicates the high-degree consistency among Model 1, Model 2, and CLS15, and the obvious differences from DTU18. The STDs of Model 1 were smaller than those of Model 2, compared with CLS15 and DTU18.

These four models were compared in terms of SSH in the open ocean and coastal region (i.e., ~10 km from land), respectively, and the results are listed in Table 5. The table shows that the STDs of Model 1 and Model 2 compared with CLS15 and DTU18 were 5~7 cm in the open ocean, whereas they rose up to the several decimeters in coastal region. Moreover, the STDs of Model 1 and Model 2 compared with DTU18 were almost twice of those compared with CLS15 in the coastal region. This indicates that DTU18 mainly differed from Model 1, Model 2, and CLS15 in coastal region, and it is assumed that this was mainly caused by the differences in the preprocessing of altimeter data. The STDs of the discrepancy between Model 1 and Model 2 were less than 2 cm in the open ocean and 3 cm in the coastal region. This indicates that these two models displayed more difference in the coastal region. In addition, the STDs of Model 1 compared with the two models CLS15 and DTU18 were lower than those of Model 2 compared with the two, indicating that the accuracy of Model 1 was superior to that of Model 2.

Table 5. Statistics on the differences between different MSS models (Model 1 and Model 2 were established using the retracked sea surface heights (SSHs) from the SSA + 50% threshold and 50% threshold retracker, respectively; M1 for Model 1; M2 for Model 2; C for CLS15; D for DTU18) in the open ocean and coastal region (in m).

Model Discrepancy		M1-C	M2-C	M1-D	M2-D	M1-M2	C-D
Coastal region	Max	1.8059	1.7862	4.3607	4.3481	0.3909	4.0920
	Min	−1.2853	−1.3391	−0.6792	−0.7294	−0.3536	−1.5820
	Mean	0.1173	0.1033	0.2125	0.1985	0.0140	0.0951
	STD	0.1521	0.1543	0.3215	0.3230	0.0262	0.3120
	RMS	0.1921	0.1857	0.3853	0.3791	0.0297	0.3262
	Number of points	31,257	31,257	31,257	31,257	31,257	31,257
Open ocean	Max	2.2095	2.1881	3.3181	3.3000	0.1403	4.5890
	Min	−1.2709	−1.2890	−0.3862	−0.4003	−0.3463	−1.4930
	Mean	0.1439	0.1308	0.1552	0.1420	0.0132	0.0112
	STD	0.0535	0.0545	0.0641	0.0651	0.0121	0.0461
	RMS	0.1536	0.1416	0.1679	0.1562	0.0179	0.0475
	Number of points	294,694	294,694	294,694	294,694	294,694	294,694

5. Conclusions

Altimeter waveforms are usually contaminated by land, island, sea reef, sea ice, seabed terrain, etc., which leads to incorrect SSHs retracked from these corrupted waveforms [30]. Moreover, the precision of MSSH model established from these poor SSHs is affected as well. SSA is a classical technique used in signal processing, and is also sometimes used in denoising signals. Therefore, it can be used to reduce the noise level in altimeter waveforms.

In this paper, the c340-p153 track of Jason-1 from 13.80° N to 21.75° N was selected as the experimental object to introduce the specific process of SSA to reduce the noise level in altimeter waveforms. All waveforms data of Jason-1 from 2002 to 2013 (including ERM1, ERM2 and GM) were processed by SSA noise reduction to obtain SSA-denoised waveforms, which were retracked by a 50% threshold retracker to obtain corresponding retracked SSHs. Then, these retracked SSHs were used to establish MSSH model over South China Sea with grid of $2' \times 2'$.

A comparison of the IMP and statistical results of crossover differences of the retracked SSHs from the SSA + 50% threshold retracker and those from the 50% threshold retracker showed that the SSA allowed a noise reduction on the Jason-1 altimeter waveforms, and can successfully improve the accuracy of retracked SSHs either in the open ocean or coastal region.

Model 1 and Model 2 were compared with CLS15 and DTU18 in the South China Sea, and the results showed that these four models had a high-degree consistency. Moreover, Model 1 showed higher accuracy than Model 2, and the main difference between these two models was found mainly in the coastal region. This result indicates that using SSA to reduce the noise level of the Jason-1 altimeter waveforms can effectively improve the accuracy of the MSSH model.

The SSA algorithm was used to denoise the waveforms of Jason-1 to establish an MSSH model over the South China Sea. It can also be employed to denoise other satellite altimeter waveforms to establish MSSH models of other regions, even a global MSSH model, which is our main aim for the next study.

Author Contributions: Formal analysis, methodology, software, validation, writing—original draft, writing—review & editing: J.Y.; Conceptualization, funding acquisition, investigation, methodology, project administration, supervision: J.G.; Formal analysis, methodology, software, validation: Y.N.; Data curation, resources, visualization: C.Z. and Z.L.; Conceptualization, investigation, methodology, supervision: X.L. All authors have read and agreed to the published version of the manuscript.

Funding: This research was funded by the National Natural Science Foundation of China, grant number 41774001.

Acknowledgments: We thank the Archiving, Validation, and Interpretation of Satellite Oceanographic Data (AVISO) for providing the Jason-1 sensor geophysical data records (SGDR) products. We also thank the Collecte Localisation Satellites (CLS) and the French Centre National d'Etudes Spatiales (CNES) for providing MSS_CNES_CLS15, and the Technical University of Denmark (DTU) for providing DTU18 MSS.

Conflicts of Interest: The authors declare no conflict of interest.

References

1. Quartly, G.; Chen, G. Introduction to the Special Issue on "Satellite Altimetry: New Sensors and New Applications". *Sensors* **2006**, *6*, 616–619. [CrossRef]
2. Guo, J.; Gao, Y.; Hwang, C.; Sun, J. A multi-subwaveform parametric retracker of the radar satellite altimetric waveform and recovery of gravity anomalies over coastal oceans. *Sci. China Earth Sci.* **2010**, *53*, 610–616. [CrossRef]
3. Hernandez, F.; Schaeffer, P. *The CLS01 Mean Sea Surface: A Validation with the GSFC00.1 Surface*; CLS: Ramonville St Agne, France, 2001.
4. Hwang, C.; Hsu, H.-Y.; Jang, R. Global mean sea surface and marine gravity anomaly from multi-satellite altimetry: Applications of deflection-geoid and inverse Vening Meinesz formulae. *J. Geodesy* **2002**, *76*, 407–418. [CrossRef]
5. Fu, L.L.; Cazenave, A. *Satellite Altimetry and Earth Sciences: A Handbook of Techniques and Applications*; Academic Press: San Diego, CA, USA, 2001.
6. Deng, X. Improvement of Geodetic Parameter Estimation in Coastal Regions from Satellite Radar Altimetry. Ph.D. Thesis, Curtin University of Technology, Perth, Australia, 2003.
7. Deng, X.; Featherstone, W.E. A coastal retracking system for satellite radar altimeter waveforms: Application to ERS-2 around Australia. *J. Geophys. Res. Oceans* **2006**, *111*. [CrossRef]
8. Gommenginger, C.; Thibaut, P.; Fenodlio-Marc, L.; Quartly, G.; Deng, X.; Gomez-Enri, J.; Challenor, P.; Gao, Y. Retracking Altimeter Waveforms Near the Coasts. In *Coastal Altimetry*; Vignudelli, S., Kostianoy, A.G., Cipollini, P., Benveniste, J., Eds.; Springer: Berlin/Heidelberg, Germany, 2011; pp. 61–102.
9. Deng, X.; Featherstone, W.E.; Hwang, C.; Berry, P.A.M. Estimation of Contamination of ERS-2 and POSEIDON Satellite Radar Altimetry Close to the Coasts of Australia. *Mar. Geod.* **2002**, *25*, 249–271. [CrossRef]
10. Yang, L.; Lin, M.; Pan, D. Retracking Jason1 Altimeter Waveform Over Chinacoastal Zone. In Proceedings of the SPIE-Microwave Remote Sensing of the Atmosphere and Environment VI, Noumea, New Caledonia, 19 November 2008; Volume 7154.
11. Idris, N.H.; Deng, X. The Retracking Technique on Multi-Peak and Quasi-Specular Waveforms for Jason-1 and Jason-2 Missions near the Coast. *Mar. Geod.* **2012**, *35*, 217–237. [CrossRef]
12. Vautard, R.; Yiou, P.; Ghil, M. Singular-spectrum analysis: A toolkit for short, noisy chaotic signals. *Physica D* **1992**, *58*, 95–126. [CrossRef]
13. Golyandina, N.; Zhigljavsky, A. *Singular Spectrum Analysis for Time Series*; Springer: New York, NY, USA, 2013.

14. Shen, Y.; Guo, J.; Liu, X.; Kong, Q.; Guo, L.; Li, W. Long-term prediction of polar motion using a combined SSA and ARMA model. *J. Geodesy* **2018**, *92*, 333–343. [CrossRef]
15. Guo, J.; Shi, K.; Liu, X.; Sun, Y.; Li, W.; Kong, Q. Singular spectrum analysis of ionospheric anomalies preceding great earthquakes: Case studies of Kaikoura and Fukushima earthquakes. *J. Geodyn.* **2019**, *124*, 1–13. [CrossRef]
16. CNES. *Jason-1 Products Handbook*; SALP-MU-M5-OP-13184-CN; Issue: 5 Rev 1; CNES: Paris, France, 2016.
17. Yang, L.; Lin, M.; Liu, Q.; Pan, D. A coastal altimetry retracking strategy based on waveform classification and sub-waveform extraction. *Int. J. Remote Sens.* **2012**, *33*, 7806–7819. [CrossRef]
18. Hwang, C.; Guo, J.; Deng, X.; Hsu, H.-Y.; Liu, Y. Coastal Gravity Anomalies from Retracked Geosat/GM Altimetry: Improvement, Limitation and the Role of Airborne Gravity Data. *J. Geodesy* **2006**, *80*, 204–216. [CrossRef]
19. Guo, J.; Hwang, C.; Chang, X.; Liu, Y. Improved threshold retracker for satellite altimeter waveform retracking over coastal sea. *Prog. Nat. Sci.* **2006**, *16*, 732–738.
20. Ganguly, D.; Chander, S.; Desai, S.; Chauhan, P. A Subwaveform-Based Retracker for Multipeak Waveforms: A Case Study over Ukai Dam/Reservoir. *Mar. Geod.* **2015**, *38*, 581–596. [CrossRef]
21. Davis, C.H. Growth of the Greenland ice sheet: A performance assessment of altimeter retracking algorithms. *IEEE Trans. Geosci. Remote* **1995**, *33*, 1108–1116. [CrossRef]
22. Davis, C.H. A robust threshold retracking algorithm for measuring ice-sheet surface elevation change from satellite radar altimeter. *IEEE Trans. Geosci. Remote* **1997**, *35*, 974–979. [CrossRef]
23. Uebbing, B.; Kusche, J.; Forootan, E. Waveform retracking for improving level estimations from topex/poseidon, Jason-1, and Jason-2 altimetry observations over african lakes. *IEEE Trans. Geosci. Remote* **2015**, *53*, 2211–2224. [CrossRef]
24. Pavlis, N.K.; Holmes, S.A.; Kenyon, S.C.; Factor, J.K. The development and evaluation of the Earth Gravitational Model 2008 (EGM2008). *J. Geophys. Res.* **2012**, *117*, B04406. [CrossRef]
25. Yuan, J.; Guo, J.; Liu, X.; Zhu, C.; Niu, Y.; Li, Z.; Ji, B.; Ouyang, Y. Mean sea surface model over China seas and its adjacent ocean established with the 19-year moving average method from multi-satellite altimeter data. *Cont. Shelf Res.* **2020**, *192*, 104009. [CrossRef]
26. Pujol, M.-I.; Schaeffer, P.; Faugère, Y.; Raynal, M.; Dibarboure, G.; Picot, N. Gauging the improvement of recent mean sea surface models: A new approach for identifying and quantifying their errors. *J. Geophys. Res. Oceans* **2018**, *123*, 5889–5911. [CrossRef]
27. Andersen, O.; Knudsen, P.; Stenseng, L. A New DTU18 MSS Mean Sea Surface—Improvement from SAR Altimetry. In Proceedings of the 25 Years of Progress in Radar Altimetry Symposium, Portugal, Duration, 24–29 September 2018.
28. Guo, J.; Wang, J.; Hu, Z.; Hwang, C.; Chen, C.; Gao, Y. Temporal-spatial variations of sea level over Chinese seas derived from altimeter data of TOPEX/Poseidon, Jason-1 and Jason-2 from 1993 to 2012. *Chin. J. Geophys.* **2015**, *58*, 3103–3120.
29. Guo, J.; Hu, Z.; Wang, J.; Chang, X.; Li, G. Sea level change of China seas and neighboring ocean based on satellite altimetry missions from 1993 to 2012. *J. Coast. Res.* **2015**, *73*, 17–21. [CrossRef]
30. Huang, Z.; Wang, H.; Luo, Z.; Shum, C.K.; Tseng, K.-H.; Zhong, B. Improving Jason-2 Sea Surface Heights within 10 km Offshore by Retracking Decontaminated Waveforms. *Remote Sens.* **2017**, *9*, 1077. [CrossRef]

Journal of
Marine Science and Engineering

MDPI

Article

A Barotropic Solver for High-Resolution Ocean General Circulation Models

Xiaodan Yang [1,2,3], Shan Zhou [4], Shengchang Zhou [1,2,3,5], Zhenya Song [1,2,3,*] and Weiguo Liu [2,5]

1 First Institute of Oceanography, and Key Laboratory of Marine Science and Numerical Modeling, Ministry of Natural Resources, Qingdao 266061, China; yangxiaodan@fio.org.cn (X.Y.); sc.zhou@foxmail.com (S.Z.)
2 Laboratory for Regional Oceanography and Numerical Modeling, Pilot National Laboratory for Marine Science and Technology, Qingdao 266237, China; weiguo.liu@sdu.edu.cn
3 Shandong Key Laboratory of Marine Science and Numerical Modeling, Qingdao 266061, China
4 Vacation BU, Trip.com Group Limited, Shanghai 200335, China; shanzhou@trip.com
5 School of Software, Shandong University, Jinan 250101, China
* Correspondence: songroy@fio.org.cn

Abstract: High-resolution global ocean general circulation models (OGCMs) play a key role in accurate ocean forecasting. However, the models of the operational forecasting systems are still not in high resolution due to the subsequent high demand for large computation, as well as the low parallel efficiency barrier. Good scalability is an important index of parallel efficiency and is still a challenge for OGCMs. We found that the communication cost in a barotropic solver, namely, the preconditioned conjugate gradient (PCG) method, is the key bottleneck for scalability due to the high frequency of the global reductions. In this work, we developed a new algorithm—a communication-avoiding Krylov subspace method with a PCG (CA-PCG)—to improve scalability and then applied it to the Nucleus for European Modelling of the Ocean (NEMO) as an example. For PCG, inner product operations with global communication were needed in every iteration, while for CA-PCG, inner product operations were only needed every eight iterations. Therefore, the global communication cost decreased from more than 94.5% of the total execution time with PCG to less than 63.4% with CA-PCG. As a result, the execution time of the barotropic modes decreased from more than 17,000 s with PCG to less than 6000 s with CA-PCG, and the total execution time decreased from more than 18,000 s with PCG to less than 6200 s with CA-PCG. Besides, the ratio of the speedup can also be increased from 3.7 to 4.6. In summary, the high process count scalability when using CA-PCG was effectively improved from that using the PCG method, providing a highly effective solution for accurate ocean simulation.

Keywords: barotropic solver; PCG; CA-PCG; ocean general circulation model; NEMO

Citation: Yang, X.; Zhou, S.; Zhou, S.; Song, Z.; Liu, W. A Barotropic Solver for High-Resolution Ocean General Circulation Models. *J. Mar. Sci. Eng.* **2021**, *9*, 421. https://doi.org/10.3390/jmse9040421

Academic Editor: Marcos G. Sotillo

Received: 8 March 2021
Accepted: 9 April 2021
Published: 14 April 2021

1. Introduction

Ocean general circulation models (OGCMs) have become increasingly important for understanding oceanic dynamic processes and ocean environment forecasting. In recent years, OGCMs have been developed with finer resolution (10–100 km for eddy-resolving ocean models) and more physical procedures due to increasing scientific requirements. Accuracy is one of the important goals of ocean simulation and forecasting. In general, OGCMs with finer horizontal resolution can provide more accurate results [1]. However, the computational cost will be almost three orders of magnitude if the horizontal resolution is fined by one order of magnitude [2,3]. For an operational forecasting system, the computation should be finished within 1∼2 h. So, the parallel efficiency becomes a great challenge for OGCMs. Due to the low parallel efficiency, current global operational forecasting systems are still in 0.1° (∼10 km) to 1° (∼100 km) resolution, which falls far short of expectations. Therefore, with the resolution finer, the demand for improving computational performance is more and more urgent.

To increase the computational efficiency of OGCMs, the hydrostatic primitive equations for momentum can be solved using a split-explicit, time-stepping scheme that requires special treatment and coupling between barotropic and baroclinic modes. The baroclinic mode is used for three-dimensional prognostic variables, which are the slow processes, such as the advective term, while the barotropic mode solves the free-surface and associated barotropic velocity equations, such as surface inertial gravity wave, which are the fast processes. To improve the simulation efficiency, a finite number of barotropic time steps were carried out within each baroclinic step (details in Appendix A). In barotropic mode, the implicit free-surface method is common because it allows a large time step to efficiently calculate the fast gravity mode [4,5]. But this method requires solving elliptic equations (details are discussed in Section 2). One of the commonly used methods of solving an elliptic equation is the conjugate gradient (CG) method [6]. To reduce the execution time of the CG method, one way is to reduce the communication costs by reducing the number of iterations to convergence. Thus, preconditioning is commonly used for the CG method to reduce the number of iterations, assuming the cost of preconditioning is reasonable [7], which has been widely applied in OGCMs [8–10]. As the Preconditioned CG method (PCG) method used in OGCMs requires the calculation of global variables, which also expects a large number of global reductions, it becomes the bottleneck when the model is run in the large-scale parallel computation. To further reduce the associated communication cost and improve scalability, [11] tried to apply a preconditioned Classical Stiefel Iteration method instead of the PCG method in the Parallel Ocean Programme to solve the elliptic system of equations in the barotropic mode, which requires no global reduction except for checking convergence. However, the PCG method is still popular and widely used due to less requirement in memory, better stability, and faster and clearer representation. Thus, the other way to reduce the number of global reductions to improve the large-scale efficiency of the PCG method is necessary and rarely done before.

The Nucleus for European Modelling of the Ocean (NEMO) is a state-of-the-art OGCM that has been widely used by ocean and climate communities for oceanographic, forecasting, and climate studies [9]. The barotropic solver recommended in NEMO is also the PCG method because it can effectively solve elliptic equations with vector computers. When using thousands of processes, the PCG method in NEMO scales poorly (Figure 1), as do other OGCMs. After analyzing, we found that the poor scaling performance of the PCG method was caused by the operation of global reductions (i.e., function MPI_Allreduce combines values from all processes and distributes the result back to all processes), which could account for more than 96% of the total execution time when using more than 8000 processes (Figure 2).

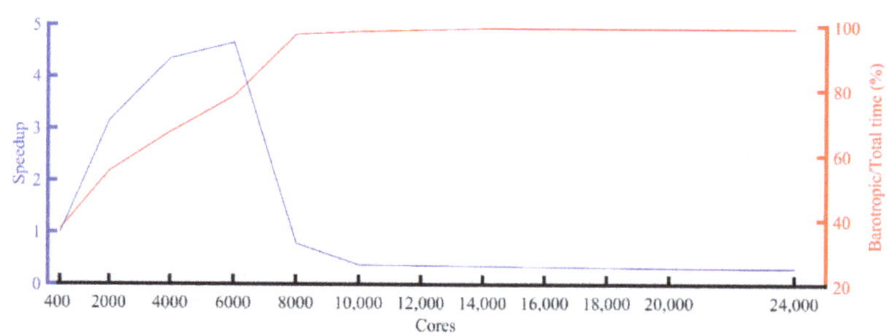

Figure 1. Speedup of the 5-km Nucleus for European Modelling of the Ocean (NEMO) experiment using the preconditioned conjugate gradient (PCG) solver (blue line) and the percentage of the barotropic time of the total execution time (red line). The baseline of the speedup is the running time corresponding to 400 processes.

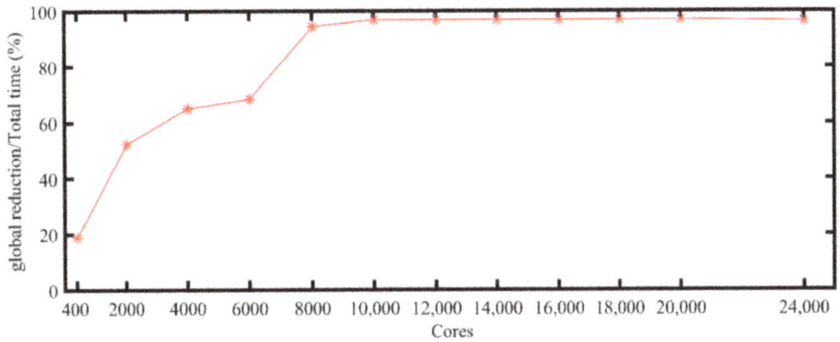

Figure 2. Percentage of the global reduction (MPI_Allreduce) time of the total execution time in the 5-km NEMO experiment with the PCG solver.

In this study, we developed a new method, namely, a communication-avoiding Krylov subspace method with PCG (CA-PCG), to reduce the number of global reductions and improve the PCG method in the barotropic mode in NEMO. The results showed a high decrease in the total execution time for the high-resolution simulation with NEMO and provided a reference for the OGCMs using PCG solver for developing effective measures to improve the scalability on large scales.

2. Materials and Methods

2.1. NEMO Model

The Nucleus for European Modelling of the Ocean (NEMO) is a framework of ocean-related engines, namely, OPA (O'cean PArall´elis´e) for the ocean dynamics and thermodynamics, LIM2 (Louvain-la-Neuve Ice Model version 2) for the sea–ice dynamics and thermodynamics, and TOP3 (Tracer in the Ocean Paradigm version 3) for the biogeochemistry [9]. The ocean component of NEMO was developed from the OPA model, as described in [12]. This model has been used for a wide range of applications, both regional and global, as a forced ocean model and as a model coupled with the sea-ice and/or the atmosphere. The barotropic mode optimized in this study is a procedure that solves the free-surface, and associated barotropic velocity equations belong to the OPA model.

2.2. Barotropic Solver

As discussed above, the solution of an elliptic equation is the most time-consuming part of the barotropic mode for large-scale clusters. Here, the elliptic equation is shown [11]:

$$\left[\nabla \cdot H\nabla - \varnothing(\tau)\right]\eta^{n+1} = \psi\left(\eta^{n}, \eta^{n-1}, \tau\right) \tag{1}$$

where H represents the depth of the ocean, τ is the time step, η^n is the sea surface height (SSH) at the n-th time step, and ψ is the function of the influence that the previous states of the SSH and forcing have on the next state.

At each time step, the time derivative of the SSH at the next time step is solved with an elliptic equation. Equation (1) is discretized into a two-dimensional orthogonal curvilinear grid using a five-point stencil in NEMO and can be reorganized into a linear symmetric system $A\mathrm{x} = \boldsymbol{b}$. The idea of the PCG method is to search for the solution of x.

2.3. NEMO PCG Solver

The PCG used in NEMO is a modified typical PCG. However, the modification does not reduce the operation of the global reductions (MPI_Allreduce). As global reductions are the main bottleneck of the NEMO model, the optimization was still based on the typical PCG for simplicity in this study. The typical PCG method (Algorithm A1) contains three major parts: computing, boundary updating, and global reduction. There are matrix-vector multiplications (MVMs) and vector–vector multiplications (VVMs) in this computation, displaying good scalability. The time cost of boundary updating is required to update the halo area after an MVM, but there is no scalability issue for large-scale clusters. The cost of global reductions required by the inner product, is the key problem in a large-core count cluster. The details of the communication costs and performance analysis are illustrated in Appendix B.

2.4. CA-PCG Solver

Optimization of NEMO scalability, which was mainly caused by the global reduction in the PCG solver, has rarely been studied until now. To solve this problem, a communication-avoiding Krylov subspace method was implemented with PCG (CA-PCG), according to Carson's work [13]. For PCG, inner product operation is needed for each iteration, which means a global reduction cost for each iteration. However, for CA-PCG, all of the iterations are divided into two layers of loops. The global reductions are only needed in the outer loop. Since we used the same preconditioner for PCG, the iteration number K should not change. If s is the inner loop count, then the outer loop count will be K/s, which means one global reduction is needed after every s iterations. After multiple tests, the inner loop count s was set as 8 in this study. Then the frequency of inner product operation was decreased to every eight iterations. Moreover, the global communication cost could decrease dramatically, and the high-core count scalability could be improved effectively. The pseudocode for the CA-PCG algorithm designed for NEMO is shown in Appendix C.

2.5. Experimental Description

The NEMO model optimized in this study was version 3.6. To assess the performance of the NEMO model and the validity of the model results after optimization using the CA-PCG method, two experiments were designed: a low-resolution experiment, named ORCA1, with a horizontal grid resolution of nearly 100 km (grid dimensions: 362×292) and a high-resolution experiment, named GYRE, with a horizontal grid resolution of nearly 5 km (grid dimensions: 7502×5002). The optimization of the NEMO model using the CA-PCG solver was mainly to reduce the communication costs. So, the performance improvement should be more effective on large-scale simulations than on a small scale. Thus, we used a small-scale experiment to validate the results' accuracy and a large-scale experiment to test the scalability of this approach. The flow chart of the total experiments is shown in Figure 3. First, a high-resolution experiment was designed to evaluate the performance of the NEMO model with a PCG solver. Second, a new solver, i.e., CA-PCG solver, was developed and applied to the NEMO model. The next step was to validate the accuracy of the model results after optimization by comparing them with those from ensembles with PCG solver using low-resolution experiments. Last, the improvement of the performance of the NEMO model with the CA-PCG solver was tested using a high-resolution experiment. The description of the parallel platform used in this study is provided in Appendix D.

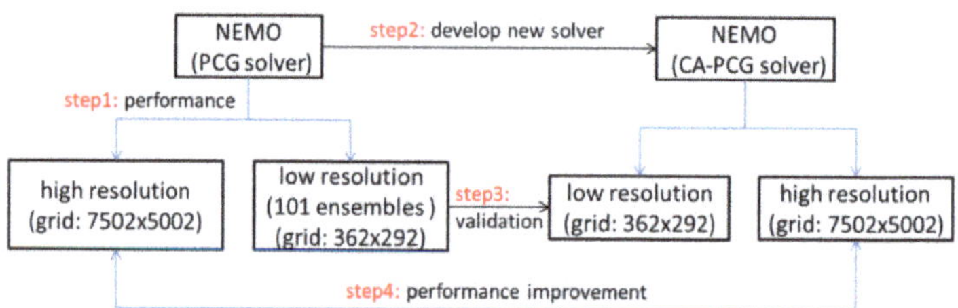

Figure 3. Flow chart of the experiments in this study. The experiments were designed in order of step1, step2, step3, and step4.

2.5.1. Low-Resolution Simulation

ORCA1 is one of the most frequently used horizontal grid resolutions in NEMO. The grid was derived from https://forge.ipsl.jussieu.fr/nemo/ (accessed on 10 April 2021), while the bathymetry file was generated by referring to http://www.noc.ac.uk/ (accessed on 10 April 2021). The initial conditions of temperature and salinity were from a combination of the World Ocean Atlas (WOA) 2009 [14,15] and the Polar Science Center Hydrographic Climatology (PHC) version 3 (updated from [16]) data. We also used the WOA 2009 and PHC v3.0 temperature data as observations to validate the model results (details in Section 2.6). The atmospheric forcing data were from the climatology Coordinated Ocean-ice Reference Experiments phase II (CORE-II) forcing data set [17,18]. The time step used in this experiment was 3600 s (i.e., the rn_rdt parameter in NEMO was set to 3600). Experiments using the PCG and CA-PCG solvers were designed, regarded as ORCA1_PCG and ORCA1_CAPCG, respectively. Besides, to validate the model results, an ensemble consisting of 101 ocean simulations using the PCG solver was also created.

2.5.2. High-Resolution Simulation

The idealized GYRE application in NEMO was presented to test and analyze the scalability of high-resolution simulation on a large scale. There was no file Input/Output other than the reading of the workload definition during initialization. This experiment used a time step of 7200 s (rn_rdt = 7200). To guarantee that the total execution time of all of the experiments was no less than 3 min, the experiments were run for 1000 steps. Corresponding experiments using the PCG and CA-PCG solvers were also designed, regarded as GYRE_PCG and GYRE_CAPCG, respectively. Notably, each experiment was repeated three times, and the mean values of the execution time were discussed.

2.6. Community Earth System Model (CESM) Port-Verification Tool

The methodology we used to validate the results was the Community Earth System Model (CESM) port-verification tool (CESM-PVT), which was developed to determine whether a change in CESM would result in distinguished biases from the natural variability of the system [19]. First, in [19], an ensemble $E = \{X_1, X_2, \ldots X_m\}$ consisting of 101 simulations, which differed only in a random perturbation of the initial temperature condition at the order of 10^{-14}, was run. At a given point j, we had a series of possible results for each variable X from the ensemble $E = \{X_1(j), X_2(j), \ldots X_m(j)\}$. Then, the mean and standard deviation of this series at a given point j was defined as $\mu(j)$ and $\delta(j)$, respectively. The root mean square z-score (RMSZ) of the ensemble data was calculated as follows.

$$RMSZ(X) = \sqrt{\frac{1}{n}\sum_{j=1}^{n}\left(\frac{X(j) - \mu(\mathrm{j})}{\delta(\mathrm{j})}\right)^2} \quad (2)$$

where n is the total number of grid points in X. Moreover, the $RMSZ$ score of the new case was also calculated. If this $RMSZ$ fell within the distribution of the ensemble's $RMSZ$, then the result was considered to have passed the accuracy test. The details could refer to [11,19].

According to [19], this methodology not only benefits the model data of the CESM, but is also applicable for evaluating the data of other simulations. To verify the accuracy of the NEMO model using the CA-PCG solver, we also considered 101 simulations, which were the same as for the ORCA1_PCG experiment, except for a random perturbation of the initial ocean temperature at the order of 10^{-14}. The ensemble runs here were 10 years in length, which is a long enough duration for an ocean model to become stable after a disturbance.

To verify the result, we did some revisions to the CESM-PVT method. First, to increase the reliability, we treated observations (WOA + PHC data) instead of the mean value of ORCA1_PCG ensemble members as the "reference value" here. This is because the only thing we cared about was whether the model results were consistent with observations, no matter before or after optimization. Second, for each ensemble member, the global mean value was calculated first, and then the mean biases and standard deviations compared to the observed global mean value were obtained at each time point. The global mean value of the CA-PCG experiment and the mean bias were calculated at the same time. Finally, the mean biases of each ensemble member and the CA-PCG experiment were divided by the corresponding standard deviations at each time point to obtain the z-score biases, respectively. We revised the CESM-PVT method by comparing the global mean value instead of the point-to-point value because the biases may be magnified at a specific point. For example, when the results between the ORCA1_PCG ensembles and observations are very close, any destabilization in the ORCA1_CAPCG results may lead to a large deviation after being divided by these extremely small standard deviations. Verification using the global mean value would avoid this issue effectively. Besides, the global mean value is an effective indicator of ocean simulation evaluation. The equation is revised as follows.

$$Z(X) = \frac{\frac{\sum_{j=1}^{n} X(j)}{n} - \mu(\mathrm{j})}{\delta(\mathrm{j})} \tag{3}$$

where $\frac{\sum_{j=1}^{n} X(j)}{n}$ is the calculation of global mean value at a given point j. The $\mu(\mathrm{j})$ represents the value in observation. The δ bs is the standard deviation of the ensemble time series.

3. Results

3.1. Validity of the Results Using the CA-PCG Solver

We verified the model results using the CA-PCG solver with the revised CESM-PVT method shown above. Figure 4 shows the distribution of the z-score biases of the temperature variable in 101 ensemble runs with the PCG solver and the CA-PCG experimental result. As the observation of temperature is more accessible than other variables and is also one of the fundamental physical quantities in ocean simulation, we chose the three-dimensional temperature field for this evaluation. To match the accuracy of CESM-PVT, the variable's z-score bias of the experiment using the CA-PCG solver should be no larger than those of the ensemble simulations. As shown in Figure 4, the CA-PCG experimental result's z-score bias fell into the distribution of the ensemble z-score biases, showing good consistency with the results simulated using the PCG solver. Above all, the model results simulated using the CA-PCG solver were accurate, and the CA-PCG solver can be used in the NEMO model in future research.

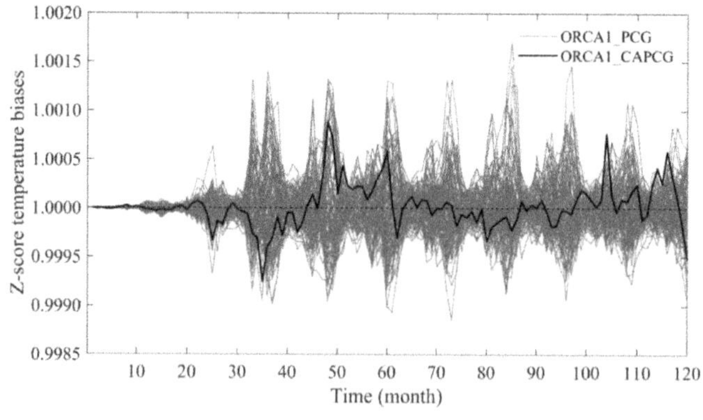

Figure 4. Z-score biases of temperature in ensemble run (grey lines) and the communication-avoiding Krylov subspace method with a preconditioned conjugate gradient (CA-PCG) solver experiment (black line). The dashed line represents the mean value of the z-score biases of all the ensemble runs.

3.2. Performance of the NEMO Model Using the CA-PCG Solver

Before investigating the performance of the model using the CA-PCG solver, it was important to compare the global reductions before and after optimization. As shown in Figure 5, the percentage of global reductions increased with the number of GYRE_PCG experiment processes. When more than 8000 processes were used, the execution time of the global reduction increased to more than 94.5% of the total execution time and could even reach 99.9% when using 24,000 processes. However, the execution time was reduced to 63.4% when using the CA-PCG solver. The ratio of the percentage of the global reductions in the two experiments was calculated (Figure 6). Almost all the ratios in the GYRE_CAPCG to GYRE_PCG experiment were less than 1, except when using 400 processes, due to the increase in the calculation time after optimization. When using more than 8000 processes, the ratio was approximately 0.70. Thus, the CA-PCG solver was useful for reducing the operation of global reduction, especially when scaled to more than 8000 processes.

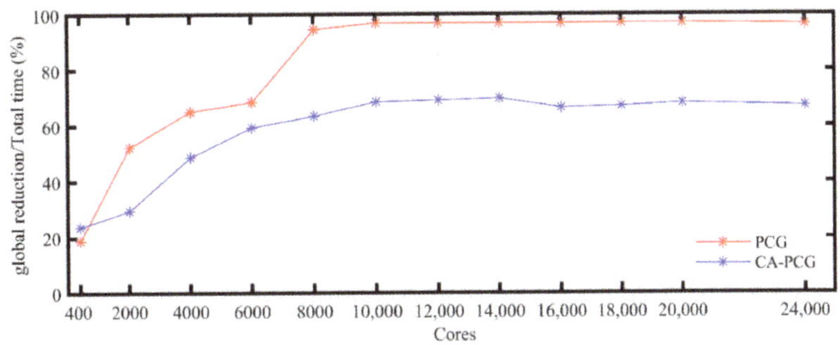

Figure 5. Percentages of the global reduction (MPI_Allreduce) time of the total execution time in 5-km Nucleus for European Modelling of the Ocean (NEMO) experiments using the preconditioned conjugate gradient (PCG) (blue line) and CA-PCG (red line) solvers.

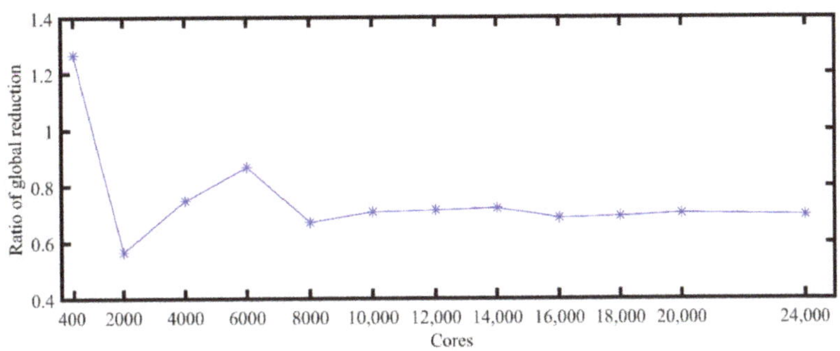

Figure 6. Ratio of the percentage of the global reduction (MPI_Allreduce) time of the total execution time determined using the CA-PCG and PCG solvers. In particular, the barotropic mode in GYRE_PCG accounted for approximately 97–99% of the total execution times when scaled to more than 8000 processes (Figure 7). After using the new CA-PCG solver, the percentage decreased slightly to approximately 92–96%. Even though the reduction in the percentage does not seem notable, the execution time of the barotropic mode decreased from more than 17,000 s to less than 6000 s (figure not shown).

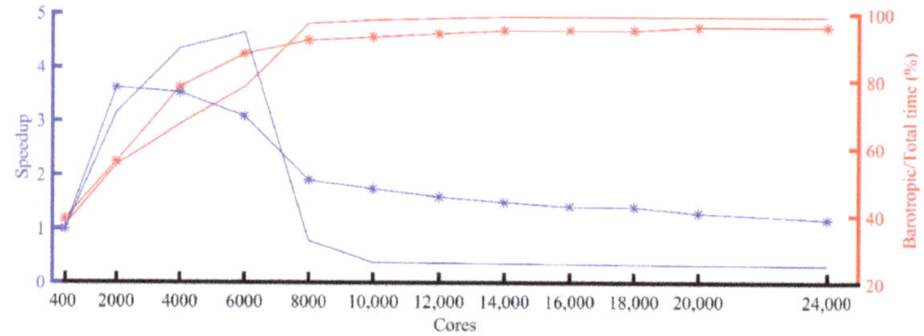

Figure 7. Speedup of 5-km NEMO experiments using the PCG (solid blue line) and CA-PCG (blue solid-star line) solvers. Percentages of the barotropic time of the total execution time using the PCG (solid red line) and CA-PCG (red solid-star line) solvers.

The total execution time considering different numbers of processes in both GYRE_PCG and GYRE_CAPCG is shown in Table 1. The total execution time decreased as the number of processes employed increased in the GYRE_PCG experiment when using less than 6000 processes, and the minimum time was 1510 s, which was due to the decrease in the cost of computation. However, it greatly increased to more than 18,000 s when using more than 10,000 processes because of the dramatic increase in global reduction. After comparing the total execution time for GYRE_PCG and GYRE_CAPCG, we found that the differences were not considered with less than 6000 processes. Sometimes the execution time in GYRE_CAPCG was longer than that in GYRE_PCG due to the increase in the required computation. However, as the model scaled to larger scales, especially more than 10,000 processes, the total execution time decreased to 4000–6000 s in GYRE_CAPCG, much faster than that in GYRE_PCG. After optimization, there was a 3.56–4.40-fold improvement at large scales, which effectively reduced the model simulation costs.

Table 1. Execution time for the high-resolution experiments.

Number of Processes	GYRE_PCG (s)	Speedup	GYRE_CAPCG (s)	Speedup
400	7025	1.00	7196	1.00
2000	2221	3.16	1987	3.62
4000	1614	4.35	2035	3.53
6000	1510	4.65	2329	3.09
8000	8841	0.79	3762	1.91
10,000	18,076	0.38	4109	1.75
12,000	18,784	0.37	4499	1.60
14,000	19,493	0.36	4789	1.50
16,000	20,005	0.35	5067	1.42
18,000	20,470	0.34	5114	1.41
20,000	20,936	0.33	5539	1.30
24,000	21,737	0.32	6104	1.18

Next, we tested the overall performance of the NEMO model. The speedup was also greatly improved after optimization, especially at large scales (Table 1). With the PCG solver, the speedup relative to 400 processes was 4.65 times when using 6000 processes. However, the speedup decreased to less than 1.0 times when using more than 8000 processes, while in GYRE_CAPCG, the speedup was still greater than 1.0 times at large scales. After comparing the speedup in GYRE_PCG and GYRE_CAPCG (Figure 8), we found that the ratio of GYRE_CAPCG to GYRE_PCG was larger than 1 when using more than 6000 processes and reached 4.61 when using more than 10,000 processes, which demonstrates an efficient improvement.

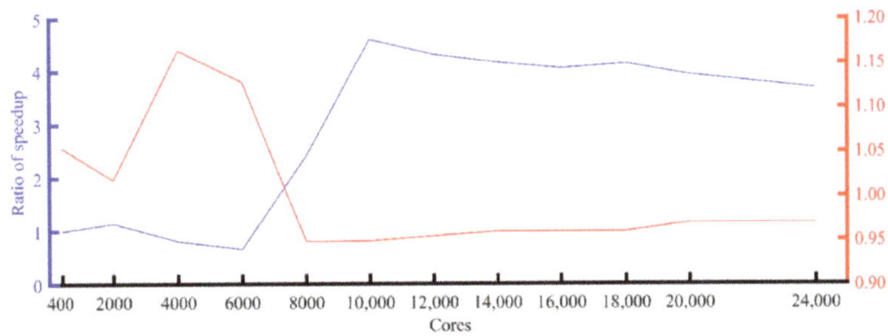

Figure 8. Ratio of the speedup of the NEMO model using the CA-PCG and PCG solvers (blue line) and the ratio of the percentage of the barotropic time of the total execution time using the CA-PCG and PCG solvers (red line).

4. Discussion and Conclusions

As the number of processors used has increased, the barotropic solver in the NEMO model has become the bottleneck for large-scale clusters. The most time-consuming part is solving elliptic equations, for which it is recommended that the PCG method is used in NEMO. In this work, we developed and implemented the CA-PCG method in the model to solve this problem. The results showed that the requirement of inner product operations decreased from each iteration to every eight iterations, which highly reduced the global reduction when using the newly developed CA-PCG solver. Thus, the execution time of the barotropic mode decreased effectively. Additionally, the speedup of NEMO after optimization was also better than that before optimization.

Above all, the optimization method could be useful in optimizing the NEMO model at large scales, as in this study. It should be noted that the main effect of the CA-PCG solver is reducing the cost of the global reduction during simulation. However, the architecture of the cluster, such as the bandwidth and the resolution of the experiment will impact the execution time of global reduction and can further influence the efficiency of the CA-PCG solver. To further evaluate the new barotropic solver's effectiveness, the optimized model should be tested on different clusters, including Heterogeneous System Architecture. Moreover, the accuracy of the results after optimization was only validated using small-scale simulation due to computation and storage limitations. Exploring different grid resolutions, particularly large-scale ones, is critical and should also be validated in the future. Besides, it should also be noted that the distribution of the ensemble z-score biases shows a seasonal variation (Figure 4), which also needs further analysis.

The experience gained from the NEMO model could benefit other OGCMs and even other ocean components in coupled models. In the Coupled Model Intercomparison Project phase 6 (CMIP6), approximately 25 climate models use NEMO as their ocean component (https://github.com/WCRP-CMIP/CMIP6_CVs/blob/master/CMIP6_source_id.json (accessed on 10 April 2021)). Climate models usually require numerous calculations and long integral times and need to be run on high-performance computing resources more than stand-alone ocean models do [20]. Therefore, the NEMO model's high performance will benefit from simple ocean process simulation and climate models that use NEMO as their ocean component.

Author Contributions: Methodology, validation, analysis, and writing–original draft preparation, X.Y.; algorithm coding and description, S.Z. (Shan Zhou); numerical experiments, S.Z. (Shengchang Zhou); conceptualization, supervision, funding acquisition, and writing–review and editing, Z.S.; algorithm and numerical experiments' suggestion, W.L. All authors discussed, read, edited, and approved the article. All authors have read and agreed to the published version of the manuscript.

Funding: This research was funded by the National Natural Science Foundation of China (nos. U1806205, 41906029, and 42022042) and the China–Korea Cooperation Project on Northwestern Pacific Climate Change and its Prediction.

Institutional Review Board Statement: Not applicable.

Informed Consent Statement: Not applicable.

Data Availability Statement: The data used within this study are the combination of the World Ocean Atlas 2009 and the Polar Science Center Hydrographic Climatology version 3 data, which can be obtained from the National Oceanographic Data Center (NODC) and Polar Science Center, respectively.

Acknowledgments: The authors are grateful to Junmin Lin and Zhe Wang for developing the algorithm.

Conflicts of Interest: The authors declare no conflict of interest.

Appendix A

The OGCMs solve the three-dimensional primitive equations, including Navier–Stokes equation, static equilibrium equation, continuity equation, temperature and salinity conservation equation, and state equation. The equations governing the dynamics of coastal circulation contain fast-moving, external gravity waves, and slow-moving, internal gravity waves. It is desirable in terms of computer economy to separate the vertically integrated equations (barotropic mode) from the vertical structure equations (baroclinic mode). This technique, known as mode splitting, permits the calculation of the free surface elevation with little sacrifice in computational time by solving the velocity transport separately from the three-dimensional calculation of the velocity and the thermodynamic properties. The split-explicit free surface formulation used in the NEMO model follows the one proposed by [21]. The general idea is to solve the free-surface equation and the associated barotropic velocity equations with a smaller time step than Δt, which is the time step used in the baroclinic mode for the three-dimensional prognostic variables (Figure A1).

The barotropic mode solves the following depth-integrated equations:

$$\frac{\partial \overline{\boldsymbol{U}}}{\partial t} = -f\mathbf{k} \times \overline{\boldsymbol{U}} - g\nabla\eta - \frac{c}{H+\eta}\overline{\boldsymbol{U}} + \overline{\mathbf{G}} \tag{A1}$$

$$\frac{\partial \eta}{\partial t} = -\nabla \cdot \left[(H+\eta)\overline{\boldsymbol{U}}\right] + P - E \tag{A2}$$

where $\overline{\boldsymbol{U}}$ is the depth-integrated barotropic velocity, η is the sea surface height, f is the Coriolis parameter, and g is the acceleration of gravity. $\overline{\mathbf{G}}$ is a forcing term held constant, containing coupling term between modes, surface atmospheric forcing, and slowly varying barotropic terms not explicitly computed to improve efficiency. The third term on the right-hand side of Equation (A1) represents the bottom stress. $\boldsymbol{H}$ represents the depth of the ocean. $\boldsymbol{P}$ and $\boldsymbol{E}$ represent the precipitation and evaporation, respectively.

Time filtering is eventually applied to barotropic quantities to avoid aliasing of fast barotropic motions into three-dimensional equations. When the filtered sea surface height option is used, the momentum equation's force is solved implicitly. Thus, an elliptic equation, which is solved using the PCG solver in the NEMO model, is formulated.

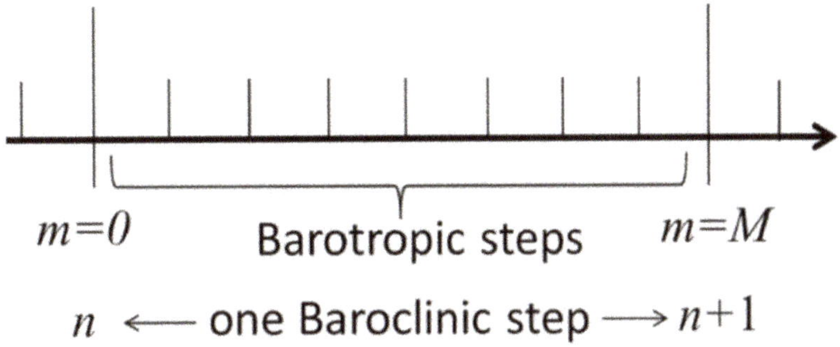

Figure A1. The split time stepping is used in the model. The m represents the time of the barotropic step, and M is the total number of steps among each baroclinic step.

Appendix B

Algorithm A1. Preconditioned conjugate gradient (PCG) [22].

Required: Coefficient matrix A, which is a definite positive symmetric matrix, preconditioner M (diagonal matrix of the diagonal elements of A), initial guess value x_1 and vector b associated with grid block $A_{i,j}$

1. Initialization: Given initial guess x_0,
2. $d_1 = 0, z_1 = 0, \sigma_1 = 0, \gamma_1 = 1, k = 0$ /* initial the value */
3. $r_1 = M^{-1}(b - Ax_1)$ /* preconditioner */
4. For $k = 1 : k_{max}$ until convergence
5. $s_k = Ar_k$ /* matrix-vector multiplication */
6. $\gamma_k = rr_k = r_k^T M r_k$ /* inner product, global reduction */
7. $\delta_k = r_k^T M s_k$ /* inner product, global reduction */
8. $\beta_k = \gamma_k / \gamma_{k-1}$
9. $d_k = r_k + \beta_k d_{k-1}$
10. $z_k = s_k + \beta_k z_{k-1}$
11. $\sigma_k = \delta_k - \beta_k^2 \sigma_{k-1}$
12. $\alpha_k = \gamma_k / \sigma_k$
13. $x_k = x_{k-1} + \alpha_k d_k$ /* update vector */
14. $r_k = r_{k-1} - \alpha_k z_k$ /* update vector */
15. Update_Halo(r_k) /* boundary communication*/
16. End for /* the result is x_k */

NEMO divides the global domain into blocks and distributes them to the processes. Each process computes the only evolution procedures related to the grid points in its block and maintains a halo region to update data with its neighbors [9]. The following assumptions were made:

- $N \times N$ is the global domain size.
- The global domain is divided into m x m blocks with a size of n x n ($n = N/m$).
- The $p = m \times m$ processes occur, and each process corresponds to exactly one grid block. Then, the total time of the barotropic mode is the execution time of the PCG solver on each block.
- K is the iteration number.

Then, for each iteration, t_{comp} is the cost of the computation, t_b is the cost of the boundary updating, and t_g is the cost of the global reduction. Below is a detailed estimation of these components.

(1) Computational cost

For Algorithm A1, the computational cost, t_{comp}, contains three parts:

- Step 5, with a total of $9n$ operations from one matrix-vector multiplication (MVM).

- Steps 6 and 7, with a total of $6n$ operations, from two vector–vector multiplication (VVMs) for inner products, two additional inverse operations of the preconditioner, which has the same cost as the preconditioner (for the diagonal preconditioner, its operation number is n), and two masking operations for land point exclusion.
- Steps 10, 11, 13, and 14, with a total of $4n$ operations, from four vector-scaling operations.

The total cost can be expressed as:

$$t_{comp} = K\left(9n^2 + 6n^2 + 4n^2\right)\sigma = 19\frac{N^2}{p}\sigma K \quad \text{(A3)}$$

where σ is the execution time per floating-point operation. If the domain size N is constant, t_{comp} decreases to a lower bound of zero when the number of processes increases.

(2) Communication cost

There are two kinds of communication in the solver. The first one is from the boundary update. After MVM and nondiagonal preconditioning, a boundary update occurs in the halo regions for each process, which requires one or more boundary layers to be updated. In NEMO, since each process covers one block and one additional halo layer is used, only one boundary update is needed per iteration. The total boundary updating cost depends on the data volume of the halo regions and network latency. When the halo size is 1, the data volume in each boundary is n, and there are four boundaries for each block (east, west, north, and south). Then, the total boundary updating cost for each iteration satisfies:

$$t_b = K(4\theta + 4n\varepsilon) = K\left(4\theta + \left(\frac{4N}{\sqrt{p}}\right)\varepsilon\right) \quad \text{(A4)}$$

where θ is the point-to-point communication latency per message and ε is the transfer time per floating-point datum. The boundary updating time also decreases as the number of processes increases but has a lower limit of 4θ. The second communication cost is due to global reduction and can be estimated with a similar method. There are two global reductions per iteration, each of which results in a reduction in computing time and communication latency. Considering a binomial tree, the communication latency cost should be $\delta \log_2 p$ [10]. Then, the global reduction execution time can be regarded as:

$$t_g = K\left(2\frac{N^2}{p}\sigma + \theta log_2 p\right) \quad \text{(A5)}$$

When the process number p increases, the cost of the computing time should decrease, and the cost of the communication latency should increase monotonically. Combining computational costs together with boundary updating and global reduction costs, the execution time of one step for NEMO PCG can be described as:

$$t_{pcg} = \left(t_{comp} + t_{comm}\right) = \left(t_{comp} + t_b + t_g\right) = K\left[21\frac{N^2}{p}\sigma + \left(\frac{4N}{\sqrt{p}}\right)\varepsilon + (4 + log_2 p)\theta\right] \quad \text{(A6)}$$

where K is constant regardless of how many processes are used [23].

Equation (A6) shows that the more processes used, the less time is required for computation and boundary updating. However, the time required for global reduction increases with the number of processes. Therefore, when the process number exceeds a certain limit, the total time of the NEMO PCG solver will increase monotonically and become dominant.

Appendix C

Algorithm A2. Communication-avoiding Krylov subspace method with a PCG (CA-PCG)

CA-PCG consists of a two-level nested loop (k,j), outer loop indexed by k, and inner loop index $j \in \{1,\ldots,s\}$, given initial PCG vectors p_{sk+1}, z_{sk+1}, and x_{sk+1}:
outer loop: prepare the required info for iterations over p_{sk+j} and z_{sk+j}, x_{sk+j}
inner loop: update the PCG vectors with their coordinates in the Krylov subspace
$\mathcal{K}_k(A,v) = \mathrm{span}\left\{v, Av, A^2v, \ldots, A^{k-1}v\right\}$

1. Initialization: Given initial guess x_0 and set residual $r_1 = b - Ax_1$, outer loop index $k = 1$,
2. $z_1 = p_1 = M^{-1}r_1$ /* preconditioner */
3. For k =1 : k_{max}/s, until convergence done
4. /* construct and compute P_k, Z_k, a polynomial of degree j with $\rho_j(\tau)$ and p_{sk+1}, z_{sk+1} */ /* matrix-vector multiplication */

$P_k = [\rho_0\left(M^{-1}A\right)p_{sk+1}, \ldots, \rho_s\left(M^{-1}A\right)p_{sk+1}]$, sat.span$(P_k) = K_{s+1}\left(M^{-1}A, p_{sk+1}\right)$
$Z_k = [\rho_0\left(M^{-1}A\right)z_{sk+1}, \ldots, \rho_{s-1}\left(M^{-1}A\right)z_{sk+1}]$, sat. span$(Z_k) = K_s\left(M^{-1}A, z_{sk+1}\right)$

5. Update_halo(V_k) /* boundary communication */
6. Compute Gram matrix $G_k = \mathcal{V}_k^T M \mathcal{V}_k$ /* inner product with global reduction */
7. Assemble B_k from $\rho_j(\tau)$, which is a change basis matrix, with all 0 in columns s + 1 and 2s + 1
8. Initial short vector $p'_{k,j} = [1, 0_{2s}]^T$,$z'_{k,j} = [0_{s+1}, 1, 0_{s+1}]^T$, $x'_{k,j} = [0_{2s+1}]^T$
9. For $j = 1, s$ do /* $s = 8$ in this paper */
10. /* inner product, G_k is ready, no global reduction */

$\alpha_{k,j} = \left(z'^T_{k,j} G_k z'_{k,j}\right) / \left(p'^T_{k,j} G_k \mathcal{B}_k p'_{k,j}\right)$

11. $x'_{k,j+1} = x'_{k,j} + \alpha_{k,j} p'_{k,j}$ /* update short vector */
12. $z'_{k,j+1} = z'_{k,j} - \alpha_{k,j} \mathcal{B}_k p'_{k,j}$ /* update short vector */
13. /* inner product, G_k is ready, no global reduction */

$\beta_{k,j} = \left(z'^T_{k,j+1} G_k z'_{k,j+1}\right) / \left(z'^T_{k,j} G_k z'_{k,j}\right)$

14. $p'_{k,j+1} = z'_{k,j+1} + \beta_{k,j} p'_{k,j}$ /* update short vector */
15. End for /* inner loop end */
16. $p_{k,j} = V_k\, p'_{k,j}$, $z_{k,j} = V_k\, z'_{k,j}$, $x_{k,j} = V_k\, x'_{k,j} + x_{k,1}$ /* update long vectors*/
17. Update_halo$\left(x_{k,j}\right)$ /* boundary communication */
18. End for /* outer loop end, the result is x_k */

Regarding B_k, $\rho_j(\tau)$ satisfies the three-term recurrence via
$\rho_0(\tau) = 1$, $\rho_1(\tau) = (\tau - \theta_0)\rho_0(\tau)/\gamma_0$, and $\rho_j(\tau) = (\left(\tau - \theta_{j-1}\right)\rho_{j-1}(\tau) - \sigma_{j-2}\rho_{j-2}(\tau)/\gamma_{j-1}$.
In this paper, the monomial bases were used, so $\gamma_j = 1$ and $\theta_j = \sigma_j = 0$.

In the CA-PCG algorithm, all of the iterations are divided into two layers of loops. In the outer loop, global reductions are still needed to obtain the Gram matrix [13], which can be used in the inner loop. Since we used the same preconditioner as that used for PCG, the convergence rate will not change, and the iteration number K should be the same. If s is the inner loop count, then the outer loop count will be K/s, which means one global reduction is needed after every s iterations. Since the inner loop count increases, the optimization efficiency will be improved, but the computational accuracy will decrease. We found that eight inner loop counts could meet both optimization efficiency and computational accuracy after multiple tests. Then the inner loop count s was set as eight in this study. However, in the PCG algorithm, one global reduction was needed after each iteration step. Thus, the amount of global reduction was reduced to 1/8 of that in the PCG algorithm. Notably, even though the global communication cost was highly reduced, the computational cost was increased compared to the PCG algorithm.

Assuming that the complexity of PCG is N_{comp}, then CA-PCG is $(2s+1)^2 N_{comp}$, which comes from steps 4, 10, 11, 12, 13, 14, and 16 in Algorithm A2, especially step 4. To reduce the computation cost of CA-PCG, we implemented optimizations to reduce the computation complexity to $6s^*N_{comp}$. The hotspots of CA-PCG were the basic operation in step 4 and the Gram matrix operation in step 6, while the costs of the inner iterations were negligible (from steps 10 to 14). The count of MVM in step 4 was $2s - 1$, approximately twice that of the PCG iterations, which could be amortized by data parallelization. The complexity of the inner production in step 6 was $(2s + 1)^2$ with a naive implementation or approximately $(2s + 1)^2/2$ by taking the symmetric property of matrix A and M^{-1}. However, the computational cost was still high and could be reduced further. With the choice of monomial basis and diagonal matrix M^{-1}, the element in the Gram matrix $G_k(i,j)$ was $(p'_{sk+1}$ or $z'_{sk+1})$ by $M(M^{-1}A)^{i-1}(M^{-1}A)^{j-1}$ by $(p_{sk+1}$ or $z_{sk+1})$. After the argument, $G_k(i_1,j_1)$ and $G_k(i_2,j_2)$ were equal, given that they were in the same submatrix (e.g., P'_kMP_k) and $i_1+j_1 = i_2+j_2$. As a result, the counts of the inner products needed for submatrices P'_kMP_k, P'_kMZ_k, and Z'_kMZ_k were $2s + 1$, $2s$, and $2s - 1$, respectively. In total, the computational complexity of the inner products was reduced to 6s at the algorithmic level compared to 2s in the corresponding PCG iterations.

The details about the computational and communication cost estimation after optimization are provided here. The communication cost comes from the boundary updating in steps 5 and 17 and the global reduction cost of step 6. Thus, we can obtain the following:

$$t_{comp} = 19K * 6s\frac{N^2}{p}\sigma \tag{A7}$$

$$t_b = \left(\frac{K}{s}\right)(8\theta + 8n\varepsilon) = \left(\frac{K}{s}\right)\left(8\theta + \left(\frac{8N}{\sqrt{p}}\right)\varepsilon\right) \tag{A8}$$

$$t_g = \left(2s\frac{N^2}{p}\sigma + \theta log_2 p\right)\left(\frac{K}{s}\right) = 2K\frac{N^2}{p}\sigma + \left(\frac{K}{s}\right)\theta log_2 p \tag{A9}$$

$$t_{capcg} = (19K * 6s + 2K)\frac{N^2}{p}\sigma + \left(\frac{K}{s}\right)\left(\frac{8N}{\sqrt{p}}\right)\varepsilon + \left(\frac{K}{s}\right)(8 + log_2 p)\theta \tag{A10}$$

According to the impact of the processor number p, we divided t_{capcg} into two parts, $F^{(1)}_{capcg}$ and $F^{(2)}_{capcg}$:

$$F^{(1)}_{capcg} = (19K * 6s + 2K)\frac{N^2}{p}\sigma + \left(\frac{K}{s}\right)\left(\frac{8N}{\sqrt{p}}\right)\varepsilon \tag{A11}$$

$$F^{(2)}_{capcg} = \left(\frac{K}{s}\right)(8 + log_2 p)\theta \tag{A12}$$

Implementing the same operation on Equation (A6), we can convert Equation (A6) into:

$$t_{pcg} = F^{(1)}_{pcg} + F^{(2)}_{pcg},$$

where:

$$F^{(1)}_{pcg} = 21K\frac{N^2}{p}\sigma + \left(K\frac{4N}{\sqrt{p}}\right)\varepsilon] \tag{A13}$$

$$F^{(2)}_{pcg} = K(4 + log_2 p)\theta] \tag{A14}$$

where s is the inner loop count. As shown in Equation (A12), the cost of global communication will decrease as s increases in CA-PCG. However, s has a limit value, which depends on the basis we choose in the algorithm. Since the s monomial basis was used in this paper, the maximum value of s should be 8, or the solver cannot achieve convergence regardless of the iteration number [24].

From Equations (A11) and (A13), we can determine that the computational cost of the CA-PCG algorithm is greater than that of the PCG algorithm. However, for a certain workload, as p increases, there will be a lower limit of zero for computational cost. From Equations (A12) and (A14), we found that as p increases, the global reduction in CA-PCG reduces to approximately $\frac{1}{s}$ of that of the PCG algorithm. Moreover, we could also predict that there is a certain value p_{inf} for the process number. When $p < p_{\text{inf}}$, the computational cost is dominant, and the performance of the CA-PCG method is even worse than that of the PCG method. However, when $p > p_{\text{inf}}$, the communication bottleneck due to the global reductions becomes considerable, and the CA-PCG method becomes very effective in reducing the communication cost, compared to the PCG method. Thus, the scalability is improved effectively with the CA-PCG solver.

Even though the CA-PCG solver can improve the performance of NEMO, correctness and accuracy should be ensured. After optimization, the solution of the barotropic solver should be the same or very close. In both the PCG and CA-PCG solver, the solution is an approximate value that can satisfy the condition of convergence (Algorithm A1, A2). If the solution of the CA-PCG solver is correct, it should also make the PCG solver achieve convergence. Thus, we used the most direct method to verify the correctness of the result. After applying the solution of the CA-PCG solver as the initial guess value of the PCG solver, we found that only one step iteration was needed to reach convergence. Then, the solution of the new solver CA-PCG can be recognized as the correct solution of this elliptic equation in the barotropic mode of NEMO.

Appendix D

Table A1. Description of the parallel platform.

CPU	2*Intel Xeon Scalable Cascade Lake 6248 (2.5 GHz, 20 Cores)
L3 cache	27.5 MB
Memory	12 × Samsung 16 GB DDR4 ECC REG 2666
Interconnect (Network)	Intel OmniPath 100 Gbps
Storage	Lustre (NVMe)
BIOS settings	4.1.05
OS/kernel	3.10.0-1062.el7.x86_64

References

1. Chassignet, E.P.; Xu, X. Impact of Horizontal Resolution (1/12° to 1/50°) on Gulf Stream Separation, Penetration, and Variability. *J. Phys. Oceanogr.* **2019**, *47*, 1999–2021. [CrossRef]
2. Qiao, F.; Zhao, W.; Yin, X.; Huang, X.; Liu, X.; Shu, Q.; Wang, G.; Song, Z.; Li, X.; Liu, H.; et al. A Highly Effective Global Surface Wave Numerical Simulation with Ultra-High Resolution. In Proceedings of the SC16: International Conference for High Performance Computing, Networking, Storage and Analysis, Salt Lake City, UT, USA, 13–18 November 2016; pp. 46–56.
3. Zhao, W.; Song, Z.Y.; Qiao, F.L.; Yin, X.Q. High efficient parallel numerical surface wave model based on an irregular quasi-rectangular domain decomposition scheme. *Sci. China Earth Sci.* **2014**, *44*, 1049–1050. [CrossRef]
4. Cohn, S.E.; Dee, D.; Marchesin, D.; Isaacson, E.; Zwas, G. A fully implicit scheme for the barotropic primitive equations. *Mon. Weather Rev.* **1985**, *113*, 641–646. [CrossRef]
5. Dukowicz, J.K.; Smith, R.D. Implicit free-surface method for the Bryan-Cox-Semtner ocean model. *J. Geophy. Res. Ocean.* **1994**, *99*, 7991–8014. [CrossRef]
6. Concus, P.; Golub, G.H.; O'Leary, D.P. Numerical solution of nonlinear elliptic partial differential equations by a generalized conjugate gradient method. *Computing* **1978**, *19*, 321–339. [CrossRef]
7. Concus, M. On computingINV block preconditionings for the conjugate gradient method. *BIT* **1986**, *26*, 493–504. [CrossRef]
8. Smith, R.; Jones, P.; Briegleb, B.; Bryan, F.; Danabasoglu, G.; Dennis, J.; Dukowicz, J.; Eden, C.; Fox-Kemper, B.; Gent, P.; et al. *The Parallel Ocean Program (POP) Reference Manual*; Tech. Rep. LAUR-10-01853; Los Alamos National Laboratory: Los Alamos, NM, USA, 2010; p. 141.
9. Madec, G. NEMO Ocean Engine. In *Note du Pole de Modelisation*; Institut Pierre-Simon Laplace: Paris, France, 2008; p. 27.
10. Xu, S.; Huang, X.; Oey, L.Y.; Xu, F.; Fu, H.; Zhang, Y.; Yang, G. POM.gpu-v1.0: A GPU-based Princeton Ocean Model. *Geosci. Model. Dev.* **2015**, *8*, 2815–2827. [CrossRef]
11. Hu, Y.; Huang, X.; Baker, A.; Tseng, Y.; Bryan, F.; Dennis, J.; Yang, G. Improving the scalability of the ocean barotropic solver in the community earth system model. In Proceedings of the International Conference for High Performance Computing, Networking, Storage and Analysis, Austin, TX, USA, 15 November 2015; pp. 1–12.
12. Madec, G.; Delecluse, P.; Imbard, M.; Levy, C. Opa 8 ocean general circulation model reference manual. *Tech. Rep.* **1998**, *11*, 91.
13. Carson, E. Communication-Avoiding Krylov Subspace Methods in Theory and Practice. Ph.D. Thesis, University of California, Berkeley, CA, USA, 2015.
14. Antonov, J.I.; Seidov, D.; Boyer, T.P.; Locarnini, R.A.; Johnson, D.R. *World Ocean. Atlas 2009 Volume 2: Salinity*; Levitus, S., Ed.; NOAA Atlas NESDIS 69; U.S. Government Printing Office: Washington, DC, USA, 2010; p. 184.

15. Locarnini, R.; Mishonov, A.; Antonov, J.; Boyer, T.; Garcia, H.; Baranova, O.; Zweng, M.; Johnson, D. *World Ocean. Atlas 2009, Volume 1: Temperature;* NOAA Atlas NESDIS 61; U.S. Government Printing Office: Washington, DC, USA, 2010; p. 182.
16. Steele, M.; Morley, R.; Ermold, W. PHC: A global ocean hydrography with a high quality Arctic Ocean. *J. Clim.* **2001**, *14*, 2079–2087. [CrossRef]
17. Griffies, S.M.; Biastoch, A.; Böning, C.; Bryan, F.; Jianjun, Y. Coordinated ocean-ice reference experiments (COREs). *Ocean. Modell.* **2009**, *26*, 1–46. [CrossRef]
18. Large, W.G.; Yeager, S.G. The global climatology of an interannually varying air–sea flux data set. *Clim. Dyn.* **2009**, *33*, 341–364. [CrossRef]
19. Baker, A.; Xu, H.; Dennis, J.; Levy, M.; Nychka, D.; Mickelson, S.; Edwards, J.; Vertenstein, M.; Wegener, A. A methodology for evaluating the impact of data compression on climate simulation data. In Proceedings of the 23rd International Symposium on High-Performance Parallel and Distributed Computing, Vancouver, BC, Canada, 23–27 June 2014; pp. 203–214.
20. Wang, Y.; Hao, H.; Zhang, J.; Jiang, J.; He, J.; Ma, Y. Performance optimization and evaluation for parallel processing of big data in earth system models. *Cluster Comput.* **2017**, *22*, 2371–2381. [CrossRef]
21. Shchepetkin, A.F.; McWilliams, J.C. The regional oceanic modeling system (roms)—a split-explicit, free-surface, topography-following-coordinate oceanic model. *Ocean. Model.* **2005**, *9*, 347–404. [CrossRef]
22. Eisenstat, S.C. Efficient Implementation of a Class of Preconditioned Conjugate Gradient Methods. *Siam J. Sci. Comput.* **1981**, *2*, 1–4. [CrossRef]
23. Hu, Y.; Huang, X.; Wang, X.; Fu, H.; Xu, S.; Ruan, H.; Xue, W.; Yang, G. A scalable barotropic mode solver for the parallel ocean program. In Proceedings of the Euro-Par 2013 Parallel Processing, Aachen, Germany, 26–30 August 2013; Springer: Berlin/Heidelberg, Germany, 2013; pp. 739–750.
24. Hoemmen, M. Communication-Avoiding Krylov Subspace Methods. Ph.D. Thesis, University of California, Berkeley, CA, USA, 2010.

Journal of *Marine Science and Engineering*

MDPI

Article

A Modelling Approach for the Assessment of Wave-Currents Interaction in the Black Sea

Salvatore Causio [1,*], Stefania A. Ciliberti [1], Emanuela Clementi [2], Giovanni Coppini [1] and Piero Lionello [3]

1 Fondazione Centro Euro-Mediterraneo sui Cambiamenti Climatici, Ocean Predictions and Applications Division, 73100 Lecce, Italy; stefania.ciliberti@cmcc.it (S.A.C.); giovanni.coppini@cmcc.it (G.C.)

2 Fondazione Centro Euro-Mediterraneo sui Cambiamenti Climatici, Ocean Modelling and Data Assimilation Division, 40127 Bologna, Italy; Emanuela.clementi@cmcc.it

3 Department of Biological and Environmental Sciences and Technologies, University of Salento—DiSTeBA, 73100 Lecce, Italy; piero.lionello@unisalento.it

* Correspondence: salvatore.causio@cmcc.it

Abstract: In this study, we investigate wave-currents interaction for the first time in the Black Sea, implementing a coupled numerical system based on the ocean circulation model NEMO v4.0 and the third-generation wave model WaveWatchIII v5.16. The scope is to evaluate how the waves impact the surface ocean dynamics, through assessment of temperature, salinity and surface currents. We provide also some evidence on the way currents may impact on sea-state. The physical processes considered here are Stokes–Coriolis force, sea-state dependent momentum flux, wave-induced vertical mixing, Doppler shift effect, and stability parameter for computation of effective wind speed. The numerical system is implemented for the Black Sea basin (the Azov Sea is not included) at a horizontal resolution of about 3 km and at 31 vertical levels for the hydrodynamics. Wave spectrum has been discretised into 30 frequencies and 24 directional bins. Extensive validation was conducted using in-situ and satellite observations over a five-year period (2015–2019). The largest positive impact of wave-currents interaction is found during Winter while the smallest is in Summer. In the uppermost 200 m of the Black Sea, the average reductions of temperature and salinity error are about −3% and −6%, respectively. Regarding waves, the coupling enhanced the model skill, reducing the simulation error, about −2%.

Keywords: Black Sea; wave-current interaction; NEMOv4; WaveWatchIII

Citation: Causio, S.; Ciliberti, S.A.; Clementi, E.; Coppini, G.; Lionello, P. A Modelling Approach for the Assessment of Wave-Currents Interaction in the Black Sea. *J. Mar. Sci. Eng.* **2021**, *9*, 893. https://doi.org/10.3390/jmse9080893

Academic Editor: Eugen Rusu

Received: 5 July 2021
Accepted: 17 August 2021
Published: 19 August 2021

1. Introduction

Wave–currents interaction has recently gained interest in the field of coastal and ocean forecasting [1,2]. Wind-wave-current processes control the momentum and energy exchange between the atmosphere and the ocean and must be better understood and resolved. The physics of wave-currents interaction depend on the kinematics and dynamics of the wave field. These include processes such as wind-wave growth, nonlinear wave-wave interactions, wave-currents interaction and wave dissipation, all of which can only be accurately represented using wave models. Reducing the uncertainties in the nowadays forecasting models that result from the non-linear feedback between the currents [3], water level variations [4,5] and wind waves [6] is therefore essential. Over the past 40 years, a considerable amount of research has been devoted to investigating the role of ocean waves in the air-sea interaction. An accurate representation of ocean surface waves has been recognised as essential in various applications, ranging from marine meteorology to ocean and coastal engineering to operational forecasting [7,8].

Important contributions to the field of wave-currents interaction include the pioneering work of [9] on what is now referred to as the Stokes–Coriolis force, the work of [10] on wave-driven ocean circulation and of [11] on upper ocean mixing by wave breaking. The impact of the oceanic wave field on upper-ocean mixing and mean properties has been examined using various model experiments [11–17]. Most find that waves appear to have a profound impact on the upper part of the ocean, but there is still considerable disagreement about which processes are more important. Waves define the mixing in the oceanic surface boundary layer (OSBL) via breaking and Langmuir turbulence. For example, ref. [18] identified Langmuir turbulence over wide areas of the global ocean and particularly in the Southern Ocean. In this region, they showed that the inclusion of the effect of surface waves on upper-ocean mixing during summertime led to a reduction in systematic bias in the OSBL depth. Jansen et al. 2013 [17] showed the positive impact of wave breaking on the daily cycle of the sea surface temperature. Babanin et al. 2012 [19] investigated the mixing process related to the orbital motion of non-breaking waves.

All of these wave-driven processes influence the vertical structure of the temperature and current fields in the mixed layer in general, and in the upper few meters in particular. This has implications for coupled models as these processes will affect the feedback between the ocean and the atmosphere [17]. However, on shorter time scales and at higher spatial resolution these processes will influence the drift of objects and pollutants on the sea surface or that are partially or wholly submerged. This has practical implications for oil spill modelling [20] and search and rescue [21–23]. Studies focusing on the Baltic Sea [24] and the North Sea [25] apply the coupling approach previously implemented by [26,27], which consists of coupling the wave model with the circulation model. Staneva et al. 2021 [28] in particular demonstrated how the Eulerian currents provided by a coupled NEMO-WAM model in the North Sea can improve the representation of particle transports, dealing with a number of possible impacts on downstream services for marine litter, oil spill and environmental water properties. This was then adapted to include the Stokes–Coriolis effect, the sea-state-dependent momentum and energy fluxes. These studies show that coupling the models has a pronounced effect on vertical temperature distribution and mesoscale events, and improves predictions of sea level and currents during storm events. In the Mediterranean Sea, Ref. [29] presents an approach based on the assumption that the currents are driven by surface wind stresses that in turn are a function of the sea state, while the sea state depends on the wind speed and the currents. These complex feedback mechanisms can be modelled by coupling hydrodynamic and wave models, which to date have been developed separately. Examples are provided by [8,30] for the North-West European Shelf system. In particular by understanding the role of ocean-wave coupling in an eddy-resolving operational configuration to improve the predictability of sea surface temperature, surface and sea bed salinity, sea surface height and currents.

The active literature production on this topic and the interest in regional modelling in particular for marginal seas induced our attention in understanding the complex mechanisms of wave-currents interaction in the Black Sea. This basin acts as a small-scale laboratory for investigating processes that are common to numerous areas of the world's oceans. Considering that the water and salt balances can be controlled, and the scales are smaller than in the global ocean, the basin is a useful test region for developing models, which can then be applied at larger scales [31].

From the physical point of view, the Black Sea is an inland sea, one of the largest and deepest basins in the world and the farthest to the east among the seas of the Atlantic Ocean. It works with a complex straits system—the Kerch Strait connects the Black Sea with the Azov Sea; the Bosphorus Strait makes the connection with the Marmara Sea which is connected in turn to the Mediterranean Sea through the Dardanelles Strait [32]. Straits system and limited water exchange with the open ocean play an important role in the water balance, influencing the ventilation, stratification and water mass formation [31,33]. Rivers are one of the strongest contributors of freshwater with impact to the general circulation in the Black Sea together with wind stress and bathymetric peculiarities: in particular, the North-West shelf region is dominated by the Danube River floodplain and outflow. The Danube, together with the Dniepr and the Dniester and the dense distribution of rivers along the overall coastline contribute to a positive water budget (~5.4 × 105 km^3), making the Black Sea a typical estuarine basin [34–36]. The Black Sea circulation is structured usually in two connected gyre systems encompassing the basin (the Rim Current) and quasi-stationary anticyclonic eddies in the coastal zone, such as Batumi, Sevastopol, Caucasian, Sakarya, Sinop ones [30,37,38]. Some scientific works focused attention on reconstructing the past ocean state in the Black Sea [39] with reference to mesoscale circulation and water mass properties [40–42]. The wind is the main driving force in creating a cyclonic general circulation. Simulation [43] and observations [44] showed that there are three major regions with different regimes of currents: (a) coastal zone of very variable flow, with currents speeds of up to 20–30 cm/s; (b) the main Black Sea currents zone, which has jet-stream character, with a width of 40–80 km and speed of 40–50 cm/s, reaching values of 1–1.5 m/s; (c) the open sea area, where the velocity of the current decreases gradually from the periphery to the centre, not exceeding 5–15 cm/s.

Another important feature of the Black Sea, strictly related to convective processes due to the thermohaline inertia in the upper layer and impact of air-sea fluxes, is the Cold Intermediate Layer (CIL) [45]. CIL is a sub-surface water mass that results from the Winter convective mixing in the centres of cyclonic gyres and shelf areas [32]. CIL formation in the Black Sea depends on the fact that Winter convection is limited by the shallow depth of halocline (such as in the Baltic Sea, the Sea of Okhotsk, the Gulf of St. Lawrence). A consequence of the vertical stratification is that the surface layer (about 0 to 50 m) is well oxygenated while the deep layer (100 m to 2000 m) is anoxic and contains high sulphide concentrations [46].

The wind regime over the Black Sea results from the cyclonic and anticyclonic activity over Europe. The Island cyclone and the Azores anticyclone acts throughout the year, while the Mediterranean cyclones and Siberian anticyclone are more effective in Winter [47]. The Black Sea is dominated by Northerly winds in the West and the North of the sea; Eastern and South-Easterly winds are typical for the East and South-East of the sea. In Spring and Summer, in the Western part of the sea, under the influence of the Azores High, the frequency of Western, South-Western and Southern winds increases [48–51]. South-eastern and Southern coasts of the sea are characterised by weak winds (average annual wind speed < 3 m/s); in the Western and Northwestern parts of the sea, as well as in the Kerch Strait area, stronger winds are observed (average annual wind speed > 4 m/s, at some stations > 5 m/s) [32].

Wind waves in the Black Sea have been widely studied by many researchers from the surrounding regions [52–56]. Brief information about Black Sea wave average climate based on long simulation (1949–2010) is here provided from one of the more recent works [57]. They showed that the average wave parameters are at a maximum in Winter, with maximal values of average Hs exceeding 0.95 m. Areas of most expressed storminess in terms of all average values correspond to the central part of the basin. The same spatial pattern is also observed in the distribution of Springtime averaged wave parameters, but their values are significantly lower (maximal average Hs 0.65 m). The calmest wave field corresponds to Summertime, where Hs > 0.5 m occurred only in small areas of the basin.

From a technical point of view, there are two ways to investigate the impact of ocean-wave interactions: offline or online. Offline coupling e.g., [27,58] implies that the model which provides the fields (sender) runs independently and before the receiving model (receiver). The output from the sender is subsequently used, as the other forcing, by the receiver. This methodology has computational and timing benefits but does not consider the mutual interaction between the two components of the forced system. In the online coupling e.g., [7,28] both the cores are integrated simultaneously, allowing feedback at a predefined exchange time. This methodology more intimately reflects the wave-currents interaction, but it requires an external coupler to drive the models' communication, and the overall running time is almost the sum of both the models running times.

The aim of this work is the assessment, for the first time, of the effects due to the reciprocal interaction between hydrodynamics and waves in the Black Sea through a forced system based on NEMO v4.0 for the hydrodynamics and Wave Watch III for the waves. The main scope is then to improve the representation of the physical processes and dynamics in the Black Sea region by accounting for the role of waves. To complete the overall investigation, this work presents also some numerical results on the impact of improved hydrodynamics on the sea-state.

The coupling strategy here proposed is based on the exchange of surface Stokes drift, wave energy to the ocean, sea-state dependent momentum flux, currents and the air-sea temperature difference, with the main purpose to improve firstly the mixing process in the hydrodynamical core model. The effects have been evaluated via comparison with a standalone run (without forcing) for both waves and hydrodynamics.

This paper is organised as follows. In Section 2 we present the model system and the coupling strategy. Section 3 illustrates the numerical experiment and validation strategy. Section 4 gives the numerical model results and a comparison with observations, and Section 5 provides conclusions and future works.

2. The Modelling System

The numerical system presented in this work is a reciprocal forced hydrodynamic circulation model with a third-generation spectral wind-wave model, implemented in the Black Sea basin. This section presents information about the horizontal numerical grid, which is common to the implementation of both wave and hydrodynamical models. The specific features of this grid, such as spectral discretization for waves and vertical discretization for hydrodynamics, are detailed in the related subsections.

2.1. Atmospheric Forcing, Numerical Grid and Bathymetry

Both hydrodynamics and waves are forced by the high-resolution atmospheric model at a horizontal resolution of 0.125° × 0.125° and 6-h frequency, as produced by the European Centre for Medium-range Weather Forecasts (ECMWF, [59]). The system uses Zonal and Meridional Components of the 10 m wind (m/s), Total Cloud Cover (%), 2 m Air Temperature (K), 2 m Dew Point Temperature (K) and mean sea-level pressure (Pa). Precipitation (Kg/m^2/s) data are provided as monthly climatologies from GPCP [60,61].

The Black Sea region has been discretised in horizontal space using a regular mesh of 395 × 215 grid points, at 0.037° (longitude) × 0.0278° (latitude) spatial resolution in spherical coordinates (Figure 1).

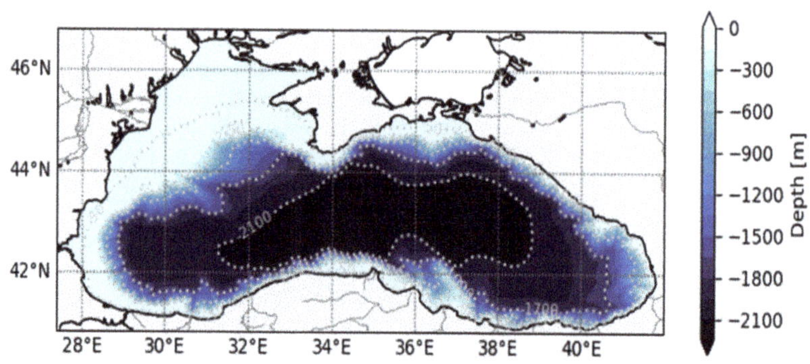

Figure 1. The Black Sea domain and bathymetry.

The bathymetric dataset used in this work is the GEBCO_14 [62] at 30″ grid resolution. The GEBCO_14 in the Black Sea basin was merged with a high-resolution dataset for the Bosphorus Strait, as described in [63].

2.2. Hydrodynamical Model

The hydrodynamical core of the forced system is based on the Nucleus for European Modelling of the Ocean (NEMO, version v4.0 [64]). The NEMO code solves primitive equations (derived through assuming hydrostatic and incompressible approximations), and we used the split-explicit free surface formulation. The vertical discretization consists of 31 unevenly spaced z-levels with partial step, with a stretching factor equal to 5, the maximum stretching at the 25th level and a minimum thickness equal to 5 m. A time step of 150 s is used, together with a barotropic time step equal to 1.5 s. The model was initialised in January 2014 using 3D temperature and salinity climatological fields, as provided in the framework of SeaDataCloud v1 for the Black Sea basin [65]. The model computes air-sea fluxes using bulk formulation as implemented for the Mediterranean Sea [66], designed to handle the ECMWF atmospheric forcing data for computing momentum, heat and water fluxes [67,68].

The model considers 72 rivers, whose discharge is estimated from monthly climatological datasets developed in the framework of the SESAME project [69]. The major rivers (e.g., the Danube, Dniepr, Dniester) have been represented as multiple distributed sources and the remainder as source points. The Danube river is represented accounting for the three main branches—the Chilia, the Sulina and the St. George (Figure 2a). Its discharge distribution is set according to [70], with the Chilia accounting for 52%, Sulina for 20% and the St. George 28%. The impact of the Bosphorus Strait on the Black Sea dynamics is accounted for in terms of a surface boundary condition, considering the barotropic transport, which has been computed to balance the freshwater fluxes on monthly basis [71,72].

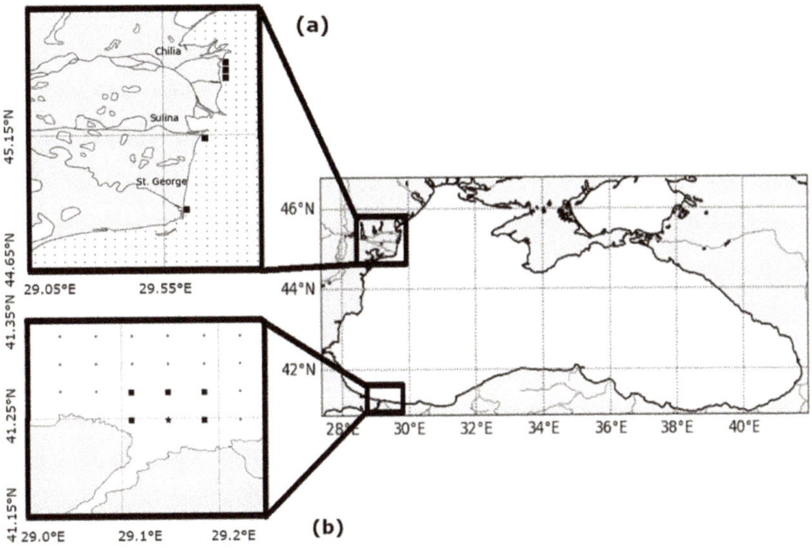

Figure 2. (**a**) The Danube Delta: the Chilia, the Sulina and St. George arms are labelled. Small grey circles show the active grid points of the numerical grid in proximity to the Danube delta. Black squares illustrate the Danube arms mouth representation in the hydrodynamical model. (**b**) The Bosphorus Box is used in the hydrodynamical model. Small grey circles show the active grid points of the numerical grid in the Bosphorus area. Black squares illustrate the points in which the temperature and salinity solution were relaxed to [73] profiles. Black star represents the location in which the T and S were relaxed as in the squares, but in addition, the surface boundary condition of "inverse river" is applied.

The representation of saltier and warmer Mediterranean waters that enter the Black Sea is performed by damping the model solution in the area of influence of the Bosphorus exit (Figure 2b) to monthly 3D temperature and salinity climatologies, as computed from [73], at hourly frequencies. The NEMO configuration uses the Turbulent Kinetic Energy vertical mixing scheme for computing the vertical diffusivity and viscosity. Both horizontal eddy viscosity and diffusivity are constant over the domain and set equal to 1.2×10^{-5} m^2/s

and 1.2×10^{-6} m^2/s, respectively. The NEMO configuration with main parameters can be found in Zenodo [74].

2.3. Wave Model

The wave model used in this work is the third-generation spectral WaveWatchIII [75] version 5.16, hereafter denoted as WW3. The model solves the wave action density balance equation for wavenumber-direction spectra $N(k, \theta, x, t)$, where, θ, is the wave direction, k is the wavenumber, $x = (x, y)$ is the position vector, and t is time. The source terms considered in this implementation concern deep water processes. The wind input source term S_{in} represents the momentum and energy transfer from air to ocean waves, the wave dissipation due to white-capping S_{ds} and the nonlinear transfer by resonant four-wave interactions S_{nl}:

$$S = S_{in} + S_{ds} + S_{nl} \tag{1}$$

WW3 has been implemented following WAM Cycle4 model physics [76]. The propagation scheme used is a third-order scheme (Ultimate Quickest) with the "Garden Sprinkler Effect" alleviation method of spatial averaging. Wind input and dissipation are based on [77], in which the wind input parameterization is adapted from Janssen's quasi-linear theory of wind-wave generation [78,79], following adjustments of [80,81]. Nonlinear wave-wave interactions have been modelled using the Discrete Interaction Approximation (DIA) [82,83].

WW3 is a spectral wave model, and therefore requires a discretization of the wave spectra in bins of frequency and direction. The spectral direction over the full cycle is selected and is divided into 24 sectors of 15° width. A first (lowest) frequency of 0.05 Hz was selected, with a frequency increment factor of 1.1. and 30 frequencies in total. The WW3 configuration parameters are provided via Zenodo [74].

2.4. Coupling Strategy

In the last few years, several works [7,8,25,27,29,84] deeply described the major wave-currents interaction processes. In this work we consider the wave-currents coupling processes implemented in a recent NEMO release (v4.0) to evaluate the impact of those processes on the Black Sea thermodynamic properties for the first time, with a primary focus on the SST, subsurface temperature and tracer advection due to vertical mixing. To a lesser extent, we gained information on the effect on water velocity, even if the exiguous number of available observations limited this analysis.

The reciprocally forced system between the hydrodynamic model (HM) NEMO and the wind-wave model (WM) WW3 is described in this section. WM and HM models share the same horizontal grid and bathymetry and are forced through a reciprocal hourly field exchange. Eight fields are exchanged between the hydrodynamic and wave models. HM sends to WM the surface currents and the Sea Surface Temperature (SST), which is used to compute the air-sea temperature difference. On the other side, WM provides to HM the following fields: significant wave height (hereafter Hs), mean wave period (Tm), wave peak frequency (Fp), Stokes drift at the surface (SD), wave energy (Φoc) and sea-state dependent momentum flux (τoc). An illustration of the coupling mechanism is given in Figure 3.

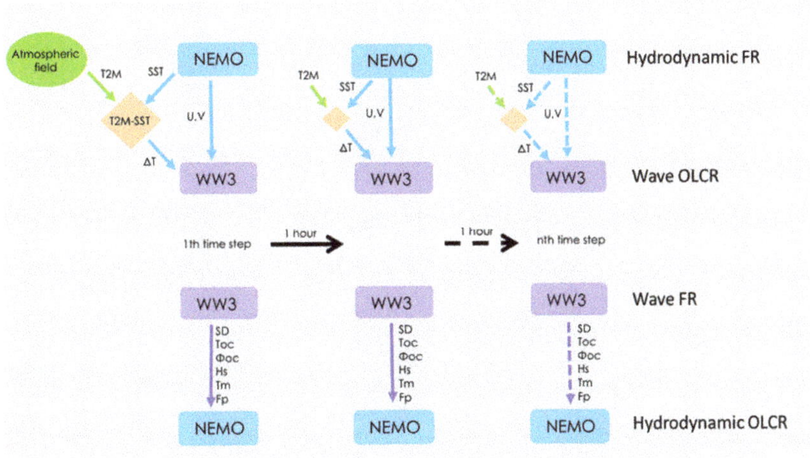

Figure 3. Sketch of the forcing mechanism used in this work. Free-runs are marked with FR, forced experiments are marked with OLCR. The fields provide occurs at a one-hour frequency. In the Wave OLCR numerical experiment, NEMO sends current fields (u, v) to WW3 and SST to a system (the light-orange coloured square in the figure), which computes air-sea temperature differences (ΔT) using 2 m temperature from the atmospheric forcing and sends ΔT to WW3. In the hydrodynamic OLCR, WW3 FR sends the Stokes Drift (SD), sea-state dependent momentum flux (τoc), wave energy to ocean energy flux (Φoc), significant wave height (Hs), mean wave period (Tm) and wave frequency peak (Fp) to NEMO.

Five physical processes related to wave-currents interaction are considered: three reflect the impact of waves on hydrodynamics (Figure 4) and two consider the effect of hydrodynamics on the wavefield:

- Sea-state dependent momentum flux;
- Stokes–Coriolis force, which requires a 3D-Stokes drift profile;
- Wave induced turbulence;
- Doppler effect and refraction due to currents;
- Effects of air stability on the growth rate of waves.

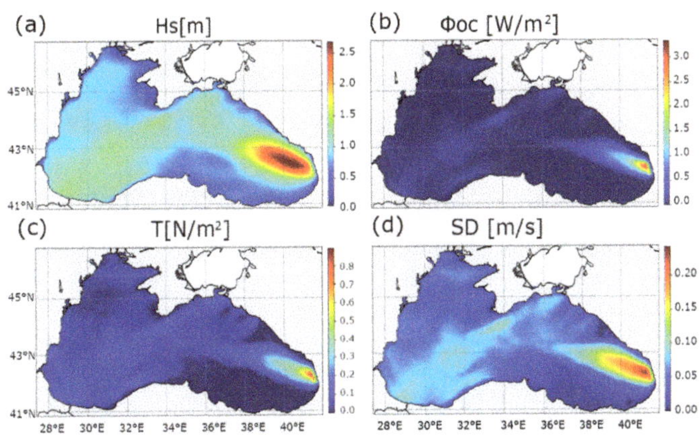

Figure 4. Example of wave fields snapshot used to force the hydrodynamic model at a given time step. (**a**) significant wave height, (**b**) wave energy to ocean turbulence, (**c**) sea-state dependent momentum flux and (**d**) Stokes Drift.

2.4.1. Sea-State Dependent Momentum Flux

In numerical ocean modelling, the momentum fluxes from the atmosphere to the ocean are traditionally calculated from the wind speed provided by an atmospheric model, using a drag coefficient relating the 10-m winds to the surface stress. With this formulation, the surface fluxes are dependent on the local wind speed only [85]. This formulation does not consider that surface stress is dependent on the sea state, that is, how the wave energy is distributed over the frequency range. Another problem related to this approach is that the net momentum flux is not necessarily conserved between the atmosphere and the ocean. For this reason, several authors continue to investigate the impact of the sea-state on wind stress for more than a half-century [25–27,77,80,86–89].

The air-sea momentum flux, or the total wind stress (τ_{tot}), is the sum of the momentum flux into both surface waves and subsurface currents. A significant proportion of the momentum lost from the atmosphere is taken up by the wind waves as increased wave momentum. The appropriate stress (τ_{oc}) (Equation (2)) to use in an Eulerian ocean model is thus the proportion of the momentum flux that is not taken up by the waves (τ_{in}) plus the momentum lost from the wind-generated waves through dissipation (τ_{dis}) [90].

$$\tau_{oc} = \tau_{tot} - \tau_w \quad (2)$$

$$\tau_w = (\tau_{in} + \tau_{dis}) \quad (3)$$

In a wave model, an important source input is the calculation of the momentum flux from the atmosphere to the ocean, which can be expressed as an integral of the wave variance spectrum multiplied by the wave growth rate (momentum-uptake rate).

$$\tau_w = \rho_w g \int_0^{2\pi} \int_0^{\omega} \frac{k}{\omega} (S_{in} + S_{dis} + S_{nl}) d\omega \, d\Theta \quad (4)$$

where ρ_w is the water density, k is the wavenumber vector, ω is the wave absolute frequency (in radians), S_{in}, S_{dis}, S_{nl} are respectively the input, dissipation and non-linear wave model source-term and Θ is the wave direction.

As in previous work e.g., [24,25,27,57,84] Equation (2), the atmospheric wind stress was corrected as follows

$$\tau_{oc} = \tau_{tot}\left(\frac{\tau_w}{\tau_{tot}}\right) - \tau_w \quad (5)$$

2.4.2. Stokes–Coriolis Force and 3D-Stokes Drift Profile

In [9], it is demonstrated that the Stokes drift yields a force on the mean currents. In a rotating ocean, the along-wave crest velocity component of the wave motion is correlated with the vertical component, thus inducing wave stress on the mean Eulerian currents. This stress is proportional to $f \cdot v_s$ where f is the Coriolis force and v_s is the Stokes drift. Following the notation in [91], this term is known as Coriolis–Stokes forcing (Equation (6)).

$$\frac{Du}{Dt} = -\frac{1}{\rho}\Delta p + (u + v_s) \times f\,\hat{Z} + \frac{1}{\rho}\frac{\partial \tau}{\partial Z} \quad (6)$$

where $\hat{Z}$ is the upward vector and τ is the stress.

Computing the Stokes drift profile is expensive as it involves evaluating an integral with the two-dimensional (2D) wave spectrum at every desired vertical level. It is also often impractical or impossible as the full 2D wave spectrum may not be available. Thus, the full Stokes drift velocity profile is commonly replaced by a monochromatic profile, matched to the transport and the surface Stokes velocity e.g., [86,92–96]. This is problematic, as it is clear that the shear under a broad spectrum is much stronger than that of a monochromatic wave of intermediate wavenumber, due to the presence of short waves, whose associated Stokes drift quickly vanishes with depth. The deep Stokes drift profile will also be stronger than that of a monochromatic wave, as the low-wavenumber components penetrate much deeper.

Staneva et al. 2017 [25] developed an alternative approximate Stokes drift profile, which has a lower mean-square error deviation than the monochromatic profile for all tested spectra. It has a stronger shear in the upper part and does not tend to zero as rapidly as the monochromatic profile in the deeper part. It mimics the effect of a broader spectrum where the low-wavenumber components penetrate deeper than the mean wavenumber component, while the shorter waves (higher wavenumbers) only affect the upper part of the water column. In this study, the computation of 3D Stokes drift profiles is conducted using the approach in [26], which is included in NEMO v4.0. The Stokes Drift contribution to the water velocity components has been integrated into tracers and momentum equations as in [7,97].

2.4.3. Wave Induced Turbulence

Following [11], the effects of breaking waves on upper ocean mixing are explicitly considered by the introduction of the energy flux from waves into the ocean water column. Thus, the transport of turbulent kinetic energy (TKE) must also be introduced. In [11], it is assumed that there is a direct conversion of mechanical energy to turbulent energy at the surface and therefore the turbulent energy flux is assumed to be given by the energy flux from waves to the ocean water column Φoc (Equation (7)), which follows from the dissipation term in the energy balance equation.

$$\Phi_{\text{oc}} = -\rho_w g \int_0^{2\pi} \int_0^{\infty} S_{ds}\, d\Theta d\omega \tag{7}$$

meaning that the injection of TKE at the surface is given by the dissipation of the wavefield via the sink term in the wave model energy balance equation (usually dominated by wave breaking) and converted into an ocean turbulence source term. In the absence of specific information on the sea state, the energy flux is parametrised in NEMO by considering a typical old wind-sea value. NEMO v4.0 does not include the wave-dependent TKE surface boundary condition, and thus we applied the code modification according to [9].

2.4.4. Doppler Effect and Refraction Due to Currents

When surface currents interact with waves, phenomena such as the Doppler shift or refraction occur. The way the Doppler shift modifies the surface waves depends on the current speed relative to the wave propagation speed; thus, slow propagating waves are mainly affected by currents. The effects of currents and waves can merge constructively, creating single exceptionally large waves (rogue waves), or if waves are in opposition to strong currents, they become shorter and steeper (potentially hazardous for navigation) and they can also break. The propagation velocity in the various phase spaces of waves interacting with currents (Equations (8)–(10)) can be written as follows.

$$\dot{x} = c_g + U_c \tag{8}$$

$$\dot{k} = -\left(\frac{\partial}{\partial d}\sigma\right)\left(\frac{\partial}{\partial s}\sigma\right) - K\cdot\frac{\partial}{\partial s}U_c \tag{9}$$

$$\dot{\Theta} = -\frac{1}{k}\left[\left(\frac{\partial}{\partial d}\sigma\right)\left(\frac{\partial}{\partial m}\sigma\right) - K\cdot\frac{\partial}{\partial m}U_c\right] \tag{10}$$

where c_g is the wave propagation velocity vector, U_c is the velocity of the current, d is the water depth, s and m are the directions along and perpendicular, respectively, to the wave direction.

2.4.5. Effects of Air Stability on the Growth Rate of Waves

The difference between the sea surface temperature (SST) and the air temperature affects the stability of the lower atmosphere and thus the wind velocity structure. Tolman, 2002 [98] formulated a stability correction by replacing the wind speed with an effective wind speed so that the wave growth reproduces [99] stable and unstable wave growth curves.

The air-sea temperature difference is used to evaluate a stability parameter, ST (Equation (11)), which is written as follows:

$$ST = \frac{hg}{Uh^2}\frac{T_a - T_s}{T_0} \tag{11}$$

where Uh is the wind speed at h height, T_a is air temperature at h height, T_s is the surface temperature and T_0 is the reference temperature. ST is used to compute effective wind speed, U_e:

$$U_e = U_{10}\left(\frac{c_0}{1 \pm c_1 tanh[c_2(ST - ST_0)]}\right)^{\frac{1}{2}} \tag{12}$$

where U_{10} is the wind speed at 10 m, values c_0, c_1, c_2 and ST_0 are set to 1.4, 0.1, 150 and −0.01, respectively.

3. Numerical Experiments Design and Validation Strategy

Nine numerical experiments were conducted from January 2014 to December 2019, according to the numerical setups explained in the previous section. Five experiments concern hydrodynamics (labelled H) and 4 examine waves (labelled W). All of the numerical experiments are listed in Table 1.

Table 1. List of numerical experiments carried out from 2014 to 2019. H refers to hydrodynamics, W refers to waves.

ID	Experiment	Description
H0	NEMO standalone	NEMO free-run. The hydrodynamic model is a standalone model
H1	NEMO forced via SD	NEMO single field-forced experiment. It uses Stokes Drift at the surface from the WM. Stokes–Coriolis Force (SCF) based on the 3-D reconstruction of the Stokes velocity profile has been computed by the HM
H2	NEMO forced via τoc	NEMO single field-forced experiment. It uses sea-state dependent momentum flux (τoc) from WM
H3	NEMO forced via Φoc	NEMO single field-forced experiment. It uses wave-induced vertical mixing (Φoc) from WM
H4	NEMO forced via SD + τoc + Φoc	NEMO fully-forced experiment. It uses SCF, τoc and Φoc
W0	WW3 standalone	WW3 free-run. The wave model is a standalone model
W1	WW3 forced via u, v	WW3 single field-forced experiment. It uses *currents* from the HM
W2	WW3 forced via ΔT	WW3 single field-forced experiment. It uses ΔT from the HM and atmospheric forcings
W3	WW3 forced via u, v + ΔT	WW3 fully-forced experiment. It uses *currents* and ΔT

Validation Strategy and Observational Data

In this work, the model evaluation was based on GODAE/Oceanpredict [100]. Standard statistics such as BIAS, Root-Mean Squared Error (RMSE), Scatter Index (SI), Pearson's correlation index (ϱ) and Standard Deviation (SDev) are used to evaluate the performance of the ocean models by comparing the numerical results against observations (in-situ and/or satellite observed data):

$$\text{BIAS} = \frac{1}{n}\sum_{i=1}^{n}(m_i - o_i) \tag{13}$$

$$\mathrm{RMSE} = \sqrt{\frac{1}{n}\sum_{i=1}^{n}(m_i - o_i)^2} \tag{14}$$

$$\mathrm{SI} = \frac{\sqrt{\frac{1}{n-1}\sum_{i=1}^{n}(m_i - o_i - BIAS)^2}}{\overline{o}} \tag{15}$$

$$\rho = \frac{\frac{1}{n-1}\sum_{i=1}^{n}(o_i - \overline{o})(m_i - \overline{m})}{\sqrt{\frac{1}{n-1}\sum_{i=1}^{n}(o_i - \overline{o})^2}\sqrt{\frac{1}{n-1}\sum_{i=1}^{n}(m_i - \overline{m})^2}} \tag{16}$$

$$\mathrm{SDev} = \sqrt{\frac{1}{n}\sum_{i=1}^{n}(x_i - \overline{x})^2} \tag{17}$$

where o and m stand for observed and modelled data, respectively. x applies for both m or o; the overbar over a variable denotes the temporal averaged value derived from the sample of length n.

Three sources of data, organised by platform, were used to evaluate the accuracy of the model results and described in the following.

- Satellite: including Hs and Sea Surface Temperature (SST). Jason-2 (J2) along-track and quality-controlled altimetric measurements of Hs at 1 Hz sampling frequency (represented in Figure 5a), from the "Archiving, Validation and Interpretation of Satellite Oceanographic data" (AVISO+) have been used for the wave validation.
- SST data are provided by the CMEMS SST Thematic Assembly Center [101]. Night-time L3 satellite data from different space missions are filtered according to quality check, bias-corrected, merged and provided at 1/16° of horizontal resolution. Dataset also provides an error estimate from the optimal interpolation. The operational maintenance of SST data is guaranteed by Consiglio Nazionale delle Ricerche—Istituto di Scienze Marine (CNR ISMAR, Venice, Italy).
- Argo: quality-controlled temperature and salinity in situ vertical profiles used in this work are provided by the CMEMS In Situ TAC [102]. The spatial distribution of almost 1400 Argo floats in the Black Sea basin over the period 2015–2019 is shown in Figure 5b. The operational maintenance of such data is coordinated by the Institute of Oceanology—Bulgarian Academy of Science (IO-BAS, Varna, Bulgaria).
- Moorings: water currents speed and direction data measured from currentometers, shown in Figure 5a, are used to evaluate surface currents from the numerical runs. Data are provided by the CMEMS In Situ TAC [102]. Operational maintenance is guaranteed by the GeoEcoMar Institute (Bucarest, Romania).

Table 2 summarises data type, producers, variables and reference links of the observations used for validating our results.

Table 2. List of observation datasets used in this work.

Dataset	Producer	Variable	Product Name	DOI/URL/Reference
SATELLITE	AVISO+	Hs	Jason-2 Geophysical Data Records (GDR) from precise orbit	https://www.aviso.altimetry.fr/en/data/products/wind/wave-products/gdr-ogdr-osdr-ra2-wwv.html#c6705 (accessed on 1 July 2021)
ARGO	CMEMS	T and S vertical profiles	INSITU_GLO_TS_REP_OBSERVATIONS_013_001_b	[102]
SATELLITE	CMEMS	SST	SST_BLK_SST_L4_NRT_OBSERVATIONS_010_006	[101]
Mooring EUXRo (01, 02, 03)	CMEMS	Currents speed and direction	INSITU_GLO_TS_REP_OBSERVATIONS_013_001_b	[102]

Due to the lack of systematic in-situ measurements in the Black Sea (buoys or moorings), wave simulations of Hs are validated by comparing them with the J2 radar altimeter. Satellite Hs measurements [2016–2018] are used to compare the wave model results. Observations for more than 30 min in time or 2 km in geographical space are rejected. Validation was conducted using scatter plots, comparing observed against modelled Hs.

The bivariate probability density function was estimated using a 2D-Gaussian kernel on a squared grid in the variable space provided in [103]. The plots include summary statistics to describe the WW3 skills to evaluate the Hs, such as BIAS, RMSE, scatter index, correlation index, slope and standard error. Salinity (S) and Temperature (T) vertical profiles from the hydrodynamics simulation were validated against ARGO profiles, while SST validation was conducted using L4 satellite data provided in the framework of the Copernicus Marine Environment and Monitoring Service. BIAS and RMSE were computed using numerical results from January 2015 to December 2019. The Hovmöller diagram allows the evaluation of the model capability to reconstruct interannual-annual and seasonal cycles of analysed scalar quantities (e.g., temperature or salinity).

In this work Hovmöller T (S) diagram considered the uppermost 300 m to evaluate the impact of coupling on the temporal variability of water masses formation in the Black Sea, from 2016 to 2019. All the grid points with depth lower than 100 m have been excluded from the computation of the basin mean to avoid contamination from the coastal zones, then, a basin averaged daily mean has been computed.

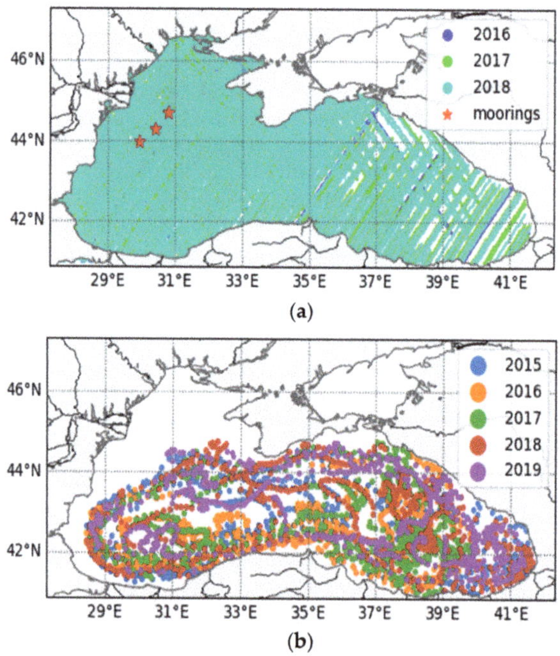

Figure 5. Location of the observations used for validating this work. (**a**) Map of Jason-2 satellite tracks from 2016 to 2018 over the Black Sea, and moorings (red stars) location. (**b**) ARGO floats available between 2015 and 2019.

4. Results and Discussion

4.1. Validation of Hydrodynamical Component

4.1.1. T/S Profiles

Sub-surface temperatures and salinity from HM simulations, as reported in Table 1, were validated using available ARGO profiles from 7.5 m to 1000 m depth as described in Table 2.

Figure 6 shows the relevant metrics for temperature and salinity as given by the set of performed experiments, averaged over the basin for the period 2015–2019. Vertical averaged bias profile for temperature (Figure 6a, left) is negative up to 1000 m and specifically the free-run experiment (H0) is characterised by the largest cold bias in the thermocline layer, up to −0.6 °C, which is only slightly reduced when the Stokes–Coriolis term and wave breaking effects are accounted in experiments H1 and H3 respectively. Temperature bias decreases up to −0.25 °C in the layer 15–40 m when the whole sets of wave-currents interaction processes are considered (H4) and the wave stress (included in experiment H2) provides the largest positive impact in reducing the temperature cold bias in the basin thermocline. In terms of vertical averaged RMSE profile, the envelope of experiments (Figure 6a, mid) shows quite similar values, with higher error in the thermocline, exceeding 2 °C.

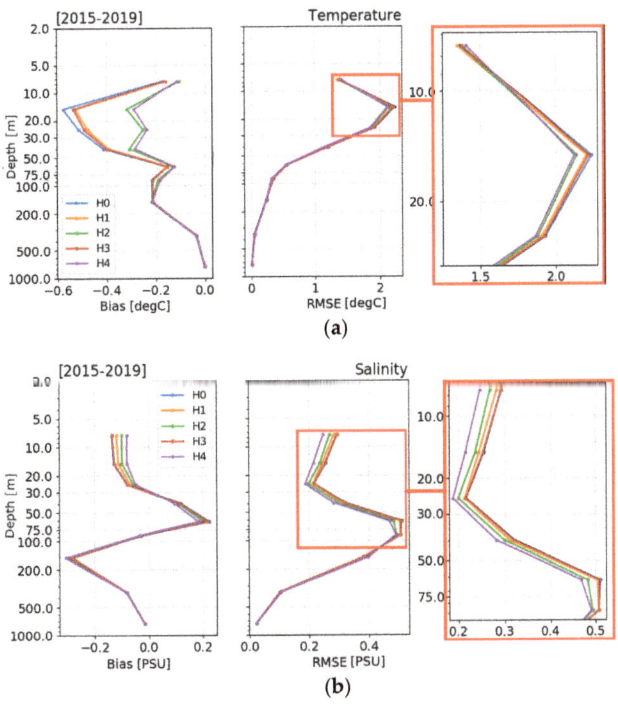

Figure 6. Domain averaged 2015–2019 validation of Temperature profiles (**a**) and Salinity (**b**) between 7.5 and 1000 m deep. For each image, the left panel shows BIAS and the right panel shows RMSE. Red boxes represent a zoomed area of the RMSE plots.

The modified ocean stress accounting for the stress absorbed/released by waves is the one providing the best skill among the "single process" experiments. H4 provides the best performance with a halving of BIAS and an RMSE reduction of about 3%, due to enhanced mixing which impacts the mixed layer positioning. Figure 6b reports the metrics for salinity: in particular, the envelope of experiments exhibits quite similar BIAS values, with a progressive reduction on the BIAS for the fully-forced experiment (H4), dealing with the lowest error at the subsurface (up to 30 m).

Tables 3 and 4 recap skills for the uppermost 200 m at basin scale for both temperature and salinity, respectively: metrics are proposed as averages over every single year and over the period 2015–2019. On a yearly basis, the mean BIAS for temperature (Table 3) is close to zero in H4, but there is no significant difference in terms of RMSE.

Table 3. Statistics evaluated by comparing temperature vertical profile measurements and model results from circulation models from 7.5 m to 200 m deep.

Metric	Experiment	Year 2015	Year 2016	Year 2017	Year 2018	Year 2019	Years 2015–2019
BIAS	H0	−0.63	−0.43	−0.15	−0.05	−0.17	−0.29 ± 0.075
	H1	−0.61	−0.41	−0.14	−0.03	−0.16	−0.27 ± 0.075
	H2	−0.54	−0.31	−0.07	0.04	−0.09	−0.19 ± 0.071
	H3	−0.61	−0.41	−0.13	−0.05	−0.17	−0.27 ± 0.077
	H4	−0.55	−0.28	−0.04	0.04	−0.1	−0.18 ± 0.073
RMSE	H0	1.22	1.04	0.86	0.95	0.88	0.99 ± 0.034
	H1	1.2	1.02	0.86	0.94	0.89	0.98 ± 0.034
	H2	1.19	0.91	0.88	0.98	0.89	0.97 ± 0.03
	H3	1.2	1.03	0.87	0.95	0.89	0.99 ± 0.032
	H4	1.18	0.91	0.88	0.97	0.88	0.96 ± 0.029
No observations		149,928	169,312	161,799	151,030	130,869	762,938

Table 4. Statistics evaluated by comparing Salinity vertical profile measurements and model results from circulation models from 7.5 to 200 m deep.

Metric	Experiment	Year 2015	Year 2016	Year 2017	Year 2018	Year 2019	Years 2015–2019
BIAS	H0	−0.09	−0.15	−0.06	0.08	−0.03	−0.05 ± 0.212
	H1	−0.09	−0.14	−0.06	0.08	−0.02	−0.05 ± 0.208
	H2	−0.08	−0.14	−0.06	0.07	−0.01	−0.04 ± 0.207
	H3	−0.09	−0.15	−0.07	0.08	−0.03	−0.05 ± 0.205
	H4	−0.09	−0.14	−0.08	0.08	−0.01	−0.04 ± 0.208
RMSE	H0	0.29	0.33	0.3	0.31	0.39	0.32 ± 0.128
	H1	0.29	0.33	0.29	0.3	0.38	0.32 ± 0.122
	H2	0.29	0.32	0.28	0.3	0.37	0.31 ± 0.113
	H3	0.29	0.33	0.3	0.31	0.38	0.32 ± 0.12
	H4	0.28	0.31	0.27	0.29	0.35	0.3 ± 0.115
No observations		149,928	169,312	161,799	151,030	130,869	762,938

Table 4 summarises the salinity validation results at the uppermost 200 m. Below 100 m depth, the profiles are very closely aligned. Over the given period, H4 shows a reduction of BIAS of about 20% and RMSE of about 6.5% up to 75 m. We speculate that the average error of 0.3 PSU given by the considered experiments, is related to weak representation of the boundary conditions at the Bosphorus Strait which is not properly representing the Mediterranean waters inflow at all.

The difference between H0 and H4 at the seasonal time scale from 2015 to 2019 was also investigated, as summarised in Table 5, and for the five-year average, the H4 experiment revealed consistently better performance in terms of both salinity and temperature BIAS and RMSE than H0. The best temperature performance in RMSE for the forced run occurs during Winter (Spring) with 0.56 °C (0.61 °C) compared to 0.60 °C (0.66 °C) for H0. During Summer and Autumn, the hydrodynamical model has reduced efficiency (both for H0 and H4 experiments) with almost double RMSE (1.17 °C and 1.18 °C, respectively), and no or negligible differences between H0 and H4.

Table 5. Seasonal evaluation of RMSE and BIAS for salinity and temperature for H0 and H4 experiments from 2015 to 2019. Bold indicates the best value between H0 and H4. DJF = Winter, MAM = Spring, JJA = Summer, SON = Autumn.

Experiment	Metric	Variable	DJF	MAM	JJA	SON
H0	BIAS	Salinity [PSU]	−0.06 ± 0.058	−0.09 ± 0.116	0 ± 0.088	−0.03 ± 0.065
		Temperature {°C}	−0.25 ± 0.245	−0.33 ± 0.179	−0.29 ± 0.243	−0.28 ± 0.255
	RMSE	Salinity [PSU]	0.3 ± 0.01	0.31 ± 0.035	0.31 ± 0.023	0.33 ± 0.08
		Temperature {°C}	0.66 ± 0.167	0.60 ± 0.123	1.17 ± 0.11	1.23 ± 0.138
H4	BIAS	Salinity [PSU]	−0.06 ± 0.56	−0.08 ± 0.107	−0.01 ± 0.085	−0.02 ± 0.06
		Temperature {°C}	−0.19 ± 0.242	−0.28 ± 0.176	−0.13 ± 0.247	−0.16 ± 0.25
	RMSE	Salinity [PSU]	0.29 ± 0.022	0.29 ± 0.027	0.29 ± 0.016	0.30 ± 0.067
		Temperature {°C}	0.61 ± 0.162	0.56 ± 0.104	1.17 ± 0.143	1.18 ± 0.099

4.1.2. SST

SST validation was conducted using L4 satellite data presents SST BIAS and RMSE for the H4 experiment over the period January 2015–December 2019. Figure 7a shows the geographic distribution of the BIAS and RMSE, Figure 7b displays basin averaged metrics on a monthly timescale. The H4–H0 RMSE difference between fully-forced H4 and free-run H0 is then shown in Figure 7c. The 2D map of computed BIAS shows colder modelled temperature than observations in the North-Western sub-basin (here the absolute minimum BIAS value of about −0.7 °C occurs in the Karkinyts'ka Gulf), in the middle of the Central sub-basin and Central-Western Turkish coasts. Warmer BIAS is shown along the Crimean coasts, in the South-Western and Eastern sub-basin (here the absolute maximum BIAS value of about +0.7 °C occurs off-shore of the area of Trebisonda and Ordu). The RMSE 2D map reveals that most of the basin has an error lower than 0.3 °C, while in the Karkinyts'ka Gulf and along the Eastern coasts the RMSE is greater than 0.5 °C. The RMSE only exceeds 0.7 °C off-shore of Trebisonda and Ordu and on the North-Eastern coast of the Black Sea. So, the Eastern basin exhibits warm waters and high error. The time series of metrics for all considered experiments shown in Figure 7b are characterised by similar BIAS values with interannual and seasonal variability—warm BIAS during summertime due to model overestimation of the measured temperature from satellite—and the lowest error provided by H2 and H4 but still very close to the reference control run H0. The H4-H0 RMSE difference map (Figure 7c) reveals that the wave coupling produces a large improvement in the South-Western area, while in the South Eastern part the coupling does not reduce the temperature error. The coupling approach also reduces the SST RMSE in the Odessa and Karkinyts'ka gulfs. Table 6 summarises SST BIAS and RMSE averaged at basin scale from 2015 to 2019 for all five hydrodynamic experiments. No significant change is found for BIAS in any of the experiments, or for RMSE in H1 and H3, which have comparable performance to H0 (≈0.88 °C). Performance is only enhanced in H2 and H4 (≈0.85 °C).

As the first conclusion, the considered single- and fully-forced experiments do not significantly ameliorate the skills at basin scale with respect to free-run experiment, apart from some specific regions in the Western basin where the effect of the air-sea interaction combined with waves affects the enhanced mixing and stress determining a general improvement of the model performances. Considering that H4 is our best implementation, all the next validations are mainly showed comparing H0 to H4.

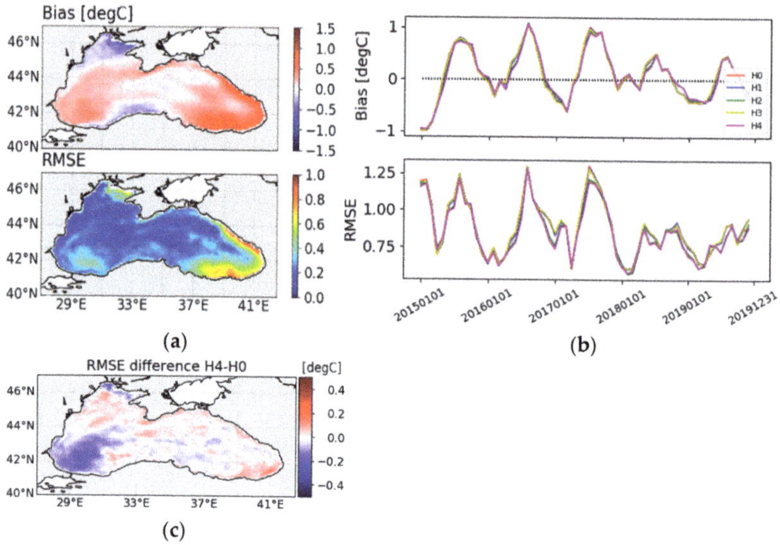

Figure 7. 2015–2019 validation of Sea Surface Temperature vs. Satellite observation. (**a**) BIAS (upper panel) and RMSE (bottom panel) for H4 numerical experiment; (**b**) 5-year BIAS (upper panel) and RMSE (lower panel) time series of SST; (**c**) 5-year SST-RMSE difference between H4 and H0 experiments.

Table 6. Statistics evaluated by comparing Sea Surface Temperature observations and results from circulation models.

Metric	Experiment	Years [2015–2019]
BIAS	H0	0.10 ± 0.129
	H1	0.10 ± 0.128
	H2	0.10 ± 0.127
	H3	0.10 ± 0.127
	H4	0.09 ± 0.129
RMSE	H0	0.882 ± 0.088
	H1	0.881 ± 0.085
	H2	0.857 ± 0.091
	H3	0.883 ± 0.087
	H4	0.854 ± 0.09

4.1.3. Water Masses

In this section, we present the Hovmöller diagrams for temperature and salinity computed by averaging daily mean fields at the basin scale. The Hovmöller diagram for temperature (Figure 8a) as computed in the free-run experiment H0 well shows that the CIL (around 50–150 m layer) is reducing between 2016 and 2019. This tendency was highlighted in [45], which described time vs. depth basin averaged properties of the Black Sea using available observations from 2005 to 2018. Winter signals, which ventilate the CIL, were weak in 2016, 2018 and 2019, becoming stronger in 2017. This phenomenon is also described in [45], in which CIL is referred to as "perforation". Figure 8b shows instead the temperature difference between H4 and H0 as Hovmöller diagram, highlighting an important mechanism activated by the fully-forced experiment, between 50–100 m, which enhances the CIL reduction as shown in [45] The coupling clearly modified the mixing processes of the basin. During Wintertime, it shows cold and homogeneous temperature differences from the surface up to almost 100 m deep, while starting from Spring, the surface temperature is slightly higher in H4 with a colder temperature core in the subsurface. The behaviour reaches the maximum at the beginning of Summer, giving a steeper seasonal thermocline for H4. This suggests that when the wave field has stronger activity, the vertical mixing is higher in the forced experiment (Winter), while if the wave activity reduces, the model has lower vertical mixing (Summer). We argued that the main reason for the mixing reduction in the fully-forced experiment might be derived from wave-dependent surface boundary conditions for the TKE. Indeed, while in H4 the amount of TKE at the surface is proportional to wave energy dissipation, in H0 the surface TKE is prescribed using a parameterization for an old wind-sea [17].

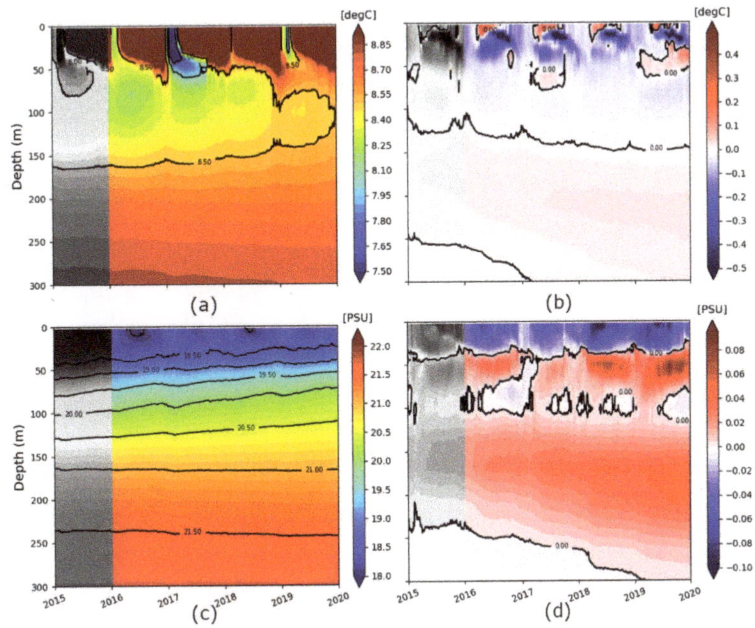

Figure 8. Time versus depth versus temperature (**a**,**b**)/salinity (**c**,**d**) diagram for H0 (**a**,**c**) and H4-H0 difference (**b**,**d**). The shaded grey area refers to the model spin-up time. The shaded area refers to the model spin-up time.

The salinity Hovmöller (Figure 8c) from the H0 model results reveals a quite stratified vertical structure. It also reveals increasing salinity in the simulated period, which is not shown in the [45] work, probably due to the closed boundary condition at the Bosphorus. Figure 8d shows the salinity difference between H4 and H0: the fully-forced experiment exhibits slightly fresher waters at the subsurface up to about 50 m and saltier ones up to 100 m. The signal of increased CIL ventilation phenomena in H4, highlighted in the temperature plot (warm water around 50–100 m) is also evident in the salinity diagram (freshwater around 50–100 m).

From Figure 8b,d it is possible to appreciate the difference in the seasonal cycle for salinity and temperature between H4 and H0.

The seasonality of the difference is more evident in 2018. During the first part of the year, warmer-fresher waters from the surface reach almost 40 m deep, while during the second part of the year becomes shallower, reaching the minimum in Autumn. This cycle is easier to observe evaluating temperature (warm core at surface, and cold-core below), while it is more hidden in salinity because the fresher core at the surface is embedded in a whole water column from 0 to 50 m with a negative BIAS. The fact that salinity is always negative in the uppermost 50 m, and saltier below is probably dependent on a higher stratification on average in H4.

4.1.4. Currents

Currents speed and direction at 2.5 m depth were validated from 2015 to 2019 using mooring data from CMEMS INS TAC as reported in Table 2 and represented in Figure 5a (red stars). Metrics are reported in Table 7. The main impact due to wave fields on currents statistics is evident when considering the velocity directions, where the coupling reduces the BIAS and RMSE by ~16% and ~10%, respectively, when compared to the control run H0. The improved skill can be mostly attributed to the use of modified stress, showing experiment H2 the lowest BIAS (36°).

Table 7. Validation of 2.5 m depth currents speed and direction: averaged statistics using data as provided by the available moorings in the 2015–2019 period.

Experiment	Variable	BIAS	RMSE
H0	Speed [m/s]	-0.055 ± 0.07	0.08 ± 0.051
	Direction [°]	-44 ± 67	110 ± 38
H4	Speed [m/s]	-0.054 ± 0.078	0.08 ± 0.056
	Direction [°]	-38 ± 66	100 ± 37

The velocity direction RMSE indicated that the H1, H2 and H3 experiments have comparable errors in the range of 107–111° and that H4 is the best implementation, with a reduced error of around 100°. Conversely to the direction, no significant differences were found for the statistics of currents speed in the forced simulation.

Figure 9 shows the five years averaged currents speed and direction at the mooring locations, and it allows us to appreciate how the currents change when hydrodynamics is forced with waves. Each column in the plot represents a specific mooring: EUXRo01, EUXRo02 and EUXRo03, from left to right.

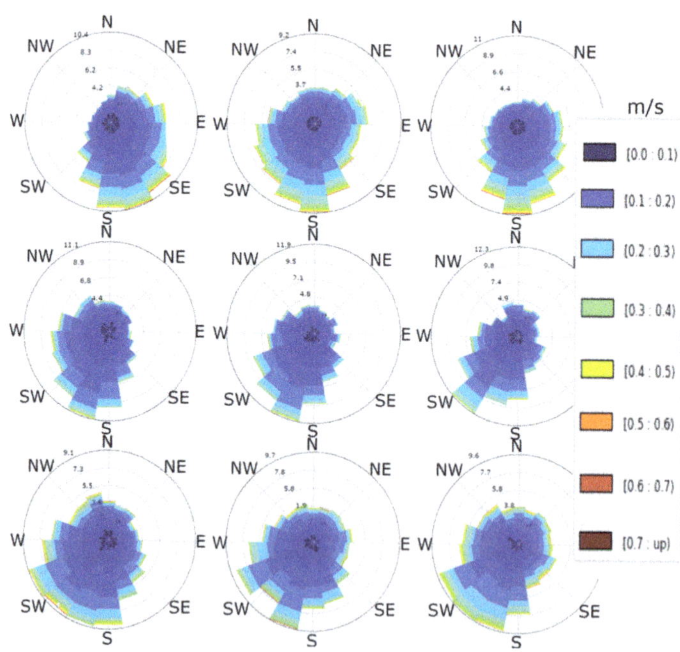

Figure 9. Currents speed and direction at 2.5 m deep for moorings (**first row**), H0 (**second row**) and H4 (**third row**). Columns represent the three different mooring locations available.

Additionally, currents speed and direction from observations (first row), from the H0 experiment (second row) and the H4 experiment (third row) is shown. From a qualitative point of view, we can see that in both forced and free-run experiments the model has a stronger western currents component than the observation, probably due to low resolution in space and time of atmospheric forcings, considering also that the observation is taken on a precise location/time.

The main wave direction in this area is aligned with that of the currents (from North-East towards South-West). H0 experiment has a prevalence of one-directional bin, while if waves are considered, the direction of the main current is described by 3–4 directional bins. This explains that the currents simulated by the H4 experiment have a wider dynamic.

Contrarily to H0, in which stronger currents are defined mainly by one direction, the forced experiment H4 has 3–4 directional bins with almost the same speeds and occurrence frequency. Despite this change in magnitude, statistics from Table 5 do not highlight improved skills in the HM simulation. Unfortunately, the very low number of available moorings in the Black Sea region prevents us from performing a more robust validation of the results, but the wider dynamics of H4, which is closer to the observation than H0, seemed to be promising.

4.1.5. RMSE vs. Significant Wave Height

To evaluate a correspondence between the improved representation of the ocean physics in the Black Sea thanks to wave-currents forced model, we propose in Figure 10 the time series of daily averaged RMSE differences for temperature (herein, RMSEdif) between fully-forced H4 and free-run H0 experiments in the layer 5–20 m and the corresponding mean, at observation location, of Hs on 2019.

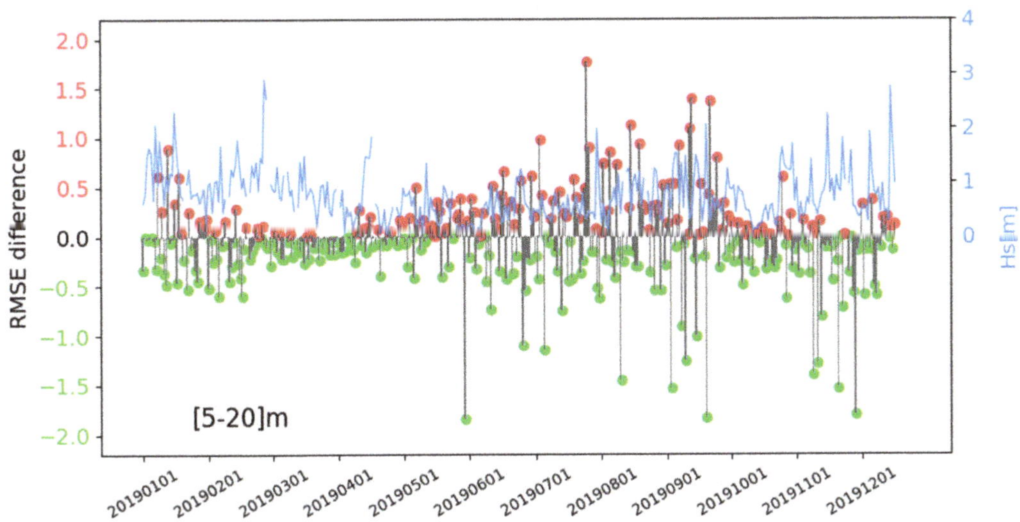

Figure 10. Timeseries of daily H4-H0 RMSE difference (RMSEdif) compared to Hs. The grey line represents RMSEdif and is associated with a green or red scatter point if the forced run is respectively better or worse than free-run. The cyan line represents the mean Hs at observation locations.

This plot helps to evaluate whether the positive or negative impact of the coupling is dependent on specific Hs values or the whole Hs spectrum. The figure shows that in general, the H4 has an error lower than H0 in most of the year. The range of variability for RMSEdif evolves seasonally, during Winter and Spring is confined to [−0.5: +1 °C], with the lowest value between mid of February—beginning of April, while during Summer and Autumn the range extends to [−2: +1.8 °C]. From March to May and from September to October, Hs rarely exceeds 1.5 m and it is always close or below 1.0 m: in this period, the fully-forced run H4 seems to have the best performances (e.g., lower error than H0). This investigation revealed that the forced experiment performs better than the free-run when the sea state has no large fluctuations, as in February–April or in November During Summer and January, in which the variability of the wavefield is high, the forced run still performs better in most of the cases, but there are several days in which the H0 is the best.

As a general indication, the forced model confirms its good performance, demonstrating that in the thermocline region the improvement can be of the order of about 0.5 °C on average. However, to better assess this conclusion, the analysis requires further dedicated investigation over a smaller time scale.

4.1.6. Validation for the Wave Component

In this study, a three-year validation (2016–2018) of Hs was conducted using J2 satellite data. The dataset, filtered according to a quality check, consisted of 10,479, 9035 and 10,447 observations for 2016, 2017 and 2018, respectively. Figure 11 compares W0 experiment (a-panel) with W3 experiment (b-panel) considering the whole dataset (29,961 observations).

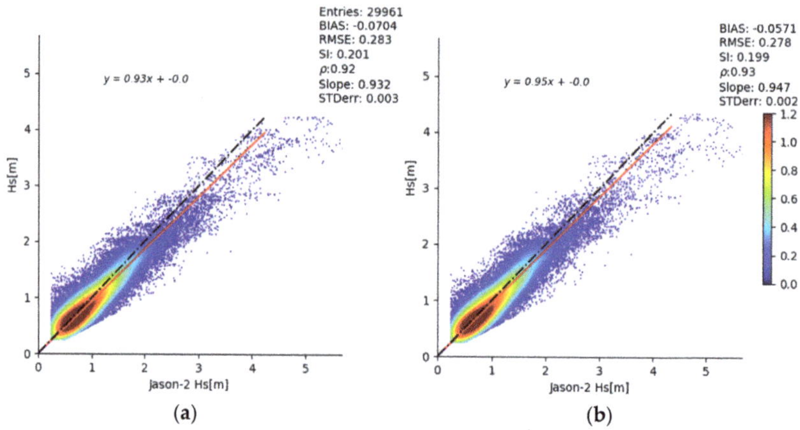

Figure 11. Significant wave height validation using Jason-2 satellite from 2016 to 2018. Scatter-plots (**a**,**b**) refer to W0 and W3 experiments, respectively.

The coupling with currents and ΔT (W3 experiment) induced a performance improvement in all the statistics: BIAS was reduced from −7 cm to −5.7 cm, RMSE from 28.3 cm to 27.8, the Scatter Index from 0.201 to 0.199, while Pearson's correlation increased from 0.92 to 0.93 and Slope from 0.93 to 0.95.

Table 8 summarises the statistics for all of the wave experiments. All three wave experiments when forced with hydrodynamic fields (W1, W2 and W3) improved model performance, albeit to different extents. The lowest and negligible impact was derived from only-currents (W1 experiment). This result could be a side-effect of the validation method here used. Again, the absence of observations, e.g., from buoy wave gauge in this case, strongly affected our capability to validate the experiments and we were obliged to use satellite data, which has two main disadvantages: it is not reliable near the coast, where currents are stronger and may impact the waves; it does not provide information about wave direction, which could be affected by refraction phenomena.

On the contrary, ΔT (W2 experiment), which acts mainly on Hs has been positively evaluated and confirmed what is demonstrated in the literature [29]. When both currents and ΔT were considered (W3 experiment), the lowest error was obtained, with a reduction of ≈−18% in BIAS, ≈−2% in RMSE. Even the precision of the simulation has been improved, with reduction of scatter index and increasing 1% in ρ in correlation.

Table 8. Significant wave height (m) validation statistics.

Metric	Experiment	Year 2016	Year 2017	Year 2018	Years 2016–2018
BIAS	W0	−0.077	−0.071	−0.06	−0.070 ± 0.007
	W1	−0.075	−0.070	−0.062	−0.069 ± 0.005
	W2	−0.062	−0.062	−0.054	−0.058 ± 0.004
	W3	−0.59	−0.059	−0.053	−0.057 ± 0.003
RMSE	W0	0.304	0.274	0.270	0.283 ± 0.015
	W1	0.302	0.271	0.269	0.282 ± 0.015
	W2	0.299	0.270	0.266	0.279 ± 0.015
	W3	0.297	0.267	0.266	0.278 ± 0.014
SI	W0	0.216	0.194	0.190	0.201 ± 0.011
	W1	0.215	0.192	0.191	0.200 ± 0.011
	W2	0.215	0.192	0.190	0.200 ± 0.011
	W3	0.214	0.190	0.190	0.199 ± 0.011
slope	W0	0.893	0.977	0.934	0.932 ± 0.034
	W1	0.895	0.978	0.933	0.933 ± 0.034
	W2	0.908	0.993	0.946	0.946 ± 0.035
	W3	0.91	0.994	0.946	0.974 ± 0.034
No observations		10,479	9035	10,447	29,961

5. Conclusions

In this study, we investigated the importance of wave-currents interaction in the Black Sea for the first time. A reciprocally forced numerical system has been implemented using the ocean circulation model NEMO v4.0, which is now including the most important wave-currents physics, and the third-generation wave model WaveWatchIII. The coupling consists of providing Sea Surface Temperature and surface currents to the wave model, computed by the hydrodynamic model and returning sea-state dependent momentum flux, Surface Stokes Drift and wave dissipated energy to ocean vertical mixing. Even if our main focus was the assessment of tracers, a positive effect has been found also on waves.

The inclusion of wave-currents interaction in the Black Sea hydrodynamics, determined reduction of the RMSE for SST (≈−3.5%) and the upper ocean from 7.5 to 200 m water depths (≈−3%). The main differences between the forced and free runs are related to the uppermost part of the water column (depth < 35 m). The strongest impact on the vertical profile caused by the for is related to the sea-state dependent momentum flux, while the Stokes–Coriolis force and the sea-state dependent vertical mixing have negligible effects. On average, the coupling produced greater benefits in Winter and Spring, which are characterised by intense wave activity and low vertical stratification. According to our validation, the forced run was found to perform better than the free run for moderate wave heights. In general, the forced run demonstrated a slightly warmer water temperature than the free one.

The forced experiment had a positive impact even on salinity, with a reduction of ≈−10% in BIAS and ≈−6.5% in RMSE on the uppermost averaged 200 m.

The main impact of waves on currents concerned the reduction of direction BIAS, without improvement/worsening for the velocity module. Anyway, we noticed a wider dynamic in direction and speed for the currents field in the fully-forced experiment.

We inferred that the physical process which has been improved in forced hydrodynamics is related to the vertical mixing, with larger mixing during Winter and lower mixing during Spring–Summer with respect to free-run experiments, as showed in Section 4.1.3.

Coupling also improved the wave model performance, which slightly better represents the Hs satellite observations, in forced configuration. The results indicate that the improvement was mainly related to the better representation of the effect of air-sea temperature differences on the wave growth, while the usage of the surface currents plays a minor role, as already shown in [28]. Unfortunately, the lack of wave buoys in the basin prevented the analysis of the coupling impact on wave period and direction.

In conclusion, this preliminary coupling configuration produced a modest but clear improvement in the simulations of temperature, salinity, currents and waves in the Black Sea. Future works could investigate this setup-up over a longer time scale to evaluate its impact on climatological time scale as in [19,104,105]. The use of an external coupler to conduct online field exchange is an important technical development that can be considered in further studies.

Author Contributions: S.C. designed the forced system and coordinated the work in collaboration with S.A.C., G.C. and P.L., E.C. participated in the scientific discussion sharing the theoretical approach as implemented in the Mediterranean Sea. All authors have read and agreed to the published version of the manuscript.

Funding: This research was funded by the Copernicus Marine Environment and Monitoring Service for the Black Sea Monitoring and Forecasting Centre, contract n. 72.

Institutional Review Board Statement: Not applicable.

Informed Consent Statement: Not applicable.

Conflicts of Interest: The authors declare no conflict of interest.

References

1. Roland, A.; Ardhuin, F. On the developments of spectral wave models: Numerics and parameterizations for the coastal ocean. *Ocean. Dyn.* **2014**, *64*, 833–846. [CrossRef]
2. Bolaños, R.; Tornfeldt Sørensen, J.V.; Benetazzo, A.; Carniel, S.; Sclavo, M. Modelling ocean currents in the northern Adriatic Sea. *Cont. Shelf Res.* **2014**, *87*, 54–72. [CrossRef]
3. Longuet-Higgins, M.S.; Stewart, R.W. The changes in amplitude of short gravity waves on steady non-uniform currents. *J. Fluid Mech.* **1961**, *10*, 529–549. [CrossRef]
4. Hsiao, S.C.; Chen, H.; Chen, W.B.; Chang, C.H.; Lin, L.Y. Quantifying the contribution of nonlinear interactions to storm tide simulations during a super typhoon event. *Ocean. Eng.* **2019**, *194*, 106661. [CrossRef]
5. Hsiao, S.C.; Wu, H.L.; Chen, W.B.; Chang, C.H.; Lin, L.Y. On the Sensitivity of Typhoon Wave Simulations to Tidal Elevation and Current. *J. Mar. Sc. Eng.* **2020**, *8*, 731. [CrossRef]
6. Janssen, P.A.E.M. Ocean wave effects on the daily cycle in SST. *J. Geophys. Res.* **2012**, *117*, C11. [CrossRef]
7. Couvelard, X.; Lemarié, F.; Samson, G.; Redelsperger, J.L.; Ardhuin, F.; Benshila, R.; Madec, G. Development of a two-way-coupled ocean–wave model: Assessment on a global NEMO (v3. 6)–WW3 (v6. 02) coupled configuration. *Geosci. Model Dev.* **2020**, *13*, 3067–3090. [CrossRef]
8. Lewis, H.W.; Castillo Sanchez, J.M.; Siddorn, J.; King, R.R.; Tonani, M.; Saulter, A.; Sykes, P.; Pequignet, A.C.; Weedon, G.P.; Palmer, T.; et al. Can wave coupling improve operational regional ocean forecasts for the north-west European Shelf? *Ocean Sci.* **2019**, *15*, 669–690. [CrossRef]
9. Hasselmann, K. Wave-driven inertial oscillations. *Geophys. Astrophys. Fluid Dyn.* **1970**, *1*, 463–502. [CrossRef]
10. McWilliams, J.C.; Restrepo, J.M.; Lane, E.M. An asymptotic theory for the interaction of waves and currents in coastal waters. *J. Fluid Mech.* **2004**, *511*, 135–178. [CrossRef]
11. Craig, P.D.; Banner, M.L. Modeling Wave-Enhanced Turbulence in the Ocean Surface Layer. *J. Phys. Oceanogr.* **1994**, *24*, 2546–2559. [CrossRef]
12. McWilliams, J.C.; Restrepo, J.M. The wave-driven ocean circulation. *J. Phys. Oceanogr.* **1991**, *29*, 2523–2540. [CrossRef]
13. McWilliams, J.C.; Sullivan, P. Vertical mixing by Langmuir circulations. *Spill Sci. Technol. Bull.* **2000**, *6*, 225–237. [CrossRef]
14. Mellor, G.; Blumberg, A. Wave Breaking and Ocean Surface Layer Thermal Response. *J. Phys. Oceanogr.* **2004**, *34*, 693–698. [CrossRef]
15. Ardhuin, F.; Jenkins, A.D. On the Interaction of Surface Waves and Upper Ocean Turbulence. *J. Phys. Oceanogr.* **2006**, *36*, 551–557. [CrossRef]
16. Huang, C.J.; Qiao, F.; Song, Z.; Ezer, T. Improving simulations of the upper ocean by inclusion of surface waves in the Mellor-Yamada turbulence scheme. *J. Geophys. Res.* **2011**, *116*. [CrossRef]
17. Janssen, P.A.E.M.; Breivik, Ø.; Mogensen, K.; Vitart, F.; Balmaseda, M.; Bidlot, J.R.; Molteni, F. Air-sea interaction and surface waves. *Eur. Cent. Medium-Range Weather. Forecast.* **2013**, *35*. [CrossRef]
18. Belcher, S.E.; Grant, A.L.; Hanley, K.E.; Fox-Kemper, B.; Van Roekel, L.; Sullivan, P.P.; Polton, J.A. A global perspective on Langmuir turbulence in the ocean surface boundary layer. *Geophys. Res. Lett.* **2012**, *39*. [CrossRef]
19. Babanin, A.V.; Haus, B.H. On the existence of water turbulence induced by nonbreaking surface waves. *J. Phys. Oceanogr.* **2009**, *39*, 2675–2679. [CrossRef]

20. Hackett, B.; Breivik, Ø.; Wettre, C. Forecasting the Drift of Objects and Substances in the Ocean. In *Ocean Weather Forecasting*; Springer: Berlin/Heidelberg, Germany, 2006; pp. 507–523.
21. Breivik, Ø.; Allen, A.A. An operational search and rescue model for the Norwegian Sea and the North Sea. *J. Mar. Syst.* **2008**, *69*, 99–113. [CrossRef]
22. Davidson, F.J.; Allen, A.; Brassington, G.B.; Breivik, Ø.; Daniel, P.; Kamachi, M.; Kaneko, H. Applications of GODAE ocean current forecasts to search and rescue and ship routing. *Oceanography* **2009**, *22*, 176–181. [CrossRef]
23. Breivik, Ø.; Allen, A.A.; Maisondieu, C.; Olagnon, M. Advances in search and rescue at sea. *Ocean Dyn.* **2013**, *63*, 83–88. [CrossRef]
24. Alari, V.; Staneva, J.; Breivik, Ø.; Bidlot, J.R.; Mogensen, K.; Janssen, P.A.E.M. Surface wave effects on water temperature in the Baltic Sea: Simulations with the coupled NEMO-WAM model. *Ocean Dyn.* **2016**, *66*, 917–930. [CrossRef]
25. Staneva, J.; Alari, V.; Breivik, Ø.; Bidlot, J.R.; Mogensen, K. Effects of wave-induced forcing on a circulation model of the North Sea. *Ocean Dyn.* **2017**, *67*, 81–101. [CrossRef]
26. Breivik, Ø.; Janssen, P.A.E.M.; Bidlot, J.R. Approximate Stokes Drift Profiles in Deep Water. *J. Phys. Oceanogr.* **2014**, *44*, 2433–2445. [CrossRef]
27. Breivik, Ø.; Mogensen, K.; Bidlot, J.R.; Balmaseda, M.; Janssen, P.A.E.M. Surface wave effects in the NEMO ocean model: Forced and coupled experiments: Waves in NEMO. en. *J. Geophys. Res. Oceans* **2015**, *120*, 2973–2992. [CrossRef]
28. Staneva, J.; Ricker, M.; Carrasco Alvarez, R.; Breivik, Ø.; Schrum, C. Effects of wave-induced processes in a coupled Wave–Ocean model on particle transport simulations. *Water* **2021**, *13*, 415. [CrossRef]
29. Clementi, E.; Oddo, P.; Drudi, M.; Pinardi, N.; Korres, G.; Grandi, A. Coupling hydrodynamic and wave models: First step and sensitivity experiments in the Mediterranean Sea. *Ocean Dyn.* **2017**, *67*, 1293–1312. [CrossRef]
30. Bruciaferri, D.; Tonani, M.; Lewis, H.W.; Siddorn, J.R.; Saulter, A.; Castillo, J.M.; McConnell, N. The impact of ocean-wave coupling on the upper ocean circulation during storm events. *J. Geophys. Res. Ocean.* **2021**, e2021JC017343. [CrossRef]
31. Stanev, E.V. Black Sea dynamics. *Oceanography* **2005**, *18*, 56. [CrossRef]
32. Ivanov, V.A.; Belokopytov, V.N. *Oceanography of the Black Sea*; National Academy of Science of Ukraine, Marine Hydrophysical Institute: Sevastopol, Russia, 2013; 210p.
33. Özsoy, E.; Di Iorio Dm Gregg, M.C.; Backhaus, J.O. Mixing in the Bosphorus Strait and the Black Sea continental shelf: Observations and a model of the dense water outflow. *J. Mar. Syst.* **2001**, *31*, 99–135. [CrossRef]
34. Mikhailov, V.N.; Mikhailova, M.V. *River Mouths in the Black Sea Environment*; Kostianoy, A.G., Kosarev, A.N., Eds.; Springer: Berlin/Heidelberg, Germany, 2008; pp. 91–133. [CrossRef]
35. Kara, A.; Birol, J.; Wallcraft, A.; Hurlburt, H.; Stanev, E.V. Air–sea fluxes and river discharges in the Black Sea with a focus on the Danube and Bosphorus. *J. Mar. Syst.* **2008**, *74*, 74–95. [CrossRef]
36. Miladinova, S.; Stips, A.; Macias Moy, D.; Garcia-Gorriz, E. Pathways and mixing of the north western river waters in the Black Sea. *Estuar. Coast. Shelf Sci.* **2020**, *236*, 106630. [CrossRef]
37. Stanev, E.V. On the mechanisms of the Black Sea circulation. *Earth-Sci. Rev.* **1990**, *28*, 285–319. [CrossRef]
38. Oguz, T.; Malanotte-Rizzoli, P. Seasonal variability of wind and thermohaline-driven circulation in the Black Sea: Modeling studies. *J. Geophys. Res. Oceans* **1996**, *101*, 16551–16569. [CrossRef]
39. Lima, L.; Ciliberti, S.A.; Aydogdu, A.; Masina, S.; Escudier, R.; Cipollone, A.; Azevedo, D.; Causio, S.; Peneva, E.; Lecci, R.; et al. Climate signals in the Black Sea from a multidecadal eddy-resolving reanalysis. *Front. Mar. Sci.* **2021**, in press. [CrossRef]
40. Gunduz, M.; Özsoy, E.; Hordoir, R. A model of Black Sea circulation with strait exchange (2008–2018). *Geosci. Model Dev.* **2020**, *13*, 121–138. [CrossRef]
41. Sadighrad, E.; Fach, B.A.; Arkin, S.S.; Salihoğlu, B.; Hüsrevoğlu, S. Mesoscale eddies in the Black Sea: Characteristics and kinematic, properties in a high-resolution ocean model. *J. Mar. Syst.* **2021**, *2021*, 103613. [CrossRef]
42. Bruciaferri, D.; Shapiro, G.; Stanichny, S.; Zatsepin, A.; Ezer, T.; Wobus, F.; Hilton, D. The development of a 3D computational mesh to improve the representation of dynamic processes: The Black Sea test case. *Ocean Model.* **2020**, *146*, 101534. [CrossRef]
43. Staneva, J.V.; Dietrich, D.E.; Stanev, E.V.; Bowman, M.J. Rim current and coastal eddy mechanisms in an eddy-resolving Black Sea general circulation model. *J. Mar. Syst.* **2001**, *31*, 137–157. [CrossRef]
44. Oguz, T.; Besiktepe, S. Observations on the Rim Current structure, CIW formation and transport in the western Black Sea. *Deep. Sea Res. Part I Oceanogr. Res. Pap.* **1999**, *46*, 1733–1753. [CrossRef]
45. Stanev, E.; Peneva, E.; Chtirkova, B. Climate Change and Regional Ocean Water Mass Disappearance: Case of the Black Sea. *J. Geophys. Res. Ocean.* **2019**, *124*, 4803–4819. [CrossRef]
46. Murray, J.W.; Jannasch, H.W.; Honjo, S.; Anderson, R.F.; Reeburgh, W.S.; Top, Z.; Friederich, G.E.; Codispoti, L.A.; Izdar, E. Unexpected changes in the oxic/anoxic interface in the Black Sea. *Nature* **1989**, *338*, 411–413. [CrossRef]
47. Rusu, L.; Bernardino, M.; Guedes Soares, C. Wind and wave modelling in the Black Sea. *J. Oper. Oceanogr.* **2014**, *7*, 5–20. [CrossRef]
48. Belokopytov, V.N.; Kudryavtseva GFLipchenko, M.M. Atmospheric pressure and wind over the Black Sea (1961–1990). *Trudy UkrNIGMI* **1998**, *246*, 174–181. (In Russian)
49. Schrum, C.; Staneva, J.; Stanev, E.; Özsoy, E. Air–sea exchange in the Black Sea estimated from atmospheric analysis for the period 1979–1993. *J. Mar. Syst.* **2001**, *31*, 3–19. [CrossRef]
50. Efimov, V.V.; Timofeev, N.A. *Heat Balance Studies of the Black Sea and the Sea of Azov*; VNIIGMI-MCD: Obninsk, Russia, 1990. Available online: https://scholar.google.com.hk/scholar?hl=zh-CN&as_sdt=0%2C5&q=Heat+balance+studies+of+the+Black+Sea+and+the+Sea+of+Azov&btnG= (accessed on 18 August 2021).

51. Zecchetto, S.; De Biasio, F. Sea Surface Winds over the Mediterranean Basin from Satellite Data (2000–04): Meso- and Local-Scale Features on Annual and Seasonal Time Scales. *J. Appl. Meteorol. Climatol.* **2007**, *46*, 814–827. [CrossRef]
52. Munk, W.H. Wind stress on water: An hypothesis. *Q. J. R. Meteorol. Soc.* **1955**, *81*, 320–332. [CrossRef]
53. Pierson, W.J., Jr.; Moskowitz, L. A proposed spectral form for fully developed wind seas based on the similarity theory of SA Kitaigorodskii. *J. Geophys. Res.* **1964**, *69*, 5181–5190. [CrossRef]
54. Rzheplinkskij, G.V. *Wind and Wave Atlas of the Black Sea*; Gidrometeoizdat: St. Petersburg, Russia, 1969. Available online: https://www.google.com.hk/search?newwindow=1&source=univ&tbm=isch&q=54.%09Rzheplinkskij,+G.V.+Wind+and+Wave+Atlas+of+the+Black+Sea&sa=X&ved=2ahUKEwii76fdkLzyAhWSMd4KHZJ3D9QQjJkEegQIIRAC&biw=1280&bih=539 (accessed on 18 August 2021). (In Russian)
55. Simonov, A.I.; Altman, E.N. *Hydrometeorology and Hydrochemistry of the USSR Seas; The Black Sea*; Gidrometeoizdat: St. Petersburg, Russia, 1991; Volume IV. Available online: https://scholar.google.com.hk/scholar?cluster=3991505303709658658&hl=zh-CN&as_sdt=2005&sciodt=0,5 (accessed on 18 August 2021).
56. Polonsky, A.B.; Popov, Y.I.; Conditions for Formation of the Cold Intermediate Layer in the Black Sea. MGI Sevastapol 2011. Available online: https://scholar.google.com.hk/scholar?hl=zh-CN&as_sdt=0%2C5&q=Conditions+for+Formation+of+the+Cold+Intermediate+Layer+in+the+Black+Sea&btnG= (accessed on 18 August 2021). (In Russian).
57. Arkhipkin, V.S.; Gippius, F.N.; Koltermann, K.P.; Surkova, G.V. Wind waves in the Black Sea: Results of a hindcast study. *Nat. Hazards Earth Syst. Sci.* **2014**, *14*, 2883. [CrossRef]
58. Law Chune, S.; Aouf, L. Wave effects in global ocean modelling: Parametrizations vs. forcing from a wave model. *Ocean Dynam.* **2018**, *68*, 1739–1758. [CrossRef]
59. Owens, R.G.; Hewson, T. ECMWF Forecast User Guide. *ECMWF Read.* **2018**. [CrossRef]
60. Adler, R.F.; Huffman, G.J.; Chang, A.; Ferraro, R.; Xie, P.; Janowiak, J.; Rudolf, B.; Schneider, U.; Curtis, S.; Bolvin, D.; et al. The Version 2 Global Precipitation Climatology Project (GPCP) Monthly Precipitation Analysis (1979-Present). *J. Hydrometeor.* **2003**, *4*, 1147–1167. [CrossRef]
61. Huffman, G.J.; Adler, R.F.; Bolvin, D.T.; Gu, G. Improving the Global Precipitation Record: GPCP Version 2.1. *Geophys. Res. Lett.* **2009**, *36*, L17808. [CrossRef]
62. Weatherall, P.; Marks, K.M.; Jakobsson, M.; Schmitt, T.; Tani, S.; Arndt, J.E.; Wigley, R. A new digital bathymetric model of the world's oceans. *Earth Space Sci.* **2015**, *2*, 331–345. [CrossRef]
63. Gürses, O.; Aydõgdu, A.; Pinardi, N.; Özsoy, E. A finite element modeling study of the Turkish Straits System. In *The Sea of Marmara—Marine Biodiversity, Fisheries, Conservations and Governance*; TUDAV: Istambul, Turkey, 2016; pp. 169–184.
64. Gürses, Ö. Dynamics of the Turkish Straits System—A Numerical Study with a Finite Element Ocean Model Based on an Unstructured Grid Approach. Ph.D. Thesis, Middle East Technical University, Ankara, Turkey, 2016.
65. Madec, G.; NEMO System Team. Scientific Notes of Climate Modelling Center, User manual. *NEMO Ocean Engine* **2017**. [CrossRef]
66. Myroshnychenko, V.; Simoncelli, S.; Troupin, C. SeaDataCloud Temperature and Salinity Climatology for the Black Sea (Version 1). *Prod. Inf. Doc.* **2019**. [CrossRef]
67. Pettenuzzo, D.; Large, W.G.; Pinardi, N. On the corrections of ERA-40 surface flux products consistent with the Mediterranean heat and water budgets and the connection between basin surface total heat flux and NAO. *J. Geophys. Res. Oceans* **2010**, *115*. [CrossRef]
68. Tonani, M.; Pinardi, N.; Dobricic, S.; Pujol, I.; Fratianni, C. A high-resolution free-surface model of the Mediterranean Sea. *Ocean Sci. Discuss.* **2008**, *4*, 213–244. [CrossRef]
69. Oddo, P.; Adani, M.; Pinardi, N.; Fratianni, C.; Tonani, M.; Pettenuzzo, D. A nested Atlantic-Mediterranean Sea general circulation model for operational forecasting. *Ocean Sci.* **2009**, *5*, 461–473. [CrossRef]
70. Ludwig, W.; Dumont, E.; Meybeck, M.; Heussner, S. River discharges of water and nutrients to the Mediterranean and Black Sea: Major drivers for ecosystem changes during past and future decades? *Prog. Oceanogr.* **2009**, *80*, 199–217. [CrossRef]
71. Panin, N.; Jipa, D. Danube River sediment input and its interaction with the north-western Black Sea. *Estuar. Coast. Shelf Sci.* **2002**, *54*, 551–562. [CrossRef]
72. Stanev, E.; Beckers, J.M. Barotropic and baroclinic oscillations in strongly stratified ocean basins: Numerical study of the Black Sea. *J. Mar. Syst.* **1999**, *19*, 65–112. [CrossRef]
73. Peneva, E.L.; Stanev, E.; Belokopytov, V.; Le Traon, P.Y. Water transport in the Bosphorus Straits estimated from hydrometeorologycal and altimeter data: Seasonal to decadal variability. *J. Mar. Sys.* **2001**, *31*, 21–35. [CrossRef]
74. Aydogdu, A.; Pinardi, N.; Özsoy, E.; Danabasoglu, G.; Gürses, O.; Karspeck, A. Circulation of the Turkish Straits System under interannual atmospheric forcing. *Ocean Sci.* **2018**, *14*, 999–1019. [CrossRef]
75. Causio, S.; Ciliberti, S.A.; Clementi, E.; Coppini, G.; Lionello, P. A Modeling Approach for the Assessment of Wave-Currents Interaction in the Black Sea. Available online: https://zenodo.org/record/5184753#.YR3VVt8RU2w (accessed on 18 August 2021).
76. Tolman, H.L. User manual and system documentation of WAVEWATCH III TM version 3.14. *Tech. Note MMAB Contrib.* **2009**, *276*, 220.
77. Günther, H.; Hasselmann, S.; Janssen, P.A.E.M. *The WAM Model Cycle 4, User Manual*; Dtsch: Klimarechenzentrum, Germany, 1992. No. DKRZ-TR-4 (REV. ED.). Available online: https://www.dkrz.de/mms/pdf/reports/ReportNo.04.pdf (accessed on 18 August 2021).

78. Ardhuin, F.; Rogers, E.; Babanin, A.V.; Filipot, J.F.; Magne, R.; Roland, A.; Collard, F. Semiempirical dissipation source functions for ocean waves. Part I: Definition, calibration, and validation. *J. Phys. Oceanogr.* **2010**, *40*, 1917–1941. [CrossRef]
79. Janssen, P.A.E.M. Quasi-linear Theory of Wind-Wave Generation Applied to Wave Forecasting. *J. Phys. Oceanogr.* **1998**, *21*, 1631–1642. [CrossRef]
80. Chalikov, D.V.; Yu Belevich, M. One-dimensional theory of the wave boundary layer. *Bound. Layer Meteorol.* **1993**, *63*, 65–96. [CrossRef]
81. Bidlot, J.; Abdalla, S.; Janssen, P.A.E.M. A revised formulation for ocean wave dissipation in CY25R1. *Intern. Memo. Res. Department ECMWF* **2005**. [CrossRef]
82. Bidlot, J.R. Intercomparison of operational wave forecasting systems against buoys: Data from ECMWF, MetOce, FNMOC, NCEP, DWD, BoM, SHOM and JMA, September 2008 to November 2008. In *Joint WMO-IOC Technical Commission for Oceanography and Marine Meteorology*; World Meteorological Organization: Geneva, Switzerland, 2008.
83. Hasselmann, S.; Hasselmann, K.; Allender, J.K.; Barnett, T.P. Computations and parameterizations of the nonlinear energy transfer in a gravity-wave specturm. Part II: Parameterizations of the nonlinear energy transfer for application in wave models. *J. Phys. Oceanogr.* **1985**, *15*, 1378–1391. [CrossRef]
84. Hasselmann, D.; Bösenberg, J.; Dunckel, M.; Richter, K.; Grünewald, M.; Carlson, C. Measurements of wave-induced pressure over surface gravity waves. In *Wave Dynamics and Radio Probing of the Ocean Surface*; Springer: Boston, MA, USA, 1986; pp. 353–368.
85. Wu, L.; Staneva, J.; Breivik, Ø.; Rutgersson, A.; Nurser, A.G.; Clementi, E.; Madec, G. Wave effects on coastal upwelling and water level. *Ocean Mod.* **2019**, *140*, 101405. [CrossRef]
86. Saetra, Ø.; Albretsen, J.; Janssen, P.A. Sea-state-dependent momentum fluxes for ocean modeling. *J. Phys. Oceanogr.* **2007**, *37*, 2714–2725. [CrossRef]
87. Charnock, H. Wind stress on a water surface. *Q. J. R. Meteorol. Soc.* **1955**, *81*, 639–640. [CrossRef]
88. Janssen, P.A. Wave-induced stress and the drag of air flow over sea waves. *J. Phys. Oceanogr.* **1989**, *19*, 745–754. [CrossRef]
89. Taylor, P.K.; Yelland, M.J. The dependence of sea surface roughness on the height and steepness of the waves. *J. Phys. Oceanogr.* **2001**, *31*, 572–590. [CrossRef]
90. Oost, W.A.; Komen, G.J.; Jacobs CM, J.; Van Oort, C. New evidence for a relation between wind stress and wave age from measurements during ASGAMAGE. *Bound. Layer Meteorol.* **2002**, *103*, 409–438. [CrossRef]
91. Weber, J.E.H.; Broström, G.; Saetra, Ø. Eulerian versus Lagrangian approaches to the wave-induced transport in the upper ocean. *J. Phys. Oceanogr.* **2006**, *36*, 2106–2118. [CrossRef]
92. Polton, J.A.; Lewis, D.M.; Belcher, S.E. The Role of Wave-Induced Coriolis–Stokes Forcing on the Wind-Driven Mixed Layer. *J. Phys. Oceanogr.* **2005**, *35*, 444–457. [CrossRef]
93. Skyllingstad, E.D.; Denbo, D.W. An ocean large-eddy simulation of Langmuir circulations and convection in the surface mixed layer. *J. Geophys. Res. Oceans* **1995**, *100*, 8501–8522. [CrossRef]
94. Carniel, S.; Sclavo, M.; Kantha, L.H.; Clayson, C.A. Langmuir cells and mixing in the upper ocean. *Il Nuovo Cim.* **2005**, *28*, 33–54.
95. Polton, J.A.; Belcher, S.E. Langmuir turbulence and deeply penetrating jets in an unstratified mixed layer. *J. Geophys. Res. Oceans* **2007**, *112*. [CrossRef]
96. Tamura, H.; Miyazawa, Y.; Oey, L.-Y. The Stokes drift and wave induced-mass flux in the North Pacific. *J. Geophys. Res. Oceans* **2012**, *117*. [CrossRef]
97. Bennis, A.C.; Ardhuin, F. Comments on "The depth-dependent current and wave interaction equations: A revision". *J. Phys. Oceanogr* **2011**, *41*, 2008–2012. [CrossRef]
98. Tolman, H.L. Validation of WAVEWATCH III Version 1.15 for a Global Domain. Technical Report. *NOAA/NWS/NCEP/OMB* **2002**, *213*, 33.
99. Kahma, K.K.; Calkoen, C.H. Reconciling Discrepancies in the Observed Growth of Wind-generated Waves. *J. Phys. Oceanogr.* **1992**, *22*, 1389–1405. [CrossRef]
100. Hernandez, F.; Blockley, E.; Brassington, G.B.; Davidson, F.; Divakaran, P.; Drévillon, M.; Zhang, A. Recent progress in performance evaluations and near real-time assessment of operational ocean products. *J. Phys. Oceanogr* **2015**, *8* (Suppl. 2), S221–S238. [CrossRef]
101. Buongiorno Nardelli, B.; Tronconi, C.; Pisano, A.; Santoleri, R. High and Ultra-High resolution processing of satellite Sea Surface Temperature data over Southern European Seas in the framework of MyOcean project. *Remote Sens. Environ.* **2013**, *129*, 1–16. [CrossRef]
102. Copernicus Marine In Situ TAC. Copernicus Marine In Situ TAC quality Information Document for Near Real Time. *Situ Products (QUID SQO)* **2021**. [CrossRef]
103. Venables, W.N.; Ripley, B.D. Random and mixed effects. In *Modern Applied Statistics with S*; Springer: New York, NY, USA, 2002; pp. 271–300.
104. Cavaleri, L.; Fox-Kemper, B.; Hemer, M. Wind Waves in the Coupled Climate System. *Bull. Am. Meteorol. Soc.* **2012**, *93*, 1651–1661. [CrossRef]
105. Fan, Y.; Griffes, S.M. Impacts of Parameterized Langmuir Turbulence and Nonbreaking Wave Mixing in Global Climate Simulations. *J. Clim.* **2014**, *27*, 4752–4775. [CrossRef]

Journal of
Marine Science and Engineering

MDPI

Article

Framework for Improving Land Boundary Conditions in Ocean Regional Products

Francisco Campuzano [1,2,*], Flávio Santos [3], Lucian Simionesei [2], Ana R. Oliveira [2], Estrella Olmedo [4], Antonio Turiel [4], Rodrigo Fernandes [5], David Brito [5], Marco Alba [6], Antonio Novellino [6] and Ramiro Neves [2]

1 +ATLANTIC CoLAB, 2520-614 Peniche, Portugal
2 MARETEC/LARSyS, Instituto Superior Técnico, Universidade de Lisboa, 1049-001 Lisbon, Portugal; lucian.simionesei@tecnico.ulisboa.pt (L.S.); anaramosoliveira@tecnico.ulisboa.pt (A.R.O.); ramiro.neves@tecnico.ulisboa.pt (R.N.)
3 Hidromod Lda., 2740-278 Porto Salvo, Portugal; flavio.santos@hidromod.com
4 Barcelona Expert Centre, E08003 Barcelona, Spain; olmedo@icm.csic.es (E.O.); turiel@icm.csic.es (A.T.)
5 Bentley Systems, 1990-208 Lisbon, Portugal; rodrigo.fernandes@bentley.com (R.F.); david.brito@bentley.com (D.B.)
6 ETT People & Technology, 16153 Genoa, Italy; marco.alba@ettsolutions.com (M.A.); antonio.novellino@ettsolutions.com (A.N.)
* Correspondence: francisco.campuzano@colabatlantic.com

Abstract: The coupling of coastal or regional ocean models to hydrological models or observed data is currently an uncommon practice in operational oceanography. Though hydrological models are regarded as a powerful and useful tool for estimating the quantity and quality of freshwater running in a watershed, they fail to provide accurate results for river flow reaching the coastal area due to water-management activities occurring within the river catchment, activities such as human consumption, irrigation, storage, etc. For this reason, many coastal and regional ocean models continue to impose surface zero-salinity discharges as land boundary conditions for representing such a dynamic boundary. Moreover, river flows are based in climatologies, thus neglecting seasonal and interannual variability. To achieve those objectives, this study proposes an integrated methodology ranging from watershed models to validation in the coastal area and passing through methods and proxies for integrating the freshwater flows into regional ocean models. The main objective of this study is to explore the results obtained by using more sophisticated land boundary conditions based on the capacities of state-of-the-art hydrologic models combined with observation networks. In addition to the evaluation of the source of river-flow data, this work also explores the use of estuarine proxies based on simple modelling grids. The estuarine proxies enable the incorporation of the mixing processes that take place in estuaries into the land fluxes and obtain the plume momentum. The watershed, estuarine proxies, and ocean were modelled using the MOHID Water modelling system and evaluated in western Iberia waters. The modelling results served to illustrate the sea surface salinity extension of the Western Iberia Buoyant Plume (WIBP) during an extreme event in March 2018.

Keywords: freshwater discharges; operational oceanography; sea surface salinity; IBI region; land boundary; estuarine proxy; water continuum; numerical models

Citation: Campuzano, F.; Santos, F.; Simionesei, L.; Oliveira, A.R.; Olmedo, E.; Turiel, A.; Fernandes, R.; Brito, D.; Alba, M.; Novellino, A.; et al. Framework for Improving Land Boundary Conditions in Ocean Regional Products. *J. Mar. Sci. Eng.* 2022, 10, 852. https://doi.org/10.3390/jmse10070852

Academic Editor: Rodger Tomlinson

Received: 16 April 2022
Accepted: 1 June 2022
Published: 22 June 2022

1. Introduction

River runoff to coastal waters strongly influences local dynamics in several ways, such as, for example, modifying water stratification [1], introducing significant fluctuations in circulation patterns, and modulating the impact of upwelling events [2,3]. In the current context of a global decline of hydrometric networks [4] that were originally designed for purposes other than fulfilling the needs of coastal end-users, it is a challenge in many locations to obtain near real time river runoff values and other associated coastal water properties. For this reason, river climatologies have been commonly used as land boundary conditions in coastal and regional ocean operational models. This type of boundary condition neglects temporal river discharge variability issuing from changes in human management that may severely impact the circulating flow. Moreover, generating climatological flows for small or poorly monitored rivers can also be a challenge. On the other hand, watershed models tend to overestimate river flows, especially during dry seasons when drained water is scarce and human management activities are more intense through dam retention, irrigation, human consumption, etc. [5].

Once the freshwater fluxes have been characterised, they need to be incorporated into regional ocean models, with a horizontal grid resolution of 5–10 km. The most common method of including river discharges in operational oceanography involves directly adding the river volume with zero salinity into one or more layers of the model [6] or through a rectangular breach in the coastal wall with uniform inflow water properties [1].

Campuzano et al. [5] proposed a methodology for the off-line extraction and analysis of water fluxes and properties extracted from full-scale estuarine models in order to be integrated into regional mesoscale ocean models, such as temporal evolution [7]. In this paper, a method to replace full-scale estuarine models by simple proxy models is proposed. The use of estuarine proxies permits extending the application of the proposed methodology with a low computational cost while including most of the local tidal signal complexity (i.e., tidal prism, range, and phases) and realistic water properties.

The presented research is based on outcomes from the Copernicus Marine Service Evolution LAMBDA (Land-Marine Boundary Development & Analysis; hereafter referred to as LAMBDA) project (http://www.cmems-lambda.eu/, accessed on 15 April 2022) that generated watershed model outputs for two Copernicus Marine Environment Monitoring Service (CMEMS) Monitoring Forecasting Centres (MFCs): the Iberia Biscay Irish (IBI-MFC) and the North West Shelf (NWS-MFC). The impact of LAMBDA watershed model inputs in the IBI-MFC was recently analysed by Sotillo et al. [8]. Here, we evaluate the differences and impacts between using direct discharges and estuarine proxies on a western Iberia regional ocean model.

2. Materials and Methods

Each section of the water continuum, from the watershed to the open ocean, is reproduced with the different components of the MOHID Water Modelling System (http://www.mohid.com [9], accessed on 15 April 2022). The MOHID Water Modelling System is an open-source modular finite volume modelling system written in ANSI FORTRAN 95 with an object-oriented programming philosophy integrating several numerical programs accessible via the GitHub repository (https://github.com/Mohid-Water-Modelling-System/Mohid, accessed on 15 April 2022). MOHID Land and MOHID Water are the core numerical models.

2.1. Study Area

The largest rivers of the Iberian Peninsula other than the Ebro River discharge to the Atlantic coastal waters, draining almost two thirds of the territory along the way. In Portugal, river mouths are mostly situated in the northern part of the country, corresponding to the wettest region of the Iberian Peninsula with around 3000 mm of rain annually (Portuguese Water Atlas, https://sniamb.apambiente.pt/content/geo-visualizador, accessed on 15 April 2022).

Combined with the rain pattern, the concentration of river mouths in the Northwest Iberian coastal zone, such as the Douro, Minho, and Mondego rivers, provides the conditions required for the generation of a significant regional feature: the Western Iberia Buoyant Plume (WIBP; [10]). This is a year-round low-salinity water lens ($S < 35.8$), and it extends along the Northwest Iberian coast, influencing, among other things, ocean productivity (i.e., [11,12]) and larva and egg dispersal (i.e., [1,13–16]).

2.2. Watershed Modelling

The MOHID Land model is the hydrological model used to estimate river flow and water temperature for the study region. It is a physically based, spatially distributed, continuous, variable time-step model for inland water property cycles [17]. The model is based on finite volumes organized into a structured grid, rectangular in the horizontal plane, and z-level geometry in the vertical direction. It includes four compartments or media (atmosphere, porous, soil surface, and river network). Water moves through the mediums based on mass- and momentum-conservation equations. The atmosphere compartment provides varying space and time surface boundary conditions (precipitation, solar radiation, wind, etc.). The surface layer is described by a 2D horizontal grid, while the porous media consists of a 3D domain of variable-layer thickness that shares the horizontal grid with the surface layer. The river network is a 1D domain defined from the digital terrain model (DTM), with watershed reaches linking surface-cell centres. Fluxes are computed through the finite volumes' sides, and state variables are computed at the cell centres to ensure transported properties are conserved. The model uses a dynamic time step that increases during dry periods and reduces as water fluxes increase (e.g., during rain events). The MOHID Land model can simulate single-catchment or multi-catchment domains. Although MOHID Land can be configured to simulate water properties such as oxygen, suspended sediment, and nutrient concentrations, in the LAMBDA project the simulations focused on obtaining river flows and water temperature.

The main objective of the watershed modelling simulations was to produce regional-scale water budgets covering the land boundary of the two CMEMS MFCs. For the LAMBDA project, the drainage basins draining the study area were divided into ten domains using regular grids with a horizontal resolution of 0.05° (Figure 1). Only the Loire (France) and the Severn (UK) rivers were simulated with higher resolutions due to developments during the Hazrunoff project (http://www.hazrunoff.eu/, accessed on 15 April 2022). In general, hydrologic models with a horizontal scale of 1–9 km resolution can obtain similar results and perform well for storm events [18]. The watershed and land-use grids were populated with the EU 30 m Digital Elevation Model (DEM) and Corine Land Cover 2012 obtained from the Copernicus Land Monitoring Service (https://land.copernicus.eu/, accessed on 15 April 2022), respectively. 3D soil hydraulic properties were obtained from Tóth et al. [19], while channel cross-sections were defined with the database from Andreadis et al. [20]. Each domain was simulated for the 2008–2019 period with meteorological conditions calculated with the ERA5 reanalysis product (horizontal resolution 31 km) from the Copernicus Climate Change Service (https://climate.copernicus.eu/, accessed on 15 April 2022).

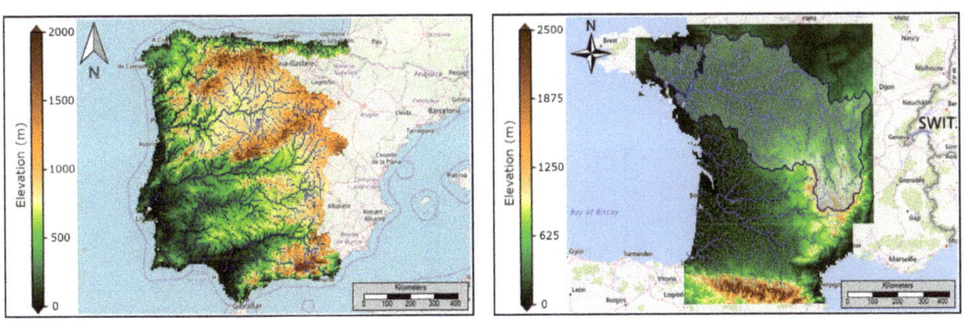

(a) (b)

Figure 1. *Cont.*

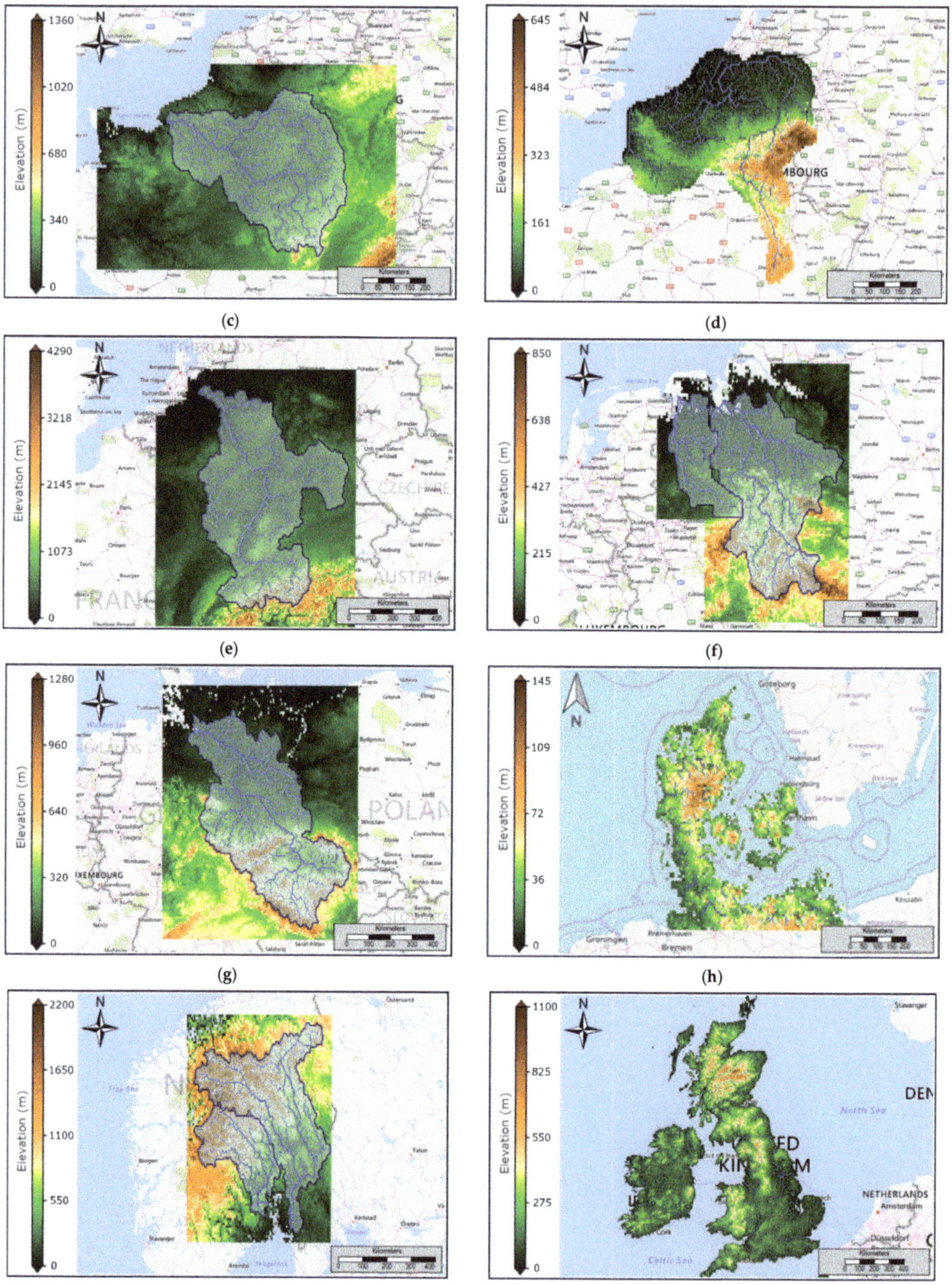

Figure 1. Topography and drainage network of the LAMBDA project watershed domains. The LAMBDA watershed domains are: (**a**) Western Iberian Peninsula domain; (**b**) Western France domain; (**c**) Seine River domain; (**d**) Somme, Scheldt, and Meuse rivers domain; (**e**) Rhine River domain; (**f**) Northwest Germany domain; (**g**) Elbe River domain; (**h**) Denmark domain; (**i**) Southern Norway domain; (**j**) United Kingdom and Ireland domain.

2.3. Estuarine Modelling

Since regional ocean models commonly impose direct river discharges into coarse grids with horizontal resolutions of several kilometres, an estuarine proxy was developed to include the tidal signal complexity, thus providing more realistic readings of salinity concentration and of discharged volumes reaching the coastal area. This proxy was adapted from a more complex approach where full scale estuarine models were used to estimate transport fluxes and related water properties [7].

A simple and scalable MOHID Water application was designed to represent estuaries schematically and to export the methodology where a full-scale estuary model is not available. The advantages of this type of application include low computing costs and the capacity to combine realistic open ocean and land boundary conditions with local tides. The proxy can calculate a realistic tidal mixing, considering each estuary's tidal prism, and different tidal cycles such as spring-neap and ebb-flow periods. Each proxy calculates cross-section water velocity that can later serve to impose momentum in the ocean model. Moreover, using a full numerical model to simulate in a simple grid, instead of tidal mixing empirical equations, allows for a future increase in complexity, i.e., adding atmospheric boundary conditions, biogeochemical modelling, etc.

The LAMBDA estuarine model proxy consists of a regular domain of 12 × 3 cells where the estuary is represented by 10 grid cells aligned in any cardinal direction plus a cell for the ocean open boundary conditions and the land boundary (Figure 2). Configuring each estuarine proxy domain requires some basic estuary geometry properties:

- Estuary depth: maximum, average, and minimum depth to populate the bathymetry grid cells;
- Estuary length: to configure the along-estuary grid cell size;
- Total area of the estuary: to estimate across-estuary cell size.
- Estuary mouth location: to define the grid origin and force of local tidal components;
- Estuary mouth orientation: to define the momentum velocity component (u or v) and its orientation;
- Ocean water properties: to force the open ocean boundary conditions.

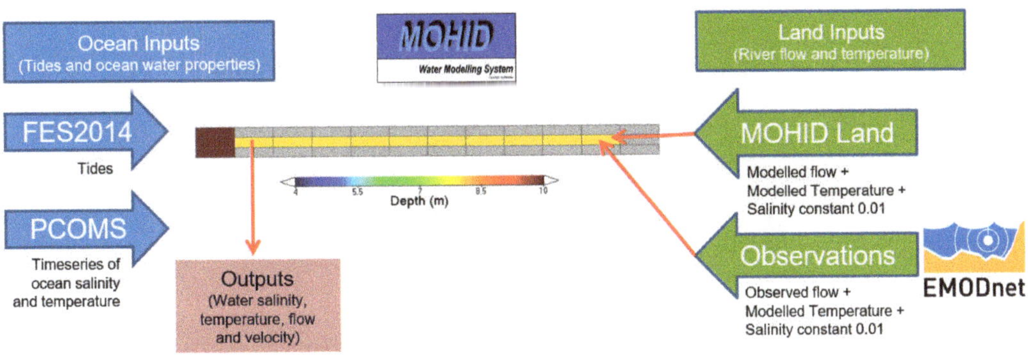

Figure 2. MOHID Water Estuarine schematic model design used for simulating an estuarine proxy. The domain consists of a 12 × 3 grid that receives land inputs (river flow, temperature and salinity) and ocean inputs from tides and regional ocean models (water level, currents, temperature, and salinity) to produce outputs at the estuarine mouth, outputs such as salinity, temperature, flow, and velocity.

At the open ocean boundary, the model generates water levels, using FES2014 tidal constituents, and receives ocean properties such as surface salinity and water temperature. Each estuarine proxy simulates its corresponding tides, supported by its georeferenced location, and using the FES2014 global tidal model [21]. Ocean-water properties at the open boundary can be defined as constant or time-varying with either timeseries or more complex 2D/3D fields from ocean global or regional models. In the innermost cell, river flow, temperature, and zero salinity is imposed.

Each estuarine proxy produces timeseries of flow, velocity (u or v according to estuary mouth orientation), and water properties estimated at its outer estuarine cell. Water velocity serves to impose momentum on the estuarine plume in the regional ocean model. In contrast with river discharges, modelled estuarine flows have positive and negative values according to the tidal phase (ebb and flood). The receiving model should be capable of handling negative discharges to fully include this property as a land boundary condition. The proxy outputs are independent of the downstream ocean model and for this reason can be applied to any regional or global ocean model.

To evaluate the performance of the estuarine proxy, simulations using this method were applied to the main six estuaries in western Iberia (from South to North): Guadalquivir, Sado, Tagus, Mondego, Douro, and Minho (Figure 3). Fluxes, temperature, salinity, and velocity timeseries were obtained for the period of May 2017–December 2018. The geomorphology information used to configure each of these estuary proxies is included in Table 1.

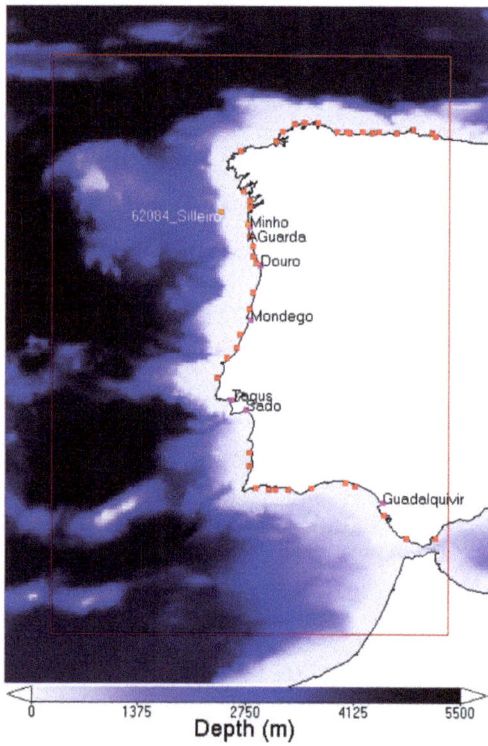

Figure 3. PCOMS regional ocean operational system domains and bathymetry. The full domain corresponds to the WestIberia 2D Domain where tide is imposed. The red box indicates the outer limit of the Portugal 3D Domain. Pink dots indicate the location of the six main estuaries in western Iberia. Red dots indicate the location of extra 45 rivers implemented with direct discharges. The locations of Silleiro and A Guarda monitoring stations are indicated by orange dots.

Table 1. Proxy configuration for the main six estuaries in western Iberia ordered from South to North. The Tagus estuary is the only proxy configured with variable depth along its channel.

Estuary	Mouth Orientation	Cell Length (Degrees)	Cell Width (Degrees)	Ocean Depth (m)	Ocean Salinity	Estuary Depth (m)	Longitude Mouth (° W)	Latitude Mouth (° N)
Guadalquivir (ES)	West	0.11000	0.00475	20	36.2	10.0	6.440	36.785
Sado (PT)	West	0.08000	0.02000	10	36.0	8.0	8.930	38.470
Tagus (PT)	West	0.05000	0.06200	25	36.0	2.0–20.0	9.420	38.620
Mondego (PT)	West	0.00150	0.04500	6	36.0	2.0	8.880	40.143
Douro (PT)	West	0.02200	0.00272	8.2	36.0	8.2	8.700	41.140
Minho (PT-ES)	West	0.04000	0.00575	6	36.0	2.0	8.885	41.860

Two proxy scenarios were set according to observed and modelled land boundary conditions to assess human management impact on river flows. Observed river flows for the Douro, Guadalquivir, Mondego, and Tagus rivers were obtained from EMODnet physics near real time data (NRT). These were the only rivers with data available for the analysed period when the simulations were made. Water temperature was forced in all proxies with watershed modelling results, since this variable is rarely monitored continuously. Salinity was considered as freshwater with zero salinity in all land scenarios. At the open ocean boundary, each proxy received surface temperature and salinity values according to the locations of its mouth from a regional ocean model without land inputs, and the values are described in the next section.

2.4. Ocean Modelling

The LAMBDA boundary products were implemented as proof of concept (PoC) in an updated version Portuguese Coast Operational Modelling System (hereafter referred as PCOMS, [22,23]) that covers the western Iberia regional ocean. Simulations were made for the period of October 2017–December 2018 that includes the extreme rain event of late March 2018.

The PCOMS system is a 3D full baroclinic hydrodynamic and biogeochemical regional ocean operational model application that uses MOHID Water as its numerical core. It is composed of two nested domains: WestIberia (2D) and Portugal (3D) covering the Iberian Atlantic coast and its contiguous ocean populated with bathymetric information derived from the EMODnet Bathymetry portal (Figure 3; https://www.emodnet-bathymetry.eu/, accessed on 15 April 2022). The WestIberia domain covers the area limited by specific latitudes (33.48° N, 45.90° N) and longitudes (4.20° W, 13.50° W), resulting in a grid of 207 × 155 cells with a maximum depth around 5600 m. The Portugal domain covers the area comprised of specific latitudes (34.38° N, 45.00° N) and longitudes (5.10° W, 12.60° W), resulting in a grid of 177 × 125 cells and a maximum depth around 5300 m. Both domains use constant horizontal spatial resolution of 0.06°, resulting in 6.7 × 5.2 km cells. The Portugal domain is located at the centre of the WestIberia grid (Figure 3). Vertically, the Portugal domain uses a hybrid discretisation with near-surface, variable-thickness sigma layer with 7 levels, increasing from 1 m at the surface to 8.68 m of depth, above a layer with 43-z-levels increasing in thickness towards the bottom; this domain is based on the Mercator-Océan PSY2V4 vertical geometry [23].

Tides in the current configuration are also forced with FES2014 harmonic components along the WestIberia domain boundary. The Portugal domain receives tides from the WestIberia domain and uses CMEMS Global Ocean 1/12° Physics Analysis and Forecast updated Daily product (hereafter referred as CMEMS-Global) as initial and boundary condition for water levels, currents, temperature, and salinity. CMEMS-Global fields are relaxed in the first 10 cells of the open boundary, while the inner area the model runs free, without assimilation; these parameters rest on the assumption that the open ocean boundary condition does not affect the river-influenced areas. At the atmospheric boundary, the Portugal domain was forced by hourly results from a MM5 model application (Meteorological Model 5; [24]) based on two nested grids with a horizontal resolution of

27 and 9 km, respectively, implemented by the IST meteorological group (http://meteo.tecnico.ulisboa.pt/ [25], accessed on 15 April 2022).

3. Modelling Results and Validation

3.1. Watershed Modelling

Each modelling domain was independently calibrated, which resulted in flow and temperature timeseries for the main 54 rivers discharging into the European Atlantic Ocean and the North Sea for the period of 2008–2019. To achieve a more comprehensive water budget entering the coastal areas, 70 extra discharges for Western Iberia and 364 extra discharges for the Ireland and UK region were obtained from the domains. However, these extra rivers could not be fully validated, since resolution may be too coarse to accurately reproduce out-river flowrates.

Observed data for each river was collected from several data sources including EMODnet physics NRT river database (https://portal.emodnet-physics.eu/, accessed on 15 April 2022) for recent data and the Global Runoff Data Base (GRDB; https://www.bafg.de/GRDC/EN/01_GRDC/13_dtbse/database_node.html, accessed on 15 April 2022) for historical data. Nevertheless, the lack of available observations for some catchments or presented incomplete datasets for the simulation period hampered efforts to validate and calibrate some domains. Moreover, the available hydrologic stations are in some cases located far from the coastal area, so that validation was only possible for the upper part of the catchment. Figure 4 shows the geographic distribution of the coefficient of determination between observed and modelled data for the stations used for validation in this study.

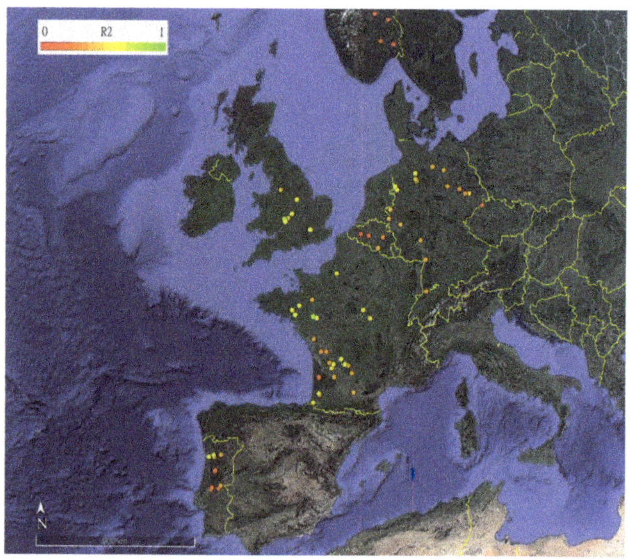

Figure 4. Coefficient of determination (R^2) between observed and modelled river flow for the stations with available data for the period 2008–2019.

A well-calibrated and -configured watershed model will face difficulties in attempting to accurately simulate river runoff, since river flow is highly controlled by reservoirs managed by human behaviour. Watershed numerical models calculate natural river flow, and as such, their results tend to be closer to the observed values when the degree of human intervention is low [17].

To illustrate the results, the analysis of two large rivers in western Iberia, the Douro and Tagus rivers, are presented. The Douro and Tagus river mouths are the longest rivers discharging in the Atlantic Ocean, at 897 and 1007 km long respectively. They receive waters from the largest drainage areas in the Iberian Peninsula, with 98,400 and 80,000 km^2 for the Douro and Tagus respectively. The larger Douro catchment is located in the wettest region of the Iberian Peninsula and with a resultant average flow of $\approx$700 m^3 s^{-1}, an average that is higher than the Tagus River, whose average flow is $\approx$440 m^3 s^{-1}. The river mouths are separated by 275 km.

Another subject relevant to the watershed validation is the precision of river flow observations, especially since many hydrometric stations use rating curves (RC) to transform water levels into river flow. This method may introduce some errors in the context of extreme events, since RC must be extrapolated outside of the observed range [26]. Such is the case for the Almourol station in the Tagus River where flow is calculated from water levels with a rating curve. On the other hand, the Douro River monitoring station closest to the coastal area (Albufeira da Crestuma) is a hydroelectric power plant, which enables more precise observation. In both rivers, the numerical model reproduces the main flow peaks in the period timeseries; however, it presents higher average flow and lower extreme peaks (Figure 5). This excess of freshwater may be caused by human management, errors, or uncertainties in the meteorological and watershed model and observations that translate into a low coefficient of determination of around 0.6 and 0.3 for the Douro and Tagus rivers, respectively.

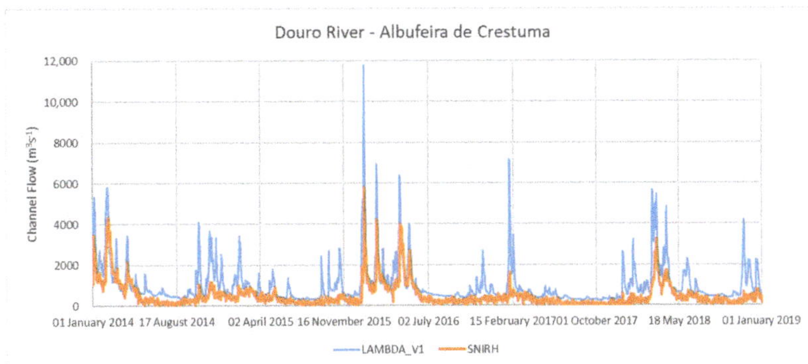

(a)

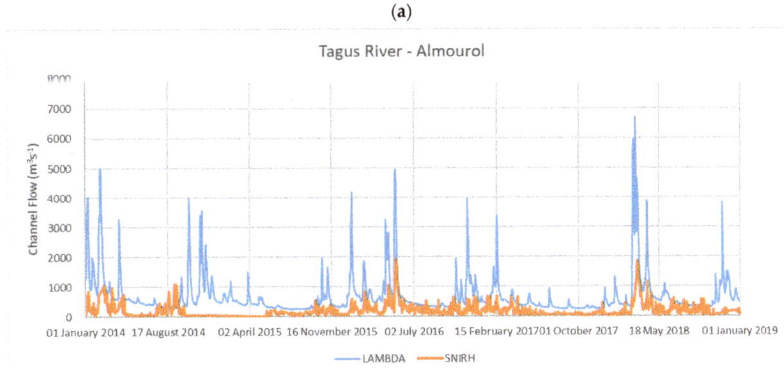

(b)

Figure 5. Observed (orange line) and modelled (blue line) river flow for the period 2014–2018 used for validation in (**a**) Douro River at Albufeira de Crestuma station; (**b**) Tagus River at Almourol station (Source for observed data: Portuguese Environmental Agency APA).

3.2. Estuary Modelling

To evaluate the differences resulting from using observed versus modelled river flow in the coastal area, the estuarine proxies defined for the Douro and the Tagus rivers were forced with both datasets for the period of May 2017–December 2018. These estuaries present large geomorphological differences (Table 1) relative to each other that affects their tidal prism size and substantially influences water volume and other properties reaching the open ocean (Figure 6). Mean maximum instant flow is around 1–1.5·10^3 m^3 s^{-1} and 40–70·10^3 m^3 s^{-1} for the Douro and Tagus estuaries, respectively. These values are similar to those found by Campuzano et al. [7] who used full-scale estuarine models. Larger volume fluxes do not imply larger momentum at the estuarine mouth. For example, the Douro estuary presents slightly larger instant velocities than the Tagus estuary that has larger estuarine flows (Table 2). During extreme events, as in March 2018, Tagus flow and velocity were barely impacted by the river discharge while the Douro estuary was highly affected by this type of event, with flows and velocities substantially affecting the plume momentum, shape, and extension.

The differing geomorphologies are also important for the water properties reaching the coastal area. When forced with the observed values of the river flow, the average salinity at the mouth of the Tagus estuary is around 33 salinity units, while in the Douro the average salinity is only 12.5 salinity units. In terms of the S-value range, the Douro has a greater variability than the Tagus estuary. During extreme conditions, freshwater conditions can almost be reached in the Douro mouth, as was also observed with the full-scale estuary model [7]. The Douro estuary temperature range is also widely impacted by the river's seasonal temperature evolution.

Proxies forced with LAMBDA watershed modelling results lead to fresher water for both estuaries. Nevertheless, salinity trends are well represented for both estuaries. The Douro estuary proxy, due to its smaller size, is more sensitive to differences in the imposed river flow.

3.3. Ocean Modelling

Six land boundary configurations were defined to evaluate the impact of river influence on the ocean waters of western Iberia. Meteorological and open ocean boundary conditions were the same for all the scenarios. The scenarios are:

- **Reference**: model running continuously, since 2017, without any land input thus considered the baseline to evaluate the rivers' impact;
- **Climatology**: direct surface river discharge of climatological values for flow and temperature with only rivers included in IBI-MFC (Douro, Guadiana, Guadalquivir, Minho, Mondego, Tagus);
- **LAMBDA**: the same rivers and methods as Climatology scenario with river flow and temperature obtained from the LAMBDA Watershed product;
- **Observed**: the same rivers and methods as Climatology but observed flow for Douro, Guadalquivir, Mondego and Tagus. River temperature from the LAMBDA watershed product;
- **Complete Observations**: six main estuaries (Minho, Douro, Mondego, Tagus, Sado and Guadalquivir) corrected by estuarine proxy. Also, an additional 45 rivers (Figure 3) discharged directly with the LAMBDA modelled flow and temperature with constant salinity S 25.
- **Complete LAMBDA**: the same as the Complete Observations scenario with all river information obtained from the LAMBDA watershed model results.

Table 2. Basic statistics for the outputs from the Tagus and Douro proxies when forced with observed flows (Obs.) and when forced with watershed models (Model). * Values were calculated with the absolute values.

		Tagus			Douro		
		Max.	Min.	Avg.	Max.	Min.	Avg.
Observations	Flow (m^3 s^{-1})	68,231.84	−68,663.64	29,264.75 *	3936.56	−1307.51	654.20 *
	Velocity (m s^{-1})	0.45	−0.46	0.19 *	0.42	−1.30	0.21 *
	Temperature (°C)	20.08	12.85	16.61	26.80	1.63	15.12
	Salinity	35.62	26.47	33.84	34.93	0.017	12.50
Model	Flow (m^3 s^{-1})	70,483.38	−68,614.38	29,255.41 *	6535.13	−1017.03	1034.52 *
	Velocity (m s^{-1})	0.45	−0.47	0.19 *	0.33	−2.21	0.33 *
	Temperature (°C)	20.51	9.68	16.53	27.23	1.22	14.81
	Salinity	35.15	5.05	30.71	30.12	0.01	3.39

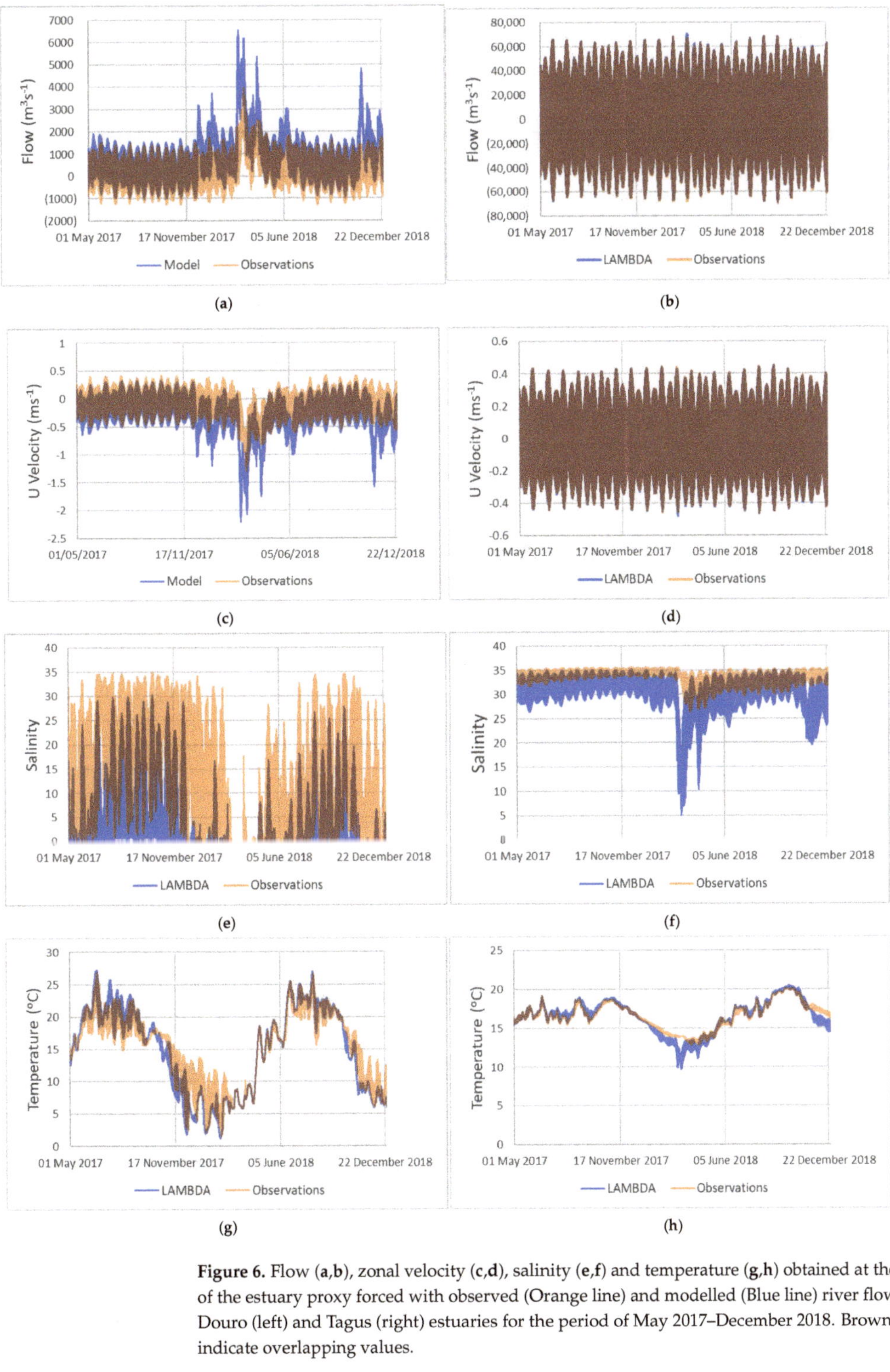

Figure 6. Flow (**a**,**b**), zonal velocity (**c**,**d**), salinity (**e**,**f**) and temperature (**g**,**h**) obtained at the mouth of the estuary proxy forced with observed (Orange line) and modelled (Blue line) river flows in the Douro (left) and Tagus (right) estuaries for the period of May 2017–December 2018. Brown colours indicate overlapping values.

3.3.1. PCOMS Validation

The PCOMS model application was forced with each land boundary condition scenario. The results were analysed in order to better understand the response of the numerical model's different land forcings. The analysis focused on the first quarter of 2018, corresponding to a wet season period, and includes an extreme rain event in late March. The obtained modelling results were compared with in-situ operational observations, remote sensing products, and state-of-the-art regional and global operational ocean models for the same study area.

Results from each land boundary scenario were compared with salinity and temperature values recorded by the Silleiro buoy. This buoy, belonging to the Puertos del Estado monitoring network, is located 50 km offshore of northwestern Iberia (coordinates: 42.119° N, 9.440° W; Figure 3). When high river discharges coincide with upwelling prevailing wind conditions, northern winds for this coastal stretch, the WIBP signature can be detected in the salinity and temperature records [23].

Observed surface salinity records remained almost constant in January 2018 at the Silleiro buoy. However, direct discharge scenarios such as Observed and LAMBDA exhibited a salinity decrease around 8 January. This was not reproduced by the scenarios using the estuarine proxy: Complete Observations and Complete LAMBDA (Figure 7). The complete scenarios receive land boundary salinity values that have been previously mixed in the estuarine proxy, while direct discharge scenarios are forced using zero salinity conditions (Figure 6). Moreover, estuarine fluxes have positive and negative fluxes while the direct discharge scenarios are releasing continuously lower salinity discharges.

To compare the present methodology results with state-of-the-art operational models, timeseries from CMEMS-Global and IBI-MFC were added to this work scenarios (Figure 7). The IBI-MFC is the regional ocean model of the European Commission Copernicus Programme based on a NEMO model application run at a horizontal resolution of 0.028° (≈2.40 km). This model implementation, operated by Puertos del Estado and Mercator-Océan, includes high frequency processes (i.e., tidal forcing, surges and high frequency atmospheric forcing, freshwater river discharge, etc.). CMEMS models use land boundary conditions drawn from monthly river climatology, conditions that are directly discharged at the surface layer.

In late February, another salinity decrease was observed but overestimated, mainly in direct discharge scenarios. The Climatology scenario presents lower salinity than the Observed scenario, due to the Douro River climatology flow for February; 1140 $m^3\ s^{-1}$ is much larger than observe values for February 2018, which only reached 264 $m^3\ s^{-1}$.

During the extreme rain event in late March, salinity values decreased by more than 2.5 salinity units and its signal lasted more than 10 days with its main peak on 22 March. The LAMBDA scenario is the direct discharge scenario that best represented this event, as the excess of freshwater estimated by the watershed model for the Douro River, two times more than observed, compensates the lack of other river sources. From this result, we concluded that it is essential to have the right freshwater budget draining in the study area to represent accurately this type of event. In the LAMBDA scenario, only two rivers, Minho and Douro, are discharging close to the buoy area, while in reality there are many other smaller watersheds that can contribute to generate this large salinity signature. CMEMS-Global results clearly underestimate the largest freshwater event, while IBI-MFC obtains good results for the extreme event. It does however overestimates other earlier events.

When the estuarine proxy is used, additional river contributions are needed to achieve similar salinity changes. In the scenarios using the proxy, Complete Observations and Complete LAMBDA, a constant salinity of 25 salinity units for the other rivers was considered as a compromise to the average salinity from different types of estuaries (Table 2). Nevertheless, this approach still seems very conservative, and salinity did not decrease as observed. It is probable that estuaries in the area have a geomorphology similar to the Douro estuary and may contribute to lower salinity values. Further analyses and experiments are needed to evaluate how minor rivers could be included and parameterised.

A possible solution to explore in future works is to implement estuarine proxies for the smaller rivers.

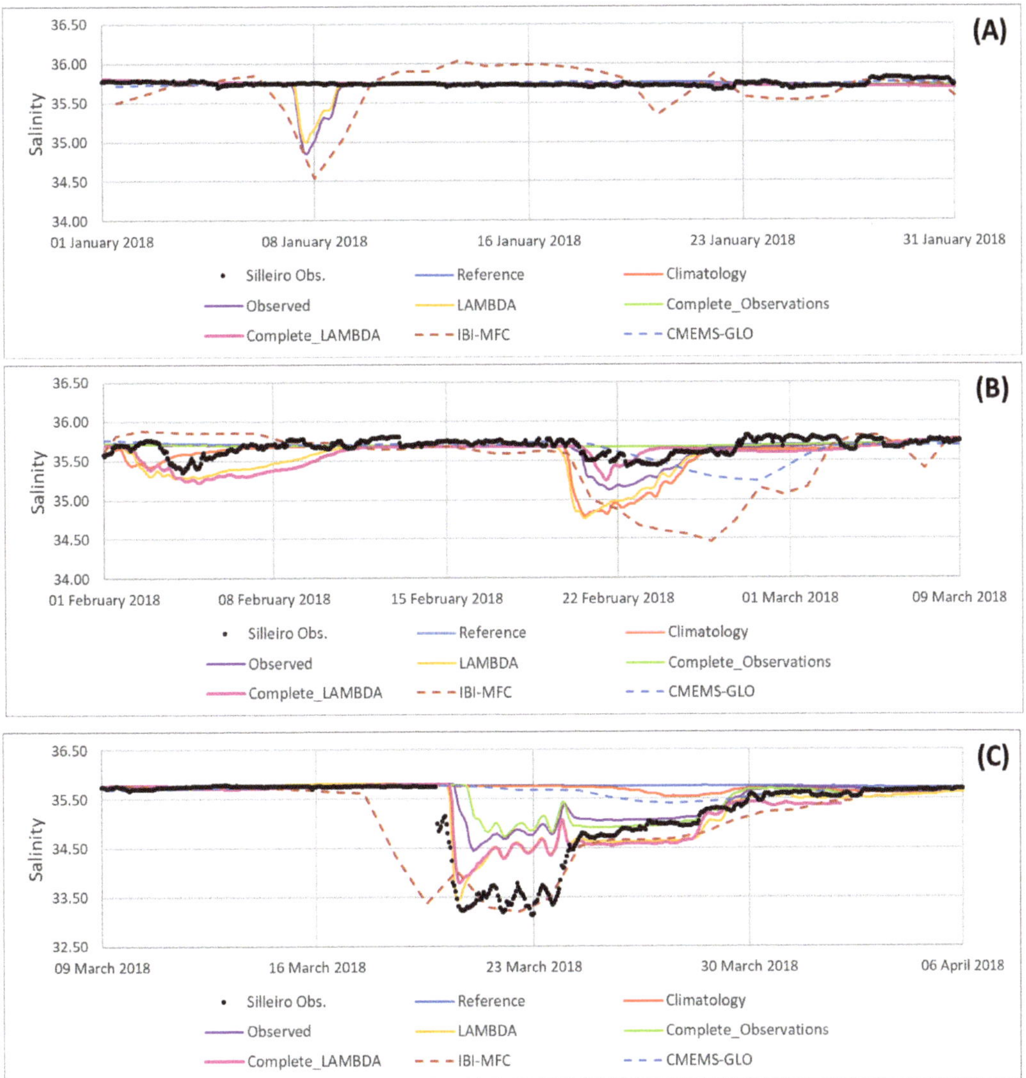

Figure 7. Sea surface salinity observed (black dots) and modelled at Silleiro buoy during the first quarter of 2018. PCOMS modelling scenarios include: no discharge simulations (Reference, Blue line), with direct discharge (Climatology (red line), Observed (purple line), and LAMBDA (yellow line)) as well as with the proxy for the main six estuaries plus 45 direct discharges (Complete Observations (green line) and Complete LAMBDA (pink line)). Surface salinity from CMEMS-Global (dashed red line) and IBI-MFC (dashed blue line) regional product are also included. The analysed was divided into three sections (**A**) (January), (**B**) (February and first week of March), and (**C**) (rest of March and first week of April) for clarity. See Section 3.3 for more configuration details.

Regarding sea surface temperature (SST), modelling results tend to represent quite accurately the main trends during the analysed period (Figure 8). As with salinity, January temperature drops a great deal in the direct discharge scenarios (Observed, LAMBDA and Climatology). However, during the extreme event, the scenarios using LAMBDA-simulated temperature were the only scenarios capable of reproducing the temperature drop. The statistical analysis of the surface temperature at this location (Table 3) showed that any scenario including rivers improved the estimation of SST with the only exception being the Climatology scenario.

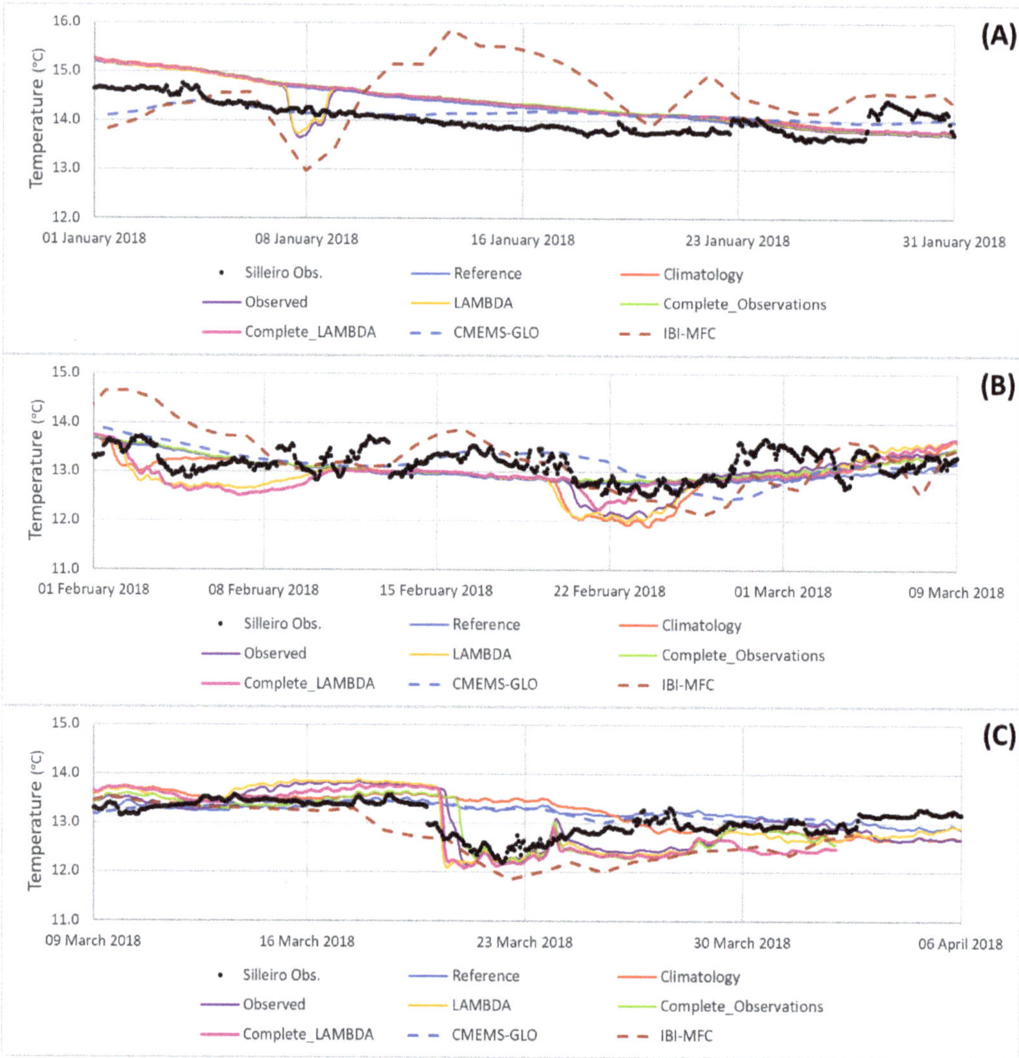

Figure 8. Sea surface temperature (black dots) and modelled at the Silleiro buoy during the first quarter of 2018. PCOMS modelling scenarios include: with no discharge (Reference, Blue line), with direct discharge (Climatology (red line), Observed (purple line), and LAMBDA (yellow line)) and with the proxy for the main six estuaries plus 45 direct discharges (Complete Observations (green line) and Complete LAMBDA (pink line)). Surface salinity from CMEMS-Global (dashed red line) and IBI-MFC (dashed blue line) regional product are also included. The analysed was divided into three sections (**A**) (January), (**B**) (February and first week of March) and (**C**) (rest of March and first week of April) for clarity. See Section 3.3 for more configuration details.

Table 3. Surface salinity and temperature coefficient of determination (R^2), Root Mean Square Error (RMSE) and Mean Absolute Error (MAE) between the observations of Silleiro buoy and each PCOMS configuration using LAMBDA boundary conditions scenario for the period of 1 January–4 January 2018. The Complete scenarios are the only ones using the proxy.

	Property	Reference	Climatology	Observed	LAMBDA	Complete Observations	Complete LAMBDA
R^2	Salinity	0.07	0.00	0.65	0.72	0.61	0.78
	Temperature	0.71	0.70	0.80	0.79	0.81	0.80
RMSE	Salinity	0.52	0.52	0.32	0.26	0.35	0.23
	Temperature	0.36	0.42	0.36	0.41	0.36	0.40
MAE	Salinity	0.27	0.27	0.10	0.07	0.12	0.06
	Temperature	0.13	0.18	0.13	0.17	0.13	0.16

To evaluate the possible impact of river discharges on the coastal hydrodynamics, the meridional velocity in the vicinity of A Guarda (Figure 3), located in the Minho estuary mouth and around 80 km north of the Douro mouth, was obtained for the Reference scenario, with no river inputs, and the Complete Observations scenario. The Complete Observations scenario achieved a meridional velocity up to three times that of the Reference scenario. Figure 9 shows that during the highest peak of river discharges (i.e., Douro River reached values around 3200 $m^3 s^{-1}$), the Reference scenario presented values around 0.5 $m s^{-1}$ while more realistic land boundary scenarios can reach maximum values around 1.5 $m s^{-1}$.

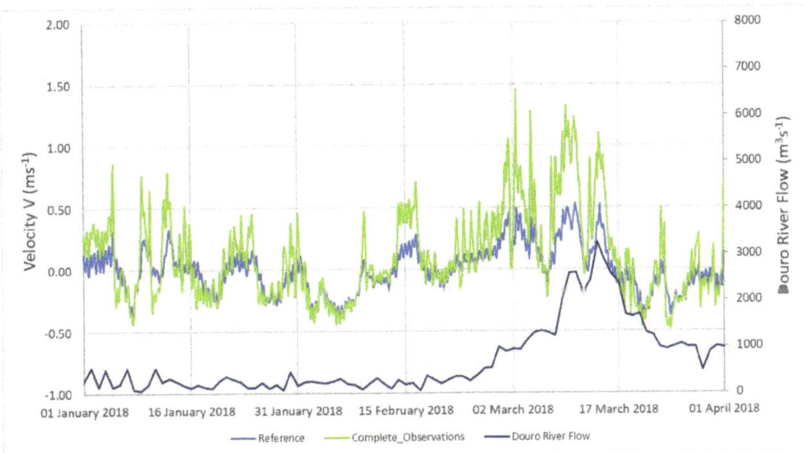

Figure 9. Surface modelled meridional velocity at the A Guarda station for the PCOMS reference scenario (no river discharge) and the Complete Observations scenario boundary conditions for the first quarter of 2018. Douro River observed flow (source: Portuguese Environmental Agency –APA) is represented in the right axis.

Earth Observations (EO) products can play an important role for model validation. They provide a spatial and temporal coverage that yields complete and verified information obtained by the numerical models. The remote sensing product chosen to evaluate the sea-surface temperature (SST) of the PCOMS modelling results is the Multi-scale Ultra-high Resolution (MUR). The MUR version 4.1 used in this study (http://dx.doi.org/10.5067/GHGMR-4FJ04, accessed on 15 April 2022) provides daily SST estimates on a global 0.01° × 0.01° grid and features the 1-km resolution MODIS retrievals, which are fused with AVHRR GAC, microwave, and in-situ SST data by applying internal correction for relative biases among the data sets [27]. In Figure 10, SST- and PCOMS-model results forced with LAMBDA Complete Observations scenario display similar structures and values for western Iberia during the extreme rain event peak. In the image, the colder temperature signature of the Douro, Tagus, and Guadalquivir rivers' individual plumes as well as other smaller rivers can be easily identified.

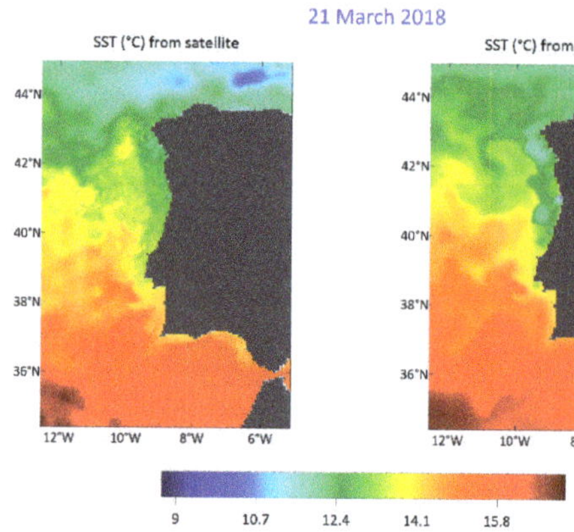

Figure 10. Sea surface temperature for 21 March 2018 from the Multi-scale Ultra-high Resolution (MUR; Left) and PCOMS with LAMBDA Complete Observations boundary conditions. Basic comparison statistics: Average PCOMS SST: 14.287; Average MUR SST: 13.992; Coef. Correlation: 0.929; Bias: 0.295 and RMSE: 0.537.

Finally, modelling results were compared with Earth Observed (EO) salinity-derived products developed in the context of the LAMBDA project [28]. This new SMOS SSS global product was developed with the main goal of capturing the coastal processes. This L4 SMOS SSS product allows downscaling the salinity by using as a template the Sea Surface Temperature (SST) at 0.05° from OSTIA. EO and numerical modelling results for the extreme event of late March 2018 present similar spatial structures and intensities (Figure 11). Though using satellite-derived SSS still has many drawbacks, i.e., low horizontal resolution to capture some coastal variability, this comparison indicates a promising evolution of these kinds of products for data analysis and model validation for coastal areas.

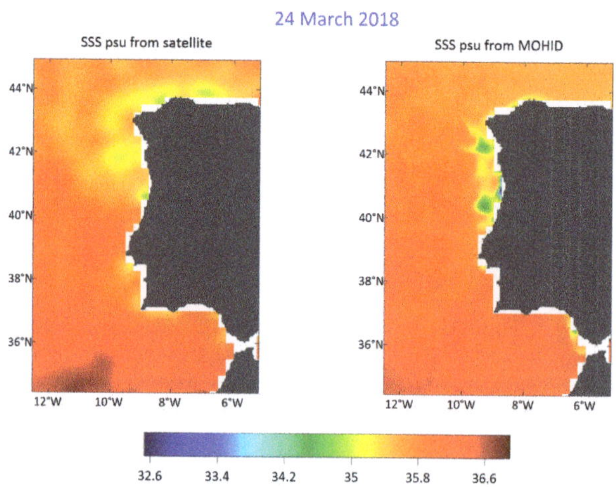

(a)

Figure 11. *Cont.*

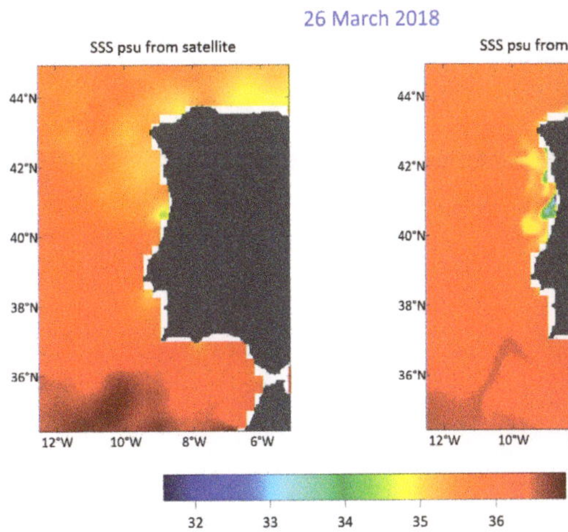

(**b**)

Figure 11. Daily Sea surface salinity for 24 March 2018 (**a**) and 26 March 2018 (**b**) from the BEC-SMOS Global L4 V2 (Left) and surface layer of PCOMS model with LAMBDA Complete Observations scenario (Right). Basic comparison statistics for (**a**) are: Average PCOMS SSS: 35.936; Average SMOS SSS: 35.892; Coef. Correlation; 0.702; Bias: 0.045 and RMSE: 0.218. Basic comparison statistics for (**b**) are: Average PCOMS SSS: 35.929; Average SMOS SSS: 35.884; Coef. Correlation; 0.676; Bias: 0.045 and RMSE: 0.275.

3.3.2. Scenario Analysis

Each of the Ocean scenarios was evaluated in terms of WIBP low salinity extension for two periods: (i) considering average values for a typical wet month (February 2018; Figure 12) and during the peak of an extreme rain event (21 March 2018; Figure 13).

During February 2018 (Figure 12), the Climatology, LAMBDA, and Complete LAMBDA scenarios exhibited the greatest extension of the West Iberia Buoyant Plume (WIBP), while scenarios using river observations (Observed and Complete Observations) had their plumes more confined near the coastal area. The latter group also exhibited a relatively reduced Tagus estuary plume (latitude 38.6). The scenarios using the estuarine proxy (Complete scenarios) reduced the extension of the plume relative to the direct discharges' scenarios, even when these included an additional 45 rivers with a constant value of 25 salinity units.

A similar pattern of results is observed during the extreme rain event for each modelling scenario. Figure 13 represents the mean SSS for the 21 March 2018, a day after the salinity minimum was registered at the Silleiro buoy. Though the Climatology scenario appears to have a similar spatial distribution during the event, it does not reach the minimum temperature and salinity values displayed in Figure 7. In addition, the Tagus estuary plume appears to be overestimated by the Climatology and LAMBDA scenarios when compared to the Observed scenario. The effect of the proxy in the Tagus estuary is clearly noticed in the Complete Observed scenarios where its impact is reduced.

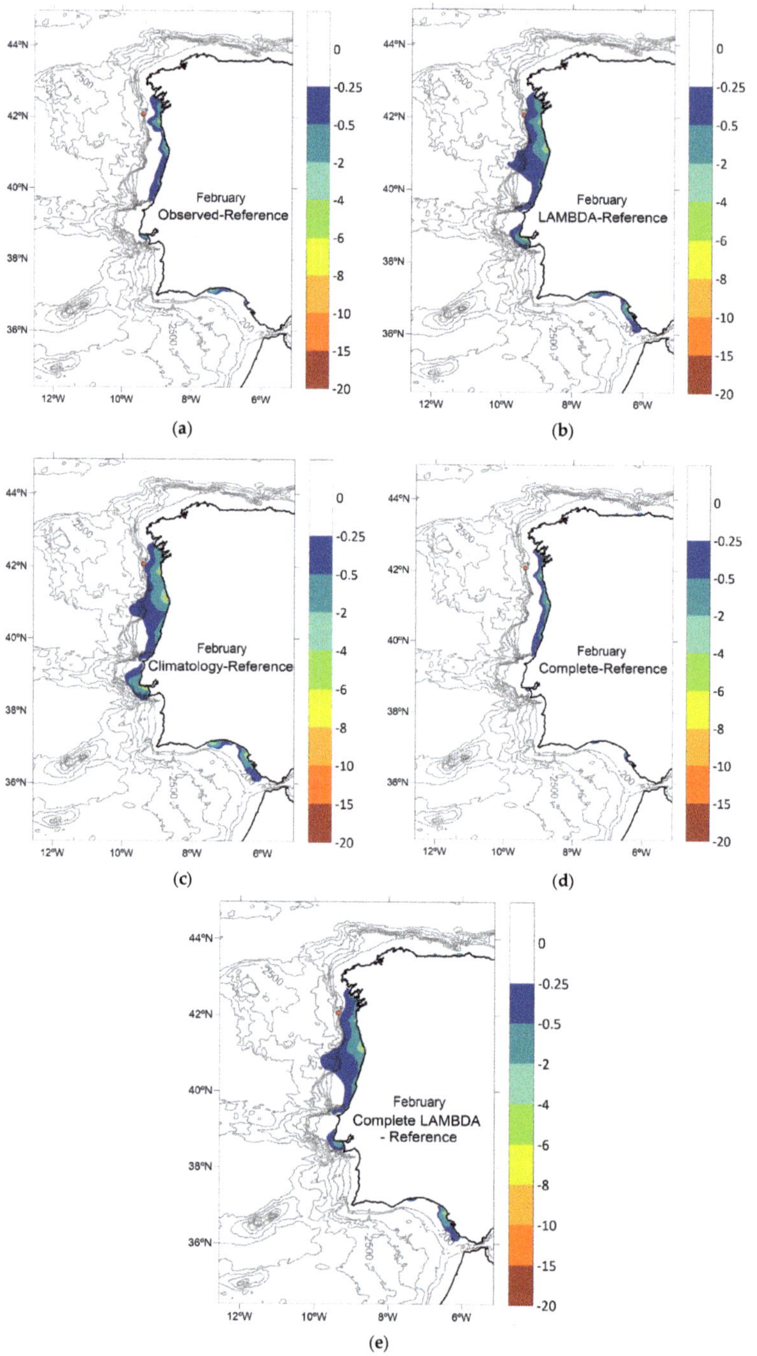

Figure 12. Surface salinity difference during February 2018 between the reference scenario with each LAMBDA land boundary scenarios (**a**) Observed; (**b**) LAMBDA; (**c**) Climatology; (**d**) Complete Observations; (**e**) Complete LAMBDA. Salinity difference between 0 and −0.5 is not represented.

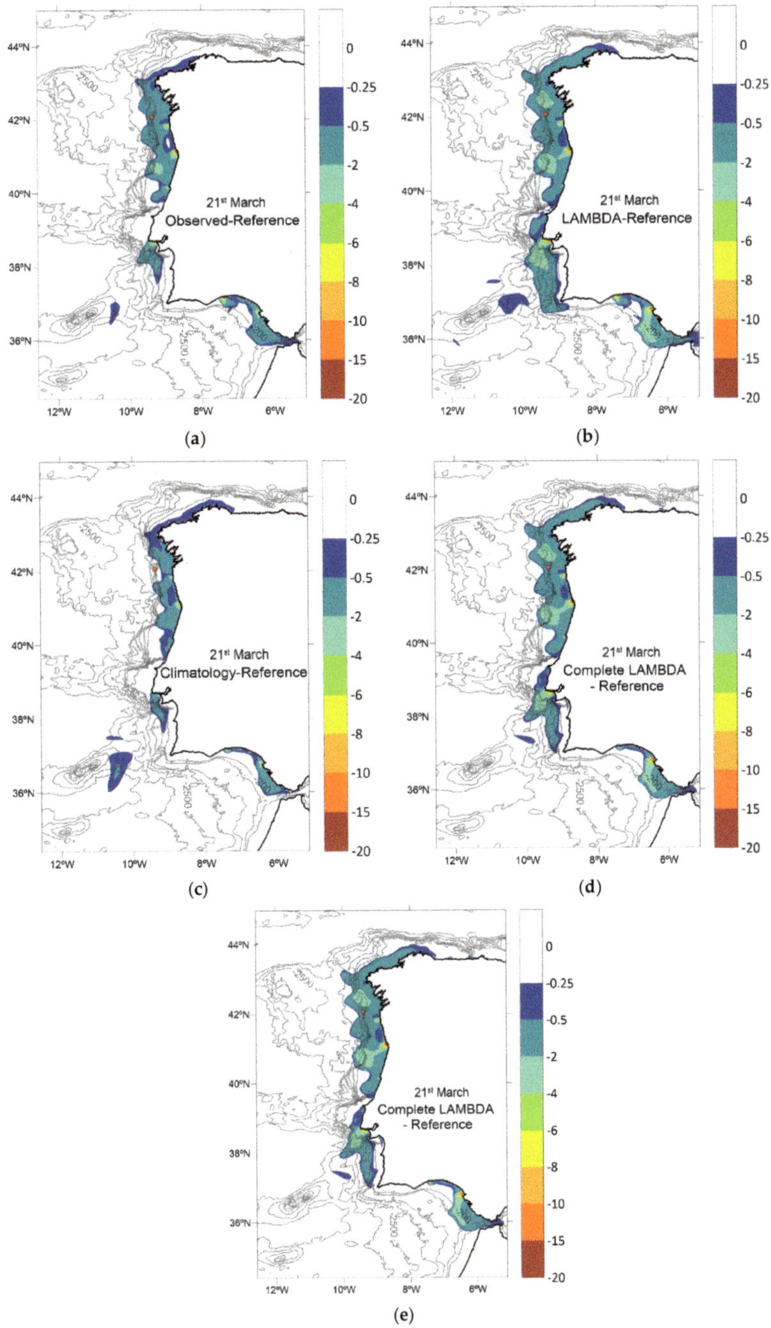

Figure 13. Surface salinity difference during the 21 March 2018 between the reference scenario with each LAMBDA land boundary scenarios (**a**) Observed; (**b**) LAMBDA; (**c**) Climatology; (**d**) Complete Observations; (**e**) Complete LAMBDA. Salinity difference between 0 and −0.5 is not represented.

4. Discussion

By improving the calculation of the freshwater quantity and associated properties reaching the coastal area, managers and scientists would be able to better reproduce the thermal and salinity fronts affecting coastal hydrodynamics and associated ecological processes. The proposed methodology for integrating the water cycle from rainwater runoff to the open ocean with numerical models was demonstrated for western Iberia by simulating the extension of the WIBP during a wet season and an extreme rain event.

The approach taken enables the continuous improvement of the solution through updates/upgrades to any upstream component such as the watershed results or through the use of a better characterisation of the estuarine proxy. The application of this methodology provides more realistic land boundary conditions than river climatology direct discharges, while also having the capacity to describe inter-annual variability between dry and wet years and effects due to extreme rain or drought events. The developed methodology is generic and could be set for any region with open-access data and open-source models.

Enhanced land boundary conditions including water properties such as temperature and salinity can provide a more realistic circulation and thermohaline fields in the coastal area. The included land boundary conditions can grow in complexity from plainly watershed modelling results that are implemented as direct discharges to fully 3D biogeochemical estuarine fluxes (positive and negative) imposed with momentum.

The main advantage of watershed modelling is to complete the hydrometric monitoring networks, providing gapless river-flow data and non-monitored variables while covering areas with monitored data. Additionally, watershed numerical modelling enables the forecasting of river flow and water properties, thereby allowing a more efficient management of the modelled systems. However, calibration and validation of many watersheds became challenging when large water reservoirs or intense human management modify their natural flow. The use of 5 km horizontal-resolution domains enabled the coverage of a large simulation area, as well as a decrease in the simulation time required, and these domains can adequately represent the channel flow behaviour of larger rivers. However, smaller rivers may not be accurately simulated under these conditions. Generally, better results are obtained from higher resolution grids and single-watershed domains, since calibration is easier to execute.

From an operational point of view, the objective is to generate a land boundary product with the best available information. For this reason, observed data can be combined with modelling results, such as river properties. In any case, the watershed-model approach has shown the ability to produce a more realistic land boundary condition than methods involving the use of river climatologies, but further development is needed for this component of the system.

The estuarine proxy has been a versatile tool that allows researchers to estimate in a simple way estuarine mixing and contributions to the open ocean. The proxy affords the inclusion of estuary time and spatial scales, due to the tides and the rivers' flow combination, in coarse regional ocean models. Its low computational cost would enable the easy construction of two-way coupled systems. Another advantage is that land boundary conditions are independent of the receiving model domain, and thus the same flow/fluxes can be used by several regional ocean models. The estuarine proxy can be regarded as a useful tool when full estuarine models do not exist.

5. Conclusions and Future Approach

Monitoring networks along the estuarine continuum from river catchment to the open ocean should be encouraged for evaluating the transfer of properties and momentum at the land-ocean interface. While open ocean network is relatively well established, operational estuarine monitoring is far from consolidated. In general, hydrometric networks should be further developed to meet the requirements of coastal region stakeholders. Numerical models can support the design of the monitoring networks, and they can provide a more comprehensive view by spatially and temporally completing the observed data. In addition, numerical models can provide non-observed variables and forecasts. Model results also permit the calculation of complex indicators such as tidal prims and the area of influence of the estuarine waters, as well as the accurate estimation of estuarine fluxes that would serve as boundary conditions for ocean regional models. The synergies between these two can be observed, and the modelled data pave the way for a more holistic view of the water continuum.

Since salinity remote sensing is starting to take its first steps in coastal areas and since in-situ monitoring has a low frequency, numerical modelling is currently the only tool able to represent and estimate the temporal and spatial scale of the WIBP and other estuarine plumes. Taking into consideration the numerical modelling limitations and assumptions, the salinity modelling results provided by the methodology described here significantly improves salinity fields and helps delimit the region of freshwater influence and salinity fronts.

Future work must include improving the incorporation of human-water management in watershed modelling in order to obtain more accurate forecasts. A possible next step in the research is to use artificial intelligence and machine learning techniques. In fact, combining machine learning and physical-based models is becoming popular in the design of predictive systems [29,30]. Including the human component is especially relevant in southern Europe where water retention is a more common practice.

In addition, efforts to identify and make available more river data are crucial for watershed-model calibration and validation. As the results show, to achieve good sea surface salinity fields in regional models, it is not enough to include only the major rivers, but a comprehensive freshwater budget is also required.

Author Contributions: Conceptualization, F.C.; methodology, F.C., F.S., L.S., A.R.O., R.F., D.B. and R.N.; validation, F.C. and F.S.; resources, E.O., A.T., M.A. and A.N.; writing—original draft preparation, F.C., F.S., E.O. and A.N.; writing—review and editing, F.C., F.S., E.O. and A.N.; project administration, F.C. and A.N. All authors have read and agreed to the published version of the manuscript.

Funding: The present work was performed within the framework of two research projects: LAMBDA project from Copernicus Marine Environment Monitoring Service (CMEMS) Service Evolution 2 (2018–2020) and the iFADO project supported with ERDF funds from the INTERREG Atlantic Area Programme under contract EAPA 165/2016.

Data Availability Statement: LAMBDA project products can be found in the LAMBDA Project Data Portal: http://www.cmems-lambda.eu/#data-portal (accessed on 8 April 2022); data from the different ocean model scenarios analysed in the study can be available on request from the corresponding author.

Acknowledgments: The authors are grateful to the following experts (and institutions) for their support to this research: Marcos G. Sotillo (Puertos del Estado), Tomasz Dabrowski (MI), Joanna Staneva (HZG), and Marina Tonani (Mercator Ocean international) for their support, evaluation, and contributions to improve the LAMBDA-project products. The authors would like to thank Angelique Melet, Isabel Garcia Hermosa, and Pierre-Yves LeTraon (Mercator-Ocean International) for their continuous advice and evaluation during the LAMBDA project. The authors would like to thank Kieran O'Driscoll for the careful English-language review. The author would like to thank the reviewers for constructive criticism of the manuscript.

Conflicts of Interest: The authors declare no conflict of interest.

References

1. Garvine, R.W.; Whitney, M.M. An estuarine box model of freshwater delivery to the coastal ocean for use in climate models. *J. Mar. Res.* **2006**, *64*, 173–194. [CrossRef]
2. Santos, A.M.P.; Chícharo, A.; Dos Santos, A.; Moita, T.; Oliveira, P.B.; Peliz, Á.; Ré, P. Physical—Biological interactions in the life history of small pelagic fish in the Western Iberia Upwelling Ecosystem. *Prog. Oceanogr.* **2007**, *74*, 192–209. [CrossRef]
3. Banas, N.S.; MacCready, P.; Hickey, B.M. The columbia river plume as cross-shelf exporter and along-coast barrier. *Cont. Shelf Res.* **2009**, *29*, 292–301. [CrossRef]
4. Mishra, A.K.; Coulibaly, P. Developments in hydrometric network design: A review. *Rev. Geophys.* **2009**, *47*, RG2001. [CrossRef]

5. Campuzano, F.; Brito, D.; Juliano, M.; Fernandes, R.; de Pablo, H.; Neves, R. Coupling watersheds, estuaries and regional ocean through numerical modelling for Western Iberia: A novel methodology. *Ocean. Dyn.* **2016**, *66*, 1745–1756. [CrossRef]
6. Schiller, R.V.; Kourafalou, V.H. Modeling river plume dynamics with the Hybrid Coordinate Ocean Model. *Ocean. Model.* **2010**, *33*, 101–117. [CrossRef]
7. Campuzano, F.J.; Juliano, M.; Sobrinho, J.; de Pablo, H.; Brito, D.; Fernandes, R.; Neves, R. *Coupling Watersheds, Estuaries and Regional Oceanography through Numerical Modelling in the Western Iberia: Thermohaline Flux Variability at the Ocean-Estuary Interface*; Froneman, E.W., Ed.; IntechOpen: London, UK, 2018; pp. 1–17. [CrossRef]
8. Sotillo, M.G.; Campuzano, F.; Guihou, K.; Lorente, P.; Olmedo, E.; Matulka, A.; Santos, F.; Amo-Baladrón, M.A.; Novellino, A. River freshwater contribution in operational ocean models along the European Atlantic Façade: Impact of a new river discharge forcing data on the CMEMS IBI Regional Model solution. *J. Mar. Sci. Eng.* **2021**, *9*, 401. [CrossRef]
9. Neves, R. The MOHID concept. In *Ocean Modelling for Coastal Management—Case Studies with MOHID*; Mateus, M., Neves, R., Eds.; IST Press: Lisbon, Portugal, 2013; pp. 1–11.
10. Peliz, Á.; Rosa, T.L.; Santos, A.M.P.; Pissarra, J.L. Fronts, jets, and counter-flows in the Western Iberian upwelling system. *J. Mar. Syst.* **2002**, *35*, 61–77. [CrossRef]
11. Ribeiro, A.C.; Peliz, Á.; Santos, A.M.P. A study of the response of chlorophyll-a biomass to a winter upwelling event off Western Iberia using SeaWiFS and in situ data. *J. Mar. Syst.* **2005**, *53*, 87–107. [CrossRef]
12. Picado, A.; Dias, J.M.; Fortunato, A. Tidal changes in estuarine systems induced by local geomorphologic modifications. *Cont. Shelf Res.* **2010**, *30*, 1854–1864. [CrossRef]
13. Santos, A.M.P.; Peliz, Á.; Dubert, J.; Oliveira, P.B.; Angélico, M.M.; Ré, P. Impact of a winter upwelling event on the distribution and transport of sardine (*Sardina pilchardus*) eggs and larvae off western Iberia: A retention mechanism. *Cont. Shelf Res.* **2004**, *24*, 149–165. [CrossRef]
14. Santos, A.M.P.; Ré, P.; dos Santos, A.; Peliz, Á. Vertical distribution of the European sardine (*Sardina pilchardus*) larvae and its implications for their survival. *J. Plankton Res.* **2006**, *28*, 523–532. [CrossRef]
15. Peliz, A.; Marchesiello, P.; Dubert, J.; Marta-Almeida, M.; Roy, C.; Queiroga, H. A study of crab larvae dispersal on the Western Iberian Shelf: Physical processes. *J. Mar. Syst.* **2007**, *68*, 215–236. [CrossRef]
16. Queiroga, H.; Cruz, T.; dos Santos, A.; Dubert, J.; González-Gordillo, J.I.; Paula, J.; Peliz, Á.; Santos, A.M.P. Oceanographic and behavioural processes affecting invertebrate larval dispersal and supply in the western Iberia upwelling ecosystem. *Prog. Oceanogr.* **2007**, *74*, 174–191. [CrossRef]
17. Brito, D.; Campuzano, F.J.; Sobrinho, J.; Fernandes, R.; Neves, R. Integrating operational watershed and coastal models for the Iberian Coast: Watershed model implementation—A first approach. *Estuar. Coast. Shelf Sci.* **2015**, *167*, 138–146. [CrossRef]
18. Tian, J.; Liu, J.; Wang, Y.; Wang, W.; Li, C.; Hu, C. A coupled atmospheric–hydrologic modeling system with variable grid sizes for rainfall—Runoff simulation in semi-humid and semi-arid watersheds: How does the coupling scale affects the results? *Hydrol. Earth Syst. Sci.* **2020**, *24*, 3933–3949. [CrossRef]
19. Tóth, B.; Weynants, M.; Pásztor, L.; Hengl, T. 3D soil hydraulic database of Europe at 250 m resolution. *Ecohydrology* **2016**, *31*, 2662–2666. [CrossRef]
20. Andreadis, K.A.; Schumann, G.J.-P.; Pavelsky, T. A simple global river bankfull width and depth database. *Water Resour. Res.* **2013**, *49*, 7164–7168. [CrossRef]
21. Carrère, L.; Lyard, F.; Cancet, M.; Guillot, A.; Picot, N. FES2014, a new tidal model—Validation results and perspectives for improvements. In Proceedings of the ESA Living Planet Conference, Prague, Czech Republic, 9–13 May 2016.
22. Mateus, M.; Riflet, G.; Chambel, P.; Fernandes, L.; Fernandes, R.; Juliano, M.; Campuzano, F.; de Pablo, H.; Neves, R. An operational model for the West Iberian coast: Products and services. *Ocean. Sci.* **2012**, *8*, 713–732. [CrossRef]
23. Campuzano, F. Coupling Watersheds, Estuaries and Regional Seas through Numerical Modelling for Western Iberia. Ph.D. Thesis, Instituto Superior Técnico, Universidade de Lisboa, Lisbon, Portugal, 2018.
24. Grell, G.A.; Dudhia, J.; Stauffer, D. *A Description of the Fifth-Generation Penn State/NCAR Mesoscale Model (MM5)*; Technical Note No.NCAR/TN-398+STR; University Corporation for Atmospheric Research: Boulder, CO, USA, 1994. [CrossRef]
25. Trancoso, A.R. Operational Modelling as a Tool in Wind Power Forecast and Meteorological Warnings. Ph.D. Thesis, Instituto Superior Técnico, Universidade Técnica de Lisboa, Lisbon, Portugal, 2012.
26. Sikorska, A.E.; Scheidegger, A.; Banasik, K.; Rieckermann, J. Considering rating curve uncertainty in water level predictions. *Hydrol. Earth Syst. Sci.* **2013**, *17*, 4415–4427. [CrossRef]
27. Chin, T.M.; Vazquez-Cuervo, J.; Armstrong, E.M. A multi-scale high-resolution analysis of global sea surface temperature. *Remote Sens. Environ.* **2017**, *200*, 154–169. [CrossRef]
28. Olmedo, E.; González-Haro, C.; Hoareau, N.; Umbert, M.; González-Gambau, V.; Martínez, J.; Gabarró, C.; Turiel, A. Nine years of SMOS sea surface salinity global maps at the Barcelona Expert Center. *Earth Syst. Sci. Data* **2021**, *13*, 857–888. [CrossRef]
29. Yaseen, Z.M.; Sulaiman, S.O.; Deo, R.C.; Chaud, K.-W. An enhanced extreme learning machine model for river flow forecasting: State-of-the-art, practical applications in water resource engineering area and future research direction. *J. Hydrol.* **2019**, *569*, 387–408. [CrossRef]
30. Sarafanov, M.; Borisova, Y.; Maslyaev, M.; Revin, I.; Maximov, G.; Nikitin, N.O. Short-term river flood forecasting using composite models and automated machine learning: The case study of Lena river. *Water* **2021**, *13*, 3482. [CrossRef]

Journal of
Marine Science and Engineering

MDPI

Article

River Freshwater Contribution in Operational Ocean Models along the European Atlantic Façade: Impact of a New River Discharge Forcing Data on the CMEMS IBI Regional Model Solution

Marcos G. Sotillo [1,2,*], Francisco Campuzano [3,4], Karen Guihou [1,5], Pablo Lorente [1,5], Estrella Olmedo [6], Ania Matulka [1], Flavio Santos [3], María Aránzazu Amo-Baladrón [1,5] and Antonio Novellino [7]

1 Puertos del Estado, 28042 Madrid, Spain; karen.guihou@nologin.es (K.G.); pablo.lorente@nologin.es (P.L.); amatulka@PUERTOS.ES (A.M.); arancha.amo@nologin.es (M.A.A.-B.)
2 Mercator Ocean International, 31520 Ramonville Saint-Agne, France
3 MARETEC/LARSyS, Instituto Superior Técnico, Universidade de Lisboa, 1649-004 Lisbon, Portugal; francisco.campuzano@colabatlantic.co (F.C.); flavio.t.santos@tecnico.ulisboa.pt (F.S.)
4 CoLAB +ATLANTIC, Molhe Leste, 2520-620 Peniche, Portugal
5 Nologin Consulting S.L., 50018 Zaragoza, Spain
6 Department of Physical Oceanography, Institute of Marine Sciences, CSIC, Barcelona Expert Center, Pg. Marítim 37-49, E-08003 Barcelona, Spain; olmedo@icm.csic.es
7 ETT S.p.A, 16153 Genoa, Italy; antonio.novellino@ettsolutions.com
* Correspondence: marcos@puertos.es

Abstract: River freshwater contribution in the European Atlantic margin and its influence on the sea salinity field are analyzed. The impacts of using a new river discharge database as part of the freshwater forcing in a regional ocean model are assessed. Ocean model scenarios, based on the CMEMS (Copernicus Marine Environment Monitoring Service) operational IBI-MFC (Iberia Biscay Ireland Monitoring Forecasting Centre) model set-up, are run to test different (observed, modeled and climatological) river and coastal freshwater forcing configurations throughout 2018. The modelled salinity fields are validated, using as a reference all known available in-situ observational data sources. The IBI model application is proven to adequately simulate the regional salinity, and the scenarios showcase the effects of varying imposed river outflows. Some model improvement is achieved using the new forcing (i.e., better capture of salinity variability and more realistic simulation of baroclinic frontal structures linked to coastal and river freshwater buoyancy plumes). Major impacts are identified in areas with bigger river discharges (i.e., the French shelf or the northwestern Iberian coast). Instead, the Portuguese shelf or the Gulf of Cadiz are less impacted by changes in the imposed river inflows, and other dynamical factors in these areas play a major role in the configuration of the regional salinity.

Keywords: river freshwater discharges; operational ocean models; sea surface salinity field; IBI region; LAMBDA river database; IBI model validation; model sensitivity tests

Citation: Sotillo, M.G.; Campuzano, F.; Guihou, K.; Lorente, P.; Olmedo, E.; Matulka, A.; Santos, F.; Amo-Baladrón, M.A.; Novellino, A. River Freshwater Contribution in Operational Ocean Models along the European Atlantic Façade: Impact of a New River Discharge Forcing Data on the CMEMS IBI Regional Model Solution. *J. Mar. Sci. Eng.* **2021**, *9*, 401. https://doi.org/10.3390/jmse9040401

Academic Editor: Alberto Ribotti

Received: 9 March 2021
Accepted: 1 April 2021
Published: 9 April 2021

1. Introduction

In the global water cycle, ocean-atmosphere surface net freshwater fluxes are balanced by ocean transport and land discharges within the oceans. The return of terrestrial runoff into oceans is mostly concentrated at the mouths of the world's major rivers, and it locally provides significant freshwater inflows, forcing changes in sea water densities. Consequently, realistic estimates of land freshwater discharges into oceans are needed to globally study water budgets [1]. At regional and local scales, and very specifically on coastal and shelf areas, the river runoff contribution is a major forcing of the ocean dynamics, having a strong influence on water stratification [2]. Besides, it also introduces significant fluctuations in local circulation patterns, generating important buoyant freshwater river plume-like flow structures [3]. Thus, noticeable on-shelf baroclinic frontal areas are caused and maintained in time by the freshwater contributions of one or various major rivers discharging in nearby coastal areas. These areas, generally located in coastal shelf seas or into estuaries where currents patterns are governed by density differences between salt sea water and fresh river water are known as Region of Fresh Water Influence (ROFI, [4]). The discharged freshwater volume, and its temporal variability, impact on the local circulation of such ROFI areas, by means of different interactions between the linked buoyancy plume structure and any existing regional ambient flow features (i.e., major current flowing along the slope, currents linked to open sea dynamics, barotropic tidal shelf currents, or ocean transports responding to prevailing wind regimes [5,6]). The coastal freshwater inputs even modulate typical coastal effects, such as upwelling events [7], having major impacts on the productivity and the whole ecological environment [8].

Due to this key role played at different scales on the ocean system, there is a growing interest in including realistic freshwater signals coming from land in operational ocean model systems.

In global ocean and basin models, which are extensively used in climate prediction and reanalysis production, this freshwater input is crucial to maintain water balances and to reproduce salinity fields with realistic latitudinal gradients that affect climate signals such as the Atlantic Meridional Overturning transports [9]. It is important to note that these global ocean model applications are not fit-for-purpose for performing on coastal areas as many processes of paramount importance to reproduce shelf dynamics are parameterized, filtered out, or simply not included. Furthermore, due to the typical global model coarse resolution, coastlines are not realistically represented. This limitation of global ocean models to deal with coastal areas has conditioned the way to force such models with punctual freshwater river discharges. Since most global models cannot accurately represent river mouths or estuaries, the freshwater river contribution has been traditionally included by means of some modification in surface ocean-atmosphere water fluxes along coastal areas (mostly modifying the evaporation-precipitation (E-P) ratio along coastal model grid points, as proposed by Bourdallé-Badie and Tréguier [10]).

The use of coastal and river freshwater forcing derived from climatological data has been a primary, and still a very commonly accepted, approach in global ocean modeling. However, available databases and services that progressively provide better estimates of coastal runoff and river discharges have started to improve ocean model freshwater coastal forcing input. Several research initiatives and established services are focused on generating at a global scale up-to-date runoff and freshwater river discharge historical datasets. Dai et al. [11] compiled for community use a global dataset with monthly streamflow at farthest downstream stations for the world's 925 largest ocean-reaching rivers; likewise, the service currently provided by the Global Runoff Data Centre [12], under the auspices of the World Meteorological Organization (WMO), delivers quality controlled "historical" databases, comprising daily and monthly mean river discharges from more than 9900 in-situ stations from 159 countries. The GloFAS-ERA5 (Global Flood Awareness System) operational global river discharge reanalysis [13] produces, consistently with the ECMWF ERA5 (European Centre for Medium-Range Weather Forecasts global reanalysis, version 5th) atmospheric reanalysis and using a river network, a global gridded data product (with a horizontal resolution of 0.1 degree and at a daily time step) that covers from 1979 to near-real-time (within a delay of 7 days).

However, when we move into coastal ocean modeling, approaches and needs are slightly different: in their review of coastal modeling science foundation, Kourafalou et al. [14] pointed out the need of applying very high-resolution models at coastal scales, able to accurately capture coastal features, and forced with realistic punctual river inflows to adequately reproduce the related dynamics occurring in ROFI areas. This demand in terms of high-resolution models is crucial when forecasting estuarine circulation. Model applications focused on estuaries show horizontal resolutions that range from kilometric scales (at coastal boundaries) to meters (within the estuary zones). These extremely high coastal model resolutions should be accompanied by a combination of ad hoc riverine and atmospheric forcing with appropriate temporal variability (being needed daily, and even higher temporal frequency data). Therefore, neither river discharge monthly climatological products nor coarse resolution watershed model estimations are fully satisfactory when used to force operational ocean coastal models [15].

Lately, the utility of dynamical model downscaling from the deep ocean, across the continental shelf and into the coastal areas, has been extensively demonstrated [16–18]. In the framework of the Copernicus Marine Environment Monitoring Service (CMEMS), a global ocean model together with a wealth of regional ocean model systems, are operationally running to deliver short-term forecast and multi-year reanalysis information on the physical and biogeochemical ocean for the global ocean and the European regional seas (CMEMS, 2020 [19]). Specifically, along the European Atlantic façade, CMEMS provides a regional service (besides its global one) for the so-called IBI area (Iberia-Biscay-Ireland regional seas [20]). These CMEMS IBI regional products, down-scaled from the CMEMS global ones, are aimed to characterize upper-ocean dynamics at finer scales and to provide better boundary conditions for very-high-resolution coastal models, embedded into IBI solution, used by other non-CMEMS downstream services (as pointed out by Le Traon et al. [21], among others, one of the goals of the CMEMS mission).

The latest review of the current European capacity in terms of operational marine and coastal modeling systems [22] maps the organizations and the operational model systems at regional and coastal scales across Europe and points out the sustained availability of CMEMS global and regional scale core products as a positive driver to favor proliferation of "downstream" services devoted to coastal monitoring and forecasting. Likewise, this review highlights the availability of different operational ocean models, covering in some cases overlapped areas.

The availability of different operational regional and coastal model solutions in some IBI areas is seen as an opportunity for on-going initiatives, such as the MyCoast Project [23]. This EU Interreg Atlantic Program project is aimed at enhancing the regional cooperation, taking benefit of synergies among existing operational oceanographic tools and services to develop improved coastal risk services to support marine safety, fight against pollution, and generally, to increase preparedness to extreme met-ocean coastal events. MyCoast develops and upgrades different risk tools, making them suitable for using different operational model solutions as inputs. In parallel to this interoperability enhancement, a multi-model intercomparison exercise was conducted to deepen the knowledge on the different model products used as forcing by the MyCoast risk tools, identifying differences and analogies between 9 operational regional and coastal model solutions on overlapping areas. This multi-model comparison exercise, operationally updated since 2018, shows general consistency among the models, especially for sea surface temperature (SST). However, noticeable differences in sea surface salinity (SSS), mostly on shelf areas, are found (see Figure 1).

These modelled salinity differences are particularly evident in ROFI zones, such as the northwestern Iberian shelf and the French shelf in the Gulf of Biscay, and seem mostly related to the diverse treatment that each operational ocean model system gives to the coastal and river freshwater discharge forcing. Indeed, Matulka et al. [24] inform that the approaches followed to include freshwater inputs in each operational forecast system, shown in Figure 1, vary from one model set-up to another, using: (1) data derived from daily observations taken in river flow stations (i.e., the MeteoGalicia MTG-ROMS set-up and the CMEMS IBI one, but this one using observations only for some of the major rivers), (2) daily freshwater discharges at river mouths produced by hydrological models (this is the case for IBI, especially for its five-days-ahead forecast run), (3) data derived from river discharge monthly climatology (a very common approach, fully or partially used in IBI, IMI-ROMS and SHOM-HyCOM operational ocean model set-ups), and finally, (4) the no river forcing option (in this case, used only by the IST-MOHID operational version, available when the study was performed). Independently of the approach used to include this freshwater signal, what seems clear is that having a non-optimal representation of this coastal and river freshwater forcing in ocean model set-ups can lead to biased simulated ocean states, jeopardizing the reliability of ocean forecasts, and diminishing forecast skill on ROFI areas.

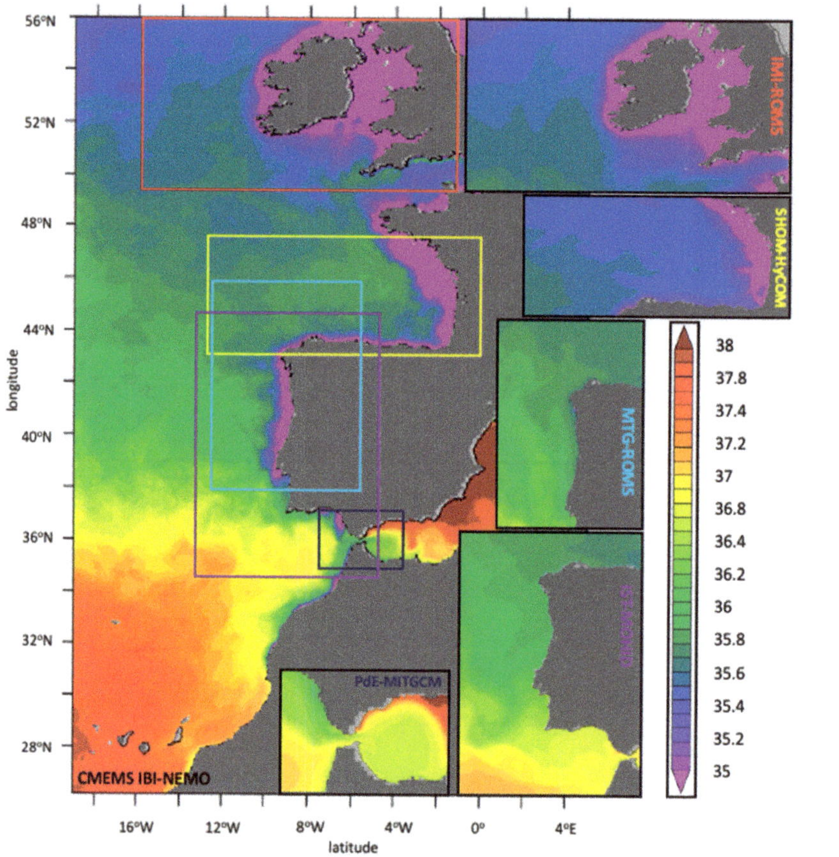

Figure 1. 2018 Spring (March-April-May) Sea Surface Salinity (SSS) field (psu) simulated by different operational ocean models in the IBI region. The following ocean model solutions, included in the MyCoast Multi-Model Exercise, are shown (each application based on a different model and covering a different geographical domain): The CMEMS IBI regional solution, based on NEMO model and covering the whole IBI region (CMEMS-IBI-NEMO), the *Puertos del Estado* model (PdE-MITGCM) I for the Strait of Gibraltar (dark blue), the *Instituto Superior Tecnico de Lisboa* (IST-MOHID) system for the western Iberian domain (purple), the *MeteoGalicia* (MTG-ROMS) forecast service for the Northwestern Iberian domain (cyan), the Irish Marine Institute (IMI-ROMS) application for Ireland (red), the *Service hydrographique et océanographique de la Marine* application (SHOM-HyCOM) for the Gulf of Biscay (yellow).

The MyCoast multi-model intercomparison exercise is further proof of the necessity to improve coastal freshwater signals in operational ocean models. This need, already identified within the operational oceanographic community [22], includes improved standardized freshwater river inputs (considering not only flow rates but also particulate and dissolved matter key for biogeochemical processes) and foresees, as a longer-term objective, to include connection and coupling with land hydrology models.

Likewise, the CMEMS Service, as part of its Evolution Strategy and scientific research priorities (CMEMS Scientific and Technical Advisory Committee, 2017 [25]), identified the enhancement of land forcing within their medium- and long-term objectives.

CMEMS has supported in the last years different research projects on cutting-edge science and technological developments, needed to ensure future state-of-the-art service evolutions. Within the last call for tenders (2018–2020), two CMEMS service evolution projects were dedicated to make the required research to improve the freshwater river discharge inputs for global and regional CMEMS models: the BRONCO (Benefits of dynamically modelled river discharge input for ocean an coupled atmosphere-land-ocena systems) Project [26] was aimed at generating a global river discharge reanalysis, using the GloFAS system driven by the ECMWF ERA5 atmospheric reanalysis. On the other hand, the LAMBDA (Land Marine Boundary Development and Analysis) Project [27], regionally focused on the European Atlantic Façade and the North Sea, had the objective of achieving better estimations, by means of high-resolution hydrological model simulations, for characterizing major, and also some minor, river contributions and their day-to-day variability (highly needed for operational ocean coastal models). The resulting new freshwater model estimates, more realistic and adjusted using observations along historical periods, reproduce not only more reliable long-term changes in continental freshwater discharges into the oceans, but also a more detailed spatial distribution. These model products and in-situ observational data are operationally updated, made available and downloadable into the LAMBDA product viewer web interface (http://www.cmems-lambda.eu/mapviewer/ (accessed on 8 April 2021)).

The present work aims to quantify the potential added value of using this new LAMBDA freshwater discharge product as part of the fresh water forcing of a regional ocean model simulation. To this aim, different model scenarios, based on the CMEMS IBI model configuration, have been designed and run. The modelled salinity fields, resulting from the different sensitivity IBI test runs, are validated by means of comparison with different in-situ observational data sources. Despite that test runs are performed over the whole IBI model domain, this study mostly focuses on ROFI areas of the Atlantic Iberia and on the French shelf in the Gulf of Biscay.

2. Materials and Methods

2.1. Description of Study Domain: In-Situ Observed Salinity Products

The region of study, hereafter referred as IBBIS, is a subset of the IBI operational configuration (Figure 2) and spans from the Strait of Gibraltar (35° N) to Bretagne (48.6° N). The warm and saline Iberian Poleward Current (IPC, [28]) flows along the steep slope of the narrow shelf of the Western Iberian Peninsula (50 km wide at its maximum, west of 8° W), then around the Galician shelf, and finally, enters the Bay of Biscay, separating the continental shelf waters from the open-ocean dynamics. This semi-enclosed Bay of Biscay features a narrow shelf on its southern part that widens towards the North (~150 km at 47° N), separated from a deep abyssal plain (more than 4000 m depth) by a steep and rugged slope, with many canyons. The main drivers on the shelf of this IBBIS domain are wind, tides and freshwater coastal runoff. The ocean dynamic varies locally, depending on the orientation of the coast, width of the shelf, freshwater discharge rates and other factors.

In this study, the IBBIS domain has been divided in three subregions, each one featuring its own characteristics: the CADIZ area, that is the Northern shelf of the Gulf of Cádiz, the Western Iberian Shelf (WISHE area), that includes the western shelf of the Iberian Peninsula, and the Gulf of Biscay (BISCA area).

The CADIZ area, at the southern tip of the Iberian Peninsula, is the place where the exchanges between the Atlantic and Mediterranean take place, across the Strait of Gibraltar. The saline MOW (Mediterranean Outflow Waters) come as a subsurface outflow from the Strait of Gibraltar, and cascade along the slope of the narrow shelf of the bay [29,30]. The circulation on the shelf is wind-driven, with a clear seasonal variability [31]: upwelling occurs in summer under the influence of northerlies, while downwelling takes place in winter under the influence of southerlies [32]. The variability on the shelf is explained by the front between the cold upwelled shelf-waters and the warmer southern waters (e.g., Reference [33]). Two main rivers, Guadalquivir and Guadiana, feed the shelf on its wider portion: in winter, the increased river runoff cools down the shelf waters [34] and contributes to the generation of a westward inshore flow [35], although this is not the main driver.

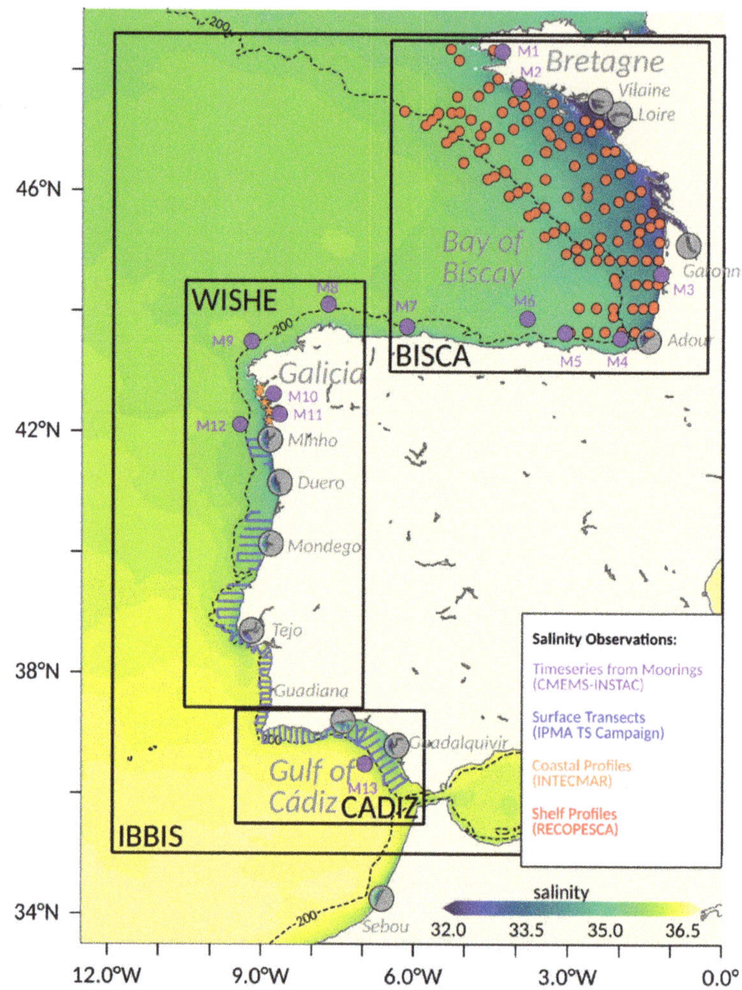

Figure 2. Study Domain. Location of the in-situ measurements used in this study: fixed moorings (purple), surface transects with thermo-salinometer (blue), coastal profiles with CTD (Conductivity, Temperature, and Depth) instrumentation (orange) and profiles with XBT (eXpendable BathyThermograph) instrumentation (red). Black rectangles show subregions in which the study is focused (IBBIS, CADIZ, WISHE, BISCA). The names of the main rivers included in the IBI model set-up and the geographical features mentioned in the paper are denoted in grey. The shaded field shows the IBI averaged surface salinity over the year 2018, from the reference simulation; the 200 m isobath, delimiting the shelf, is depicted in dashed line.

The WISHE area features a narrow shelf (30 to 90 km at its widest), and in comparison to the other proposed study subregions, this is the most sensitive to open-ocean general circulation patterns occurring in the North-eastern Atlantic. The geostrophic flow of the IPC intensifies in winter and isolates the narrow shelf's waters from the open-ocean southward circulation. Along the Galician coast, the currents are mainly wind-driven [36]. The WISHE area marks the limit of the Eastern North Atlantic Upwelling region, with a frequent development of equatorward filaments at the coast in spring and autumn [37]. The interactions and front dynamics between the IPC and the shelf waters are complex [38]. The Western Iberian Buoyant Plume extends from the Mondego river northwards and is characterized by a salinity below 35.8. It is fed by the Mondego, Douro and Minho rivers and the numerous Rias of Galicia. Under non-upwelling conditions, the plume stays confined to the inner shelf and contributes to the stratification and cooling of the shelf waters. On the other hand, under upwelling conditions, the plume extends offshore, and masks the thermal signature of the IPC.

Finally, in the BISCA area, the IPC develops instabilities and eddies along its path which propagate offshore and take part of the general anticyclonic circulation of the open-ocean section [28,39]. On the French shelf, the main drivers are the coastal density-driven jets which interact with the tidal signal [40,41], and the surface wind-driven currents that advect shelf waters southward, e.g., Solabarrieta et al. [42]. In winter, the activity is weak over the shelf, but intense mesoscale activity takes place in the open ocean. The Southwestern winds induce northeastward drift over the shelf. From spring, the wind weakens and the dynamic activity on the shelf reaches the same levels of energy as offshore [43]. The major source of freshwater on the BISCA shelf is the river outflow, with a seasonal cycle featuring a maximum of discharge in winter and spring. The main rivers are the Loire and Vilaine system (47° N), the Garonne (45° N) and the Adour, in the South-West corner of the bay. Freshwater filaments extend from the coast and participate in the cross-shelf exchange [44]. Patches of low-salinity surface waters can be pushed by wind-induced coastal upwellings and reach the shelf-break [45]. Along the northern Iberian coast, the numerous rivers have a small averaged runoff, but can have peaks of torrential discharge, principally in spring and autumn. The accumulated outflow from these coastal rivers in winter and spring induces a low-salinity front on the narrow shelf [46], that can reach the shelf-break and extend down to 50 m depth [47].

All the in-situ salinity observations available for the year 2018 were gathered and explored to assess the capacity of the IBI model configurations to reproduce salinity fronts and river plume extension on the shelf, depending on each freshwater forcing. Table 1 shows an overview of the available observations used in each of the study areas.

Table 1. Overview of the in-situ observational salinity data sources in the study domain and subregions.

	Bay of Biscay (BISCA): 9.5 W–5.8 W/ 35.5 N–37.38 N	**Western Iberian Shelf (WISHE): 10.5 W–7 W/37.4 N–44.5 N**	**Gulf of Cadiz (CADIZ): 10 W–6.5 W/43 N–48.5 N**	**Iberian-Biscay (IBBIS): 12 W–0 W/35 N–48.6 N**
Hourly SSS Timeseries at Mooring Buoy	7 stations (M1 to M7, from N to S)	5 stations (M8 to M12, from N to S)	1 station (M13)	13 stations (M1 to M13, from N to S)
(Shelf) XBT salinity profiles	106 profiles	-	-	106 profiles
Weekly (Coastal) CTD salinity profiles	-	5 stations	-	5 stations
Surface (Shelf) salinityTransects	-	1 campaign (IPMA 28 April–20 May 2018)	1 campaign (IPMA 21 May–30 May 2018)	1 campaign (IPMA 28 April–30 May 2018)

Hourly timeseries of salinity from mooring stations for the year 2018 were obtained from the CMEMS in-situ observational product (Copernicus Marine In-Situ Tac Data Management Team (2019) [48]). The buoys are mostly moored on the shelf or at the shelf-break (see their locations, represented by purple dots, in Figure 2) and measurements are collected at surface and subsurface (mostly between 1 and 3 m deep). After processing, 13 stations are available at the whole study area: 7 in the BISCA zone (numbered M1 to M7, following the coast southwards from Bretagne to Galicia), 5 in the WISHE one (numbered M8 to M12, from North to South) and 1 in the CADIZ subregion (M13). Data from all of them have been used for the model validation, however some stations provide a limited exploitable input; in some cases, there is no more than a few months of data available (e.g., M4), whereas in other cases, the buoy is moored too close to the coast, not being fully useful for a meaningful validation of a regional model such as IBI (e.g., M3 or M10). However, as salinity observational data are sparse, it has been decided not to discard them, but use them with caution.

Weekly salinity profiles from 5 conductivity-temperature-depth (CTD) stations over the year 2018 located at Rias Baixas within the WISHE area were provided by the *Instituto Tecnolóxico para o Control do medio Mariño de Galicia* (INTECMAR). These CTD measures have been running since 1992 and the current INTECMAR oceanographic network is formed by 43 stations distributed along the Galician Coast [49]. These weekly campaigns are very coastal, with most of the stations in Rias and estuaries. The depth values are not uniform over the year, depending on the devices used to measure the data [50]. In this study, only samples from the 5 most offshore INTECMAR CTD stations (identified by orange stars in Figure 2) over the year 2018 are used, and these measurements are compared with daily modelled salinity profiles extracted, at the closet model grid point, from the different IBI model runs.

Surface Transects measured every 10 min with thermo-salinometers (TSG) installed on board of the research vessel Noruega were performed during the year 2018. Data from this specific campaign were provided by the Instituto Português do Mar e da Atmosfera (IPMA). Blue transects depicted in Figure 2 show the on-shelf measurements taken from 42° N southwards. This IPMA campaign covers part of the WISHE and CADIZ zones for a short time (from 28 April until 30 May 2018). The salinity is calculated as the salinity of the water inside the TSG and it represents salinity of the oceanic surface layer. Therefore, these measurements are used to validate the surface salinity extracted from the different IBI model simulations performed.

Daily salinity expendable bathythermograph (XBT) profiles are obtained from the Coriolis Ocean Dataset for Reanalysis for the Ireland-Biscay-Iberia region [51] real-time website (www.coriolis.eu.org (accessed on 8 April 2021)). These data are located on the shelf (red dots in Figure 2) in the BISCA area and are a part of the RECOPESCA (*Réseau de mesure de l'activité de pêche spatialisé et de données environnementales à usage scientifique*) Project campaigns (IFREMER; *L'Institut Français de Recherche pour l'Exploitation de la Mer*), which collect in-situ data in the Bay of Biscay onboard fishing vessels [52]. The salinity XBT profiles are compared to the daily modelled salinity profiles extracted from IBI simulations at the closest grid point.

2.2. Description of the New River Freshwater LAMBDA Database

The CMEMS Service Evolution LAMBDA project [27] aimed to improve land boundary conditions by coupling watershed models to ocean regional models. The project study area simulated the territories draining to the European Atlantic front and the North Sea, to provide boundary conditions to the CMEMS IBI and NWS (North West Shelves) Monitoring and Forecasting Centers, using the MOHID (Water Modelling System) Land model [53] with regular grids with a 0.05° horizontal resolution. The MOHID Land model, part of the MOHID Modeling System (http://www.mohid.com (accessed on 8 April 2021); [54]), is a physically based, spatially distributed, continuous, variable time step model for the water and property cycles in inland waters that allows simulation of single and multi-catchment simulations.

Each drainage basin domain was simulated for a 11-year period (1 January 2008–1 January 2019) using ERA5 reanalysis meteorology from the Copernicus Climate Change Service (https://climate.copernicus.eu/ (accessed on 8 April 2021)). These simulations resulted in the first LAMBDA data product, used in this study, consisting of daily flow and temperature values near the river mouth for a total of 54 main European rivers. In a second phase, 70 extra discharges for minor rivers in Western Iberia and 364 extra discharges for Ireland and UK domain were extracted to obtain a more comprehensive freshwater budget in these areas. However, these extra rivers included in a second release of the LAMBDA database are not fully validated yet and the model resolution (too coarse) used may be a limitation to accurately represent flows from such minor rivers. Therefore, in this work, only used freshwater data for rivers (the ones used currently as forcing of the IBI model application) included in the first release of the LAMBDA database was used. More information about the LAMBDA project products and results can be seen in Campuzano et al. [55].

2.3. The CMEMS IBI Salinity Model Product

The CMEMS-IBI Near-Real-Time forecast and analysis operational product [56] is generated daily by means of a NEMO model application [57], run over a regional grid (see geographical coverage in Figure 1) with a horizontal resolution of 1/36° (around 3 km) and 50 unevenly distributed z vertical levels. A best estimate of the ocean state (IBI analysis generated through a model run with data assimilation on a weekly basis) and a five-day-ahead forecast (daily updated free model run) are routinely produced for a number of hydrodynamic variables, among others: temperature, salinity, mixed layer depth, zonal and meridional velocity currents and sea surface height. While hourly-averaged estimations are provided for the sea surface, daily-averaged values are computed for the rest of the three-dimensional water column in the entire IBI regional domain. Hourly 3D physical forecast data, covering exclusively the continental shelf and coastal areas, are additionally delivered to foster the dynamical downscaling and subsequent implementation of coastal downstream services.

The CMEMS IBI system uses a SAM2-based (second generation of Mercator Assimilation System) data assimilation scheme [58] to enhance its predictive skills by constraining the model in a multivariate way with a wealth of observations: Altimeter data (i.e., along-track sea-level anomalies), in-situ temperature and salinity vertical profiles (from the CMEMS CORA—Coriolis Ocean Dataset for Reanalysis- 4.1 database) and satellite-derived sea surface temperature (OSTIA—Operational Sea Surface Temperature and Sea Ice Analysis- product) are regularly assimilated to estimate periodically initial conditions for numerical forecast simulation. The resulting product from this latest IBI analysis run is what is delivered as historical best estimates.

At the time of the experiment, the IBI ocean model operational simulation is forced by 3-h daily updated high-resolution (1/8°) meteorological conditions provided by the European Center for Medium-Range Weather Forecast (ECMWF). Lateral open boundary conditions for temperature, salinity and currents, imposed from daily outputs from the parent system (the CMEMS GLOBAL ocean forecast [59]), are complemented with 11 tidal harmonics built from the FES2014 [60] tidal model solution. Land freshwater forcing is included in the IBI operational model set-up. Discharged flows for 33 of the main rivers presented in the IBI area are considered. In-depth insights into this IBI coastal and riverine freshwater forcing are provided in the next section devoted to describing the IBI model sensitivity tests performed in this study. For further technical specifications of the CMEMS-IBI model solution, the reader is referred to Sotillo et al. [20] and Aznar et al. [61].

The multi-parameter skill assessment of this CMEMS IBI ocean forecast system is operationally conducted through the NARVAL (North Atlantic Regional VALidation) toolbox, which routinely monitors its performance and prognostic capabilities by computing a variety of quality indicators [62]. Both real-time ('online mode') and regular-scheduled ('delayed-mode') comparisons are performed using a wealth of independent observational sources as a benchmark, encompassing in-situ observations (collected by moorings, tide-gauges, drifters, gliders and Argo floats networks) and remote-sensed estimations (from satellites and high-frequency radars). For salinity field, the CMEMS-IBI model performance is evaluated using a multiple spatio-temporal scale adopting a multi-platform approach (see list of observational data sources used in the IBI validation in Table 2).

Table 2. Summary of observational (Obs) platforms used in NARVAL (North Atlantic Regional VALidation) toolbox to validate IBI model salinity (* the climatology used is the World Ocean Atlas (WOA) 2013).

Obs Source	Obs Type	Vertical Coverage	Geographical Coverage	Temporal Frequency	Purpose
Mooring Buoys	In-situ	Surface	Mostly on shelf	Hourly	Validation
SMOS	Remote sensed	Surface	IBI domain	Daily	Validation
Argo profilers	In-situ	3-Dimensional	Open waters	Daily	Validation
Gliders	In-situ	3-Dimensional	Ibiza Channel	Daily	Validation
Climatology *	Gridded product	3-Dimensional	IBI domain	Monthly	Consistency

With respect to in-situ observations, the NARVAL software tool makes routine comparisons of CMEMS IBI salinity model products, extracting sea surface salinity timeseries on the grid points closest to the mooring buoy locations, and salinity profiles where Argo or glider observations are available.

Additionally, the IBI SSS field is validated using daily remotely sensed satellite estimations derived from the European Space Agency's Soil Moisture and Ocean Salinity (SMOS) mission. In recent years, the SMOS products have steadily evolved to mitigate systematic biases detected in coastal areas [63], and today, they better portray plume patterns in major river basins [64,65]. However, in the IBI area, there is still room for improvement in these SMOS-derived products, especially in the operational ones and at very coastal areas where systematic biases are identified. The operational SMOS products' limitations (not able to fully capture variability associated to minor coastal freshwater contributions) have conditioned their use in NARVAL to operationally assess IBI salinity product quality, focusing the validation only on deep open waters.

It is pointed out that recent updates of SMOS products, taking benefit of new LAMBDA Project developments, has resulted in a new 9-year (2011–2019) reprocessing of global SMOS SSS daily maps. This new dataset introduces significant improvements with respect to the operational and previous SMOS versions, tending to improve salinity gradients, preserving the small-scale ones close to the coast [66], and minimizing latitudinal and seasonal biases [67]. These improvements support the possibility of using this new SMOS satellite data as a salinity reference in model assessments, and the IBI-MFC has introduced it in NARVAL as an extra salinity data source for IBI delayed mode validations.

Highlights from the operational validation of IBI salinity products with in-situ observations and comparisons with the new SMOS satellite product are shown in Section 3.1.

2.4. Description of the IBI Model Sensitivity Tests Performed Using LAMBDA Product as Forcing

In the CMEMS IBI operational NEMO model set-up, inputs from 33 rivers are prescribed as a daily flux of freshwater (0.1 psu. The main rivers of the European Atlantic Shelf are considered (e.g., Garonne or Rhine, which have mean outflows larger than 1000 m^3s^{-1}), but also smaller rivers which can present a strong temporal variability (e.g., Guadalquivir, whose flow is usually below 200 m^3s^{-1}, but has peaks up to 6000 m^3s^{-1}). To compensate the input from missing rivers, whose individual debits can be negligible but in total have a significant impact on the coastal salinity, a monthly climatological runoff (hereafter FWF_CRF) is applied all along the coast. It is derived from the same climatology used in the CMEMS GLOBAL configuration. The overall freshwater input, surface runoff and rivers, in the operational CMEMS IBI configuration during 2018, had a daily mean flow of 16×10^3 m^3s^{-1}, where river discharge accounted for 67%. The river discharge flows used in the reference CMEMS IBI configuration consist of a composite of observations and climatology (hereafter referred to as FWF_REF) provided by the GRDC (Global Runoff Data Centre; http://www.bafg.de/GRDC (accessed on 8 April 2021)), the French "Banque Hydro" dataset (http://www.hydro.eaufrance.fr/ (accessed on 8 April 2021)), simulations from the Swedish Meteorological Service (SMHI) E-HYPE ((HYdrological Predictions for the Environment)) hydrological model (http://e-hypeweb.smhi.se (accessed on 8 April 2021)) and observations gathered by the Previmer project. For further technical details on how freshwater forcing is included in the IBI operational forecast service, the reader is referred to Amo et al. [56].

Ten rivers discharge in the IBBIS area (Figure 1). Four of them are along the French coast (BISCA area), four in the Western Iberian shelf (WISHE area) and two in the Gulf of Cádiz (CADIZ area). Since April 2018, most of the river discharges were forced by a monthly climatology (FWF_CLM) or by an annual mean flow (in the IBBIS region, only the Mondego river), calculated from the reference forcing. The new forcing set that will be assessed in this study comes from the LAMBDA project watershed product (FWF_LAM), described above. Table 3 shows the mean daily and cumulated annual flow for each of the forcing datasets in the four areas. Mean daily flow of the LAMBDA forcing for year 2018 in the IBBIS area reaches 8910 m^3s^{-1}, that is 70% more than the river input used for the IBI reference run (5223 m^3s^{-1}) and 128% more than the climatological forcing (3893 m^3s^{-1}).

Table 3. Mean daily river discharge (m^3s^{-1}) and cumulated discharge (10^3 m^3s^{-1}) over 2018 for each forcing set, in the full study domain (IBBIS) and the three subregions (BISCA, WISHE, CADIZ).

IBI Freshwater Discharge Forcing Data	Mean Daily (Cumulated) Freshwater Discharge in m^3s^{-1} (in 10^3 m^3s^{-1})			
	BISCA [1]	WISHE [2]	CADIZ [3]	IBBIS [1,2,3]
FWF_REF (IBI_OP Reference Forcing)	**1031** (2826)	**624** (1709)	**251** (687)	**1906** (5223)
FWF_LAM (LAMBDA Forcing)	**1876** (5141)	**1017** (2787)	**358** (982)	**3252** (8910)
FWF_CLM (Climatology)	**806** (2207)	**486** (1330)	**130** (355)	**1421** (3893)
FWF_CRF (Coastal runoff climatology)	**6** (463)	**3** (261)	**1** (57)	**10** (843)

The rivers prescribed as forcing in the subregions are: [1] Vilaine, Loire, Garonne, Adour. [2] Minho, Douro, Mondego, Tagus. [3] Guadiana, Guadalquivir.

Figure 3 shows the daily river discharge of the reference forcing (FWF_REF, blue line), LAMBDA river database (FWF_LAM, red line), climatology (FWF_CLM, cyan dashed line) and the extra coastal runoff (FWF_CRF, grey line), in the IBBIS domain and the three subdomains (BISCA, WISHE, CADIZ). For the IBBIS region, in the first months of 2018, the variability and intensity are equivalent in both the reference and the new LAMBDA datasets, with peaks of freshwater in January and March, though peaks are not always simultaneous (Figure 3a). From April onwards, the FWF_REF is mainly based on the climatology and shows a smooth decrease in total river debit from Spring up to August, when it reaches a minimum average daily flow of ≈2000 m^3s^{-1}. LAMBDA features a minimum in summer too, associated with a reduced variability, but shows two important peaks in June and November, that do not exist in the IBI reference forcing. The difference between the two datasets is particularly important in BISCA (Figure 3b), where the mean river discharge is 1.8 times more important in the new forcing (Table 3).

The impact of using new river discharge forcing in CMEMS-IBI, and more specifically on the simulated salinity fields and frontal features linked to river plumes, will be evaluated through specific IBI-like model scenarios. To assess the performance of the numerical model and identify the main sources of uncertainty linked to the river runoff forcing in the model simulations, four scenarios forced by the previously described freshwater discharge parametrization were performed over the year 2018.

Table 4 shows an overview of the four IBI model land boundary scenarios: IBI_REF is the reference simulation forced by the reference river discharge forcing, and its parameterization is identical to the one used in the operational IBI configuration, IBI_LAM differs from IBI_REF by using the LAMBDA river database forcing, IBI_NOR is like IBI_LAM, but without adding the extra coastal runoff (i.e., the river discharge forcing is the only input of freshwater), and finally, IBI_CLM is a run forced by climatology. The latter scenario was simulated only over the first 6 months of 2018, because the reference run from April is using as forcing the same climatology.

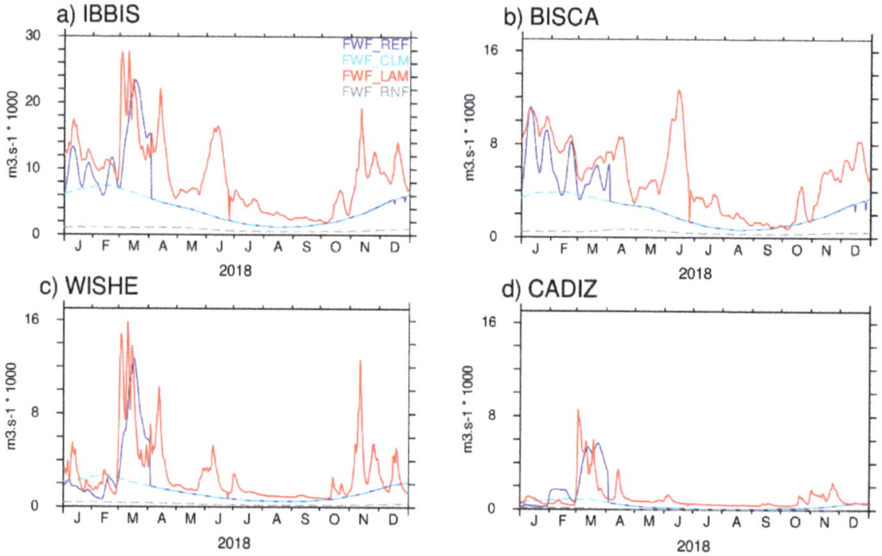

Figure 3. Average daily freshwater discharges imposed as forcing in different IBI sensitivity runs in (**a**) IBBIS, (**b**) BISCA, (**c**) WISHE and (**d**) CADIZ areas, in 2018. The freshwater forcing FWF_REF, FWF_LAM, FWF_CLM and FWF_CRF data sources are respectively depicted in blue, red, dashed blue and grey.

Table 4. IBI model scenarios: Overview of model runs with each specific treatment of river and coastal freshwater inputs.

IBI Model Scenario	River Discharge Forcing	Use of Extra Coastal Runoff	IBI Model Run Type and Main Features
IBI_REF	FWF_REF	Yes	Free run (from 1 January 2018 to 31 December 2018). Control run.
IBI_LAM	FWF_LAM	Yes	Free run (from 1 January 2018 to 26 December 2018). LAMBDA river forcing scenario
IBI_NOR	FWF_LAM	No	Free run (from 1 January 2018 to 26 December 2018). LAMBDA river forcing, but no Coastal runoff
IBI_CLM	FWF_CLM	Yes	Free run (from 1 January 2018 to 19 June 2018). River climatological forcing scenario
IBI_OP	FWF_REF	Yes	CMEMS IBI Operational product (from 1 January 2018 to 31 December 2018) Data Assimilation scheme included (weekly update from IBI Analysis). Used as reference.

All the IBI scenarios were performed with the same model parametrization and forcing datasets, excluding the freshwater forcing, as the CMEMS-IBI operational configuration. A spin-up run of 110 days was performed to generate the same initial condition for all model scenarios. The model started from rest (the 6 September 2017) with initial three-dimensional temperature and salinity fields taken from the CMEMS-IBI operational outputs. After this spin-up period, the model solution was evaluated as stable, showing on the 1 January 2018 velocity fields similar in magnitude and pattern to the CMEMS-IBI operational solution.

All the outputs from these IBI model scenario runs are delivered at hourly frequency for the sea surface, and at a daily frequency for the rest of the water column. Thus, to validate model salinity outputs, surface hourly model data have been used for comparison with mooring observed timeseries to take advantage of the higher temporal frequency of these available in-situ salinity observations. On the other hand, to compare with other observational data sources, such as the ones based on CTD or XBT profiles, daily model salinity outputs at the closest grid point and depth are collocated.

3. Results

3.1. How Does the Operational CMEMS IBI Forecast Model Reproduce Salinity Fields?

In this section, and before analyzing specific results from the IBI model sensitivity tests performed, it is shown how the CMEMS IBI operational model system reproduced salinity fields for the study year (2018). NARVAL toolbox provided a wealth of outcomes and quality indicators on a monthly basis that allowed to quantify CMEMS-IBI prognostic capabilities to reproduce the salinity field during such year. Equally, "parent-son" model intercomparisons were regularly conducted with the purpose of checking the consistency of the CMEMS global solution and the regional downscaled IBI model solution, identifying any potential problem that might be inherited from the coarser global system. In this context, Lorente et al. [17] previously showcased the discrepancies in coastal areas during 2018 between the CMEMS IBI and its parent solution (CMEMS GLOBAL) due to the differences in both the horizontal resolution and the freshwater forcing implemented in their respective operational chains.

The CMEMS operational IBI performance appeared to be rather consistent, especially in open deep waters (Figure 4). The qualitative resemblance of yearly averaged maps of SSS for 2018 between the IBI model product and SMOS estimations was significantly high, with slight differences arising in nearshore shallower areas, especially in the African coastal upwelling region (Figure 4a,b).

With the aim of assessing IBI accuracy in the entire water column, quality-controlled salinity profiles collected by Argo floats were regularly used as a benchmark (Figure 4c–f). Monthly maps of skill metrics were computed by NARVAL toolbox for several depth layers and also for the entire full profiles, as exposed in Figure 4c,d. As it can be observed, moderate RMSE and high correlation values were obtained for March 2018, with most of the values emerging in the range [0–0.5] PSU and [0.71–0.93], respectively. Complementarily, monthly qualitative comparisons were conducted with a focus on specific subregions (Figure 4e,f). Here, the salinity profiles observed and modeled in the Gulf of Biscay are shown (the BISCA subregion displayed in Figure 2). Model outputs were vertically interpolated into the observation depth levels to facilitate the visual comparison. Likewise, averaged salinity profiles (solid black lines) were calculated to infer the main features of the water column in this area. According to the values for specific levels (in blue), the model-observations accordance was relevant for March 2018. Reduced differences were detected in the upper layer, where higher spatial-temporal variability of the salinity field is observed, lying between 33 and 36 PSU. Monthly statistical results for the rest of the year (not shown) illustrated that CMEMS IBI performance in off-shelf waters was consistent, operating within acceptable ranges all the year-long.

Since accuracy of SMOS remote-sensed estimations is higher in open offshore waters (these products showing a more limited precision in coastal areas mostly due to the lack of valid observations—this is so especially in the operational products and somehow minimized in the reprocessed ones) and Argo observations were constrained to offshore waters beyond the continental shelf, ancillary validation exercises with coastal buoys were required to objectively evaluate CMEMS IBI performance in the periphery of ROFI areas, where impulsive-type freshwater discharges might be noticeable. Using surface salinity observations from the coastal moorings available along the northwestern Iberian shelf (M5–M12 mooring buoys, depicted in Figure 2), a contingency table is served, not only as a summary of the most relevant SSS drop events occurred during 2018, but also as a qualitative overview of the CMEMS IBI model ability to adequately reproduce them (Figure 5a).

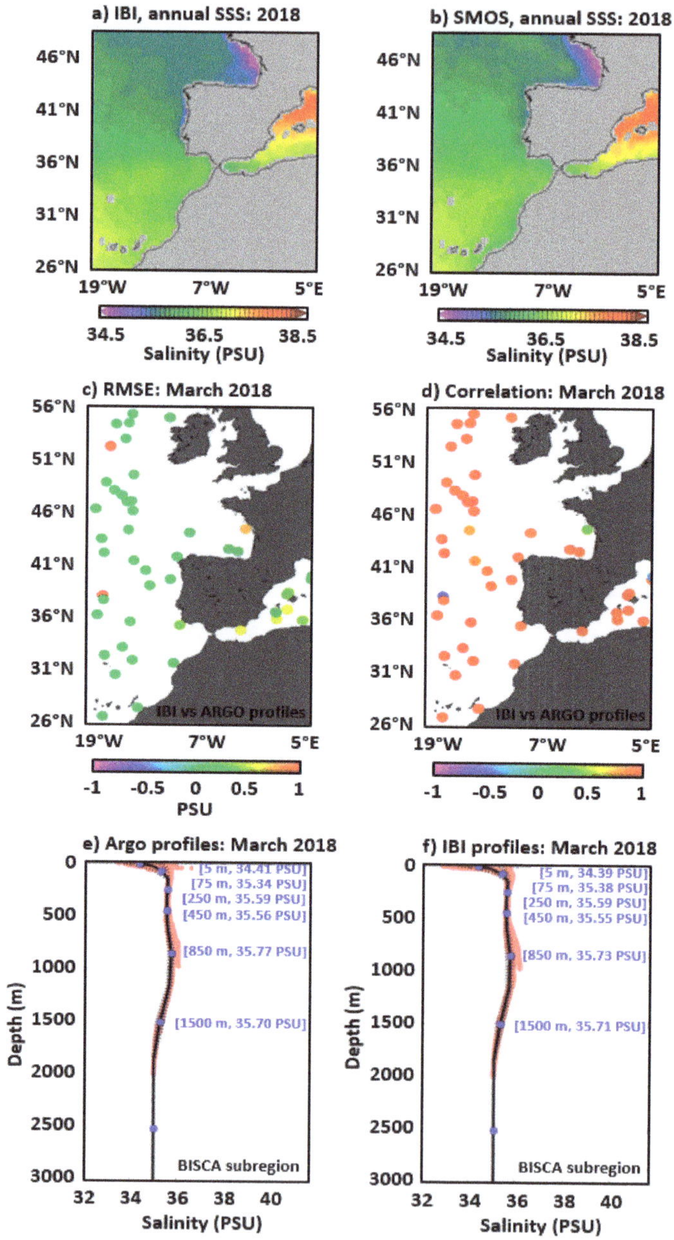

Figure 4. 2018 yearly averaged maps of sea surface salinity (SSS) simulated by CMEMS IBI (**a**) and derived from remotely sensed SMOS data (**b**). Monthly maps (March 2018) of skill metrics derived from the comparison of entire salinity profiles observed by Argo floats and modelled by CMEMS IBI: root mean squared error (**c**) and correlation (**d**). Daily salinity profiles collected by Argo floats (**e**) and IBI (**f**) in the Gulf of Biscay (BISCA zone) during March 2018. Averaged profiles exposed in solid black lines. Mean depth values gathered in blue.

a) Contingency table: relevant SSS drop events in NW Iberian waters during 2018

	EVENTS OBSERVED		
EVENTS FORECASTED	YES	NO	TOTAL EVENTS
YES	24 (hit)	22 (false alarm)	46
NO	11 (miss)	------	11
TOTAL EVENTS	35	22	57

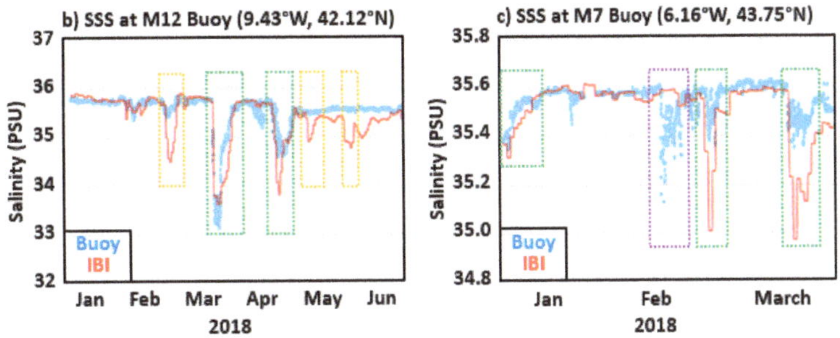

Figure 5. (**a**) Contingency table focused on significant sea surface salinity (SSS) drops detected during 2018 at coastal buoys moored around the Iberian Peninsula (buoys M2–M12 used, see Figure 2 for location). Hit, miss and false alarm events are highlighted in green, purple and yellow colors, respectively. (**b**) Six-month observed (blue dots) and IBI modeled (red line) SSS timeseries at Silleiro buoy (M12). Relevant SSS drops marked with colored dashed squares. (**c**) Three-month timeseries comparison of SSS observed at Cabo de Peñas buoy (M7) and modelled by CMEMS-IBI at the closest grid point.

As it can be deduced from the contingency table, CMEMS IBI was able to represent 69% (24 of 35) of the episodes categorized here, with variable degreed of precision. By contrast, almost 50% (22 of 46) of the forecasted SSS drop events constituted false alarms, which were not confirmed by the in-situ observation. To better illustrate this validation approach, two examples at 2 buoys' locations are provided (Figure 5b,c). In the first case, the 6-month SSS timeseries at Silleiro buoy (M12) location revealed a series of events, categorized as modelled hit and miss event (Figure 5b), chronologically described hereinafter:

- 22–28 February 2018: False alarm. IBI showed a SSS drop (more than 1 PSU), not so intense in the observed data.
- 20–30 March 2018: Hit. Both IBI and observed data exhibited a SSS drop (of more than 2 PSUs), with a very similar time evolution.
- 21 April–1 May 2018: Hit. Both IBI and Silleiro Buoy observation exhibited a SSS drop (about 2 PSUs), and the model overestimates the observed decrease.
- 10–14 May 2018: False alarm. IBI showed a non-observed SSS drop (around 1 PSU).
- 30 May–10 June 2018: False alarm. IBI showed a SSS drop around 1 PSU.

In the second case (Figure 5c), the 3-month comparison of SSS at Cabo de Peñas buoy (M7) location showed the following timeline features (hits and a miss event):

- 1–9 January 2019: Hit. Both IBI and observed data exhibited a moderate SSS drop of 0.5 PSU and a steady recovery during the following days.
- 13–17 February 2019: Miss. Observed drop of 0.5 PSU not reproduced by IBI.
- 25–28 February 2019: Hit. IBI clearly overestimated the observed SSS drop.
- 20–26 March 2019: Hit. IBI clearly overestimated the observed SSS drop.

3.2. IBI Model Sensitivity to Changes in Freshwater Coastal and River Forcing

This section shows the main results of the IBI modeling scenarios performed to evaluate the regional impact on salinity after using different river/coastal freshwater forcing. In order to assess the four different IBI scenarios performed and to identify the main sources of uncertainty linked to the use of different river runoff forcing, the modelled outcomes have been validated with different in-situ salinity observations.

Figure 6 shows the seasonal SSS of IBI_REF, and the mean difference with the other simulations. In winter, that is the first 3 months after the start of the runs, the differences in the surface salinity field between the various simulations are limited to the shelf, and more specifically to the ROFI areas. Due to the higher river discharge of LAMBDA, IBI_LAM and IBI_NOR are fresher than IBI_REF. The same patterns are found in autumn. The small differences observed offshore at this time come from the expected propagation of the differences within the domain after 12 months of simulation but confirm that there is no noticeable bias appearing with any of the forcing sets. In spring and summer, the river discharge of LAMBDA is stronger than the other sources (as seen in Figure 3) and the fresher water masses of the two simulations forced by LAMBDA (IBI_LAM and IBI_NOR) extend to the open ocean, especially in the BISCA area. The Vilaine and Loire river system (at 47° N) impacts the surface and subsurface layers all year-long. The southernmost section of BISCA is the most sensible to the impact of the coastal runoff, as shown by the higher salinity of IBI_NOR in this area.

The simulation forced by a climatology (IBI_CLM) is saltier than the other simulations (+0.05 PSU over the IBBIS domain compared to IBI_REF) in winter. This emphasizes the fact that using a climatology vs. realistic daily river discharge significatively changes the salinity budget on the shelf.

These mean SSS seasonal snapshots from the test simulations show differences in the salinity patterns. Furthermore, it is needed to validate the simulated salinity fields against observations in order to assess the impact of these changes in the freshwater forcing.

The comparison with observed salinity timeseries from moorings shows that all the simulations globally manage to accurately reproduce the spatial-temporal variability of salinity at the coast and at the shelf-break, within the IBBIS domain (Figure 7). The correlations between mooring measurements and model scenarios outputs are most of the time above 0.6, and the standard deviation below 0.5 PSU. The capacity of the IBI model to reproduce the salinity is more explained by the geography and resolution of the local dynamics than by the parametrization of the river forcing: it can indeed be seen on the Taylor diagram (Figure 7a) that points are gathered by station, rather than being gathered by simulations. The simulation forced 6 months by the climatology differs most of the time from the others: it may be explained by the impact of a diurnal vs. climatological freshwater discharge forcing or by the shorter length of the simulation (6 months vs. 1 year).

As each subregion features its proper dynamics and local processes, the assessment of the salinity variability and fronts location, and the assessment of the impact of the river discharge forcing on the IBI configuration, have been conducted separately in BISCA, WISHE and CADIZ areas.

3.2.1. BISCA Region

The BISCA area is sensible to the open-ocean dynamics in its Southern part, where the shelf is narrower and the poleward slope-current is close to the coast. Where the shelf is wider, in the French shelf area, the tides and the freshwater runoff are the dominant factors. This local variability shows in the mooring's measurements: in the northern part (M1 and M2, in Bretagne), the salinity is steady in summer and autumn, when stratification occurs. In the southern section (M5 to M7), the variability is, on the contrary, strong in spring/summer, when the mesoscale activity on the shelf is at its maximum (such as M6, Figure 7b). The simulations are all coherent with these local seasonal patterns. In the case of the station M6, IBI_CLM features an unrealistic variability in winter, that tends to support the hypothesis that a climatological river forcing degrades the solution in terms of variability, compared to a diurnal one.

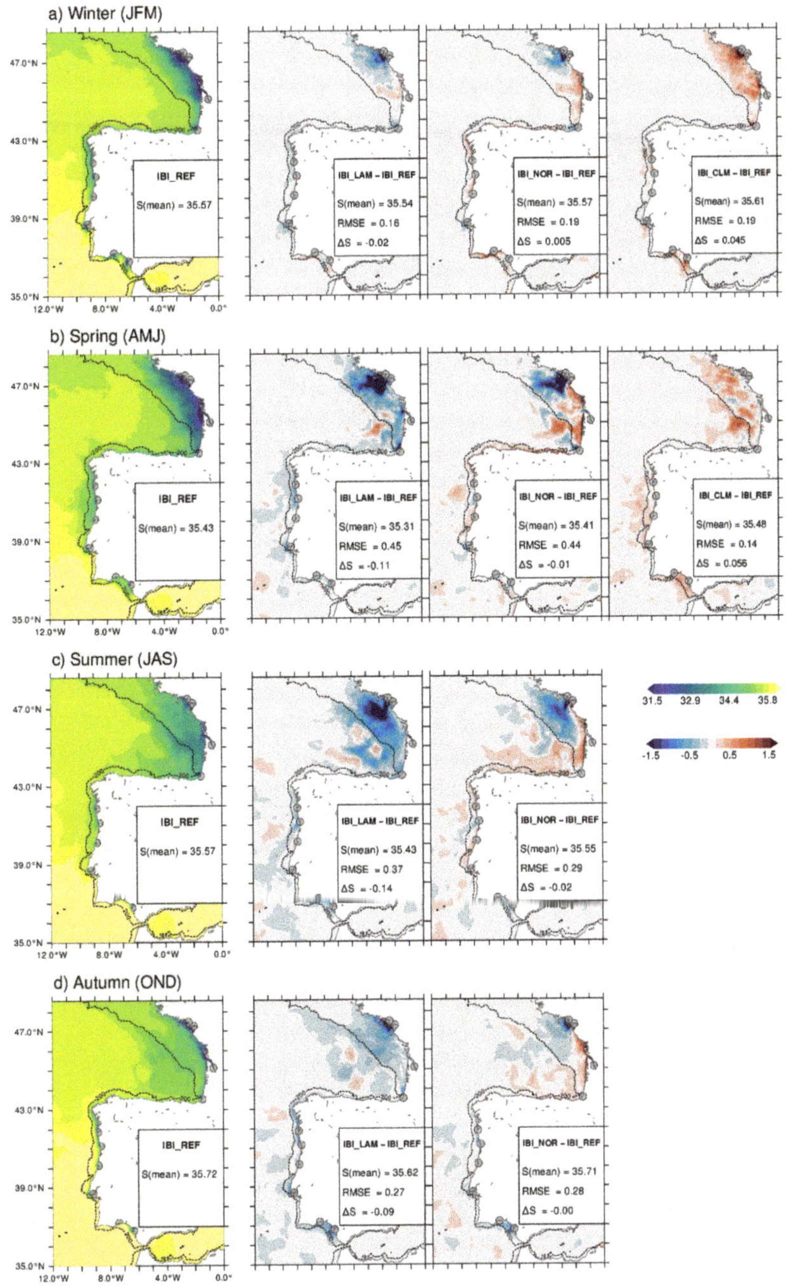

Figure 6. (**a**) Seasonal surface salinity in winter 2018 from IBI_REF (left panel) and difference of seasonal surface salinity between IBI_LAM, IBI_NOR, IBI_CLM (respectively in panels from left to right) and IBI_REF. (**b**) Same as (**a**) but in spring. (**c**) Same as (**a**) but in summer. (**d**) Same as (**a**) but in autumn. Red (blue) shading in the panels (**a**) to (**c**) means that the simulation is saltier (fresher) than IBI_REF. Salinity mean values and statistical metrics (root mean square error and bias computed when comparing each scenario with the reference one) are shown in each panel. Grey dots show the river mouth locations of the rivers considered in the IBI model set-up.

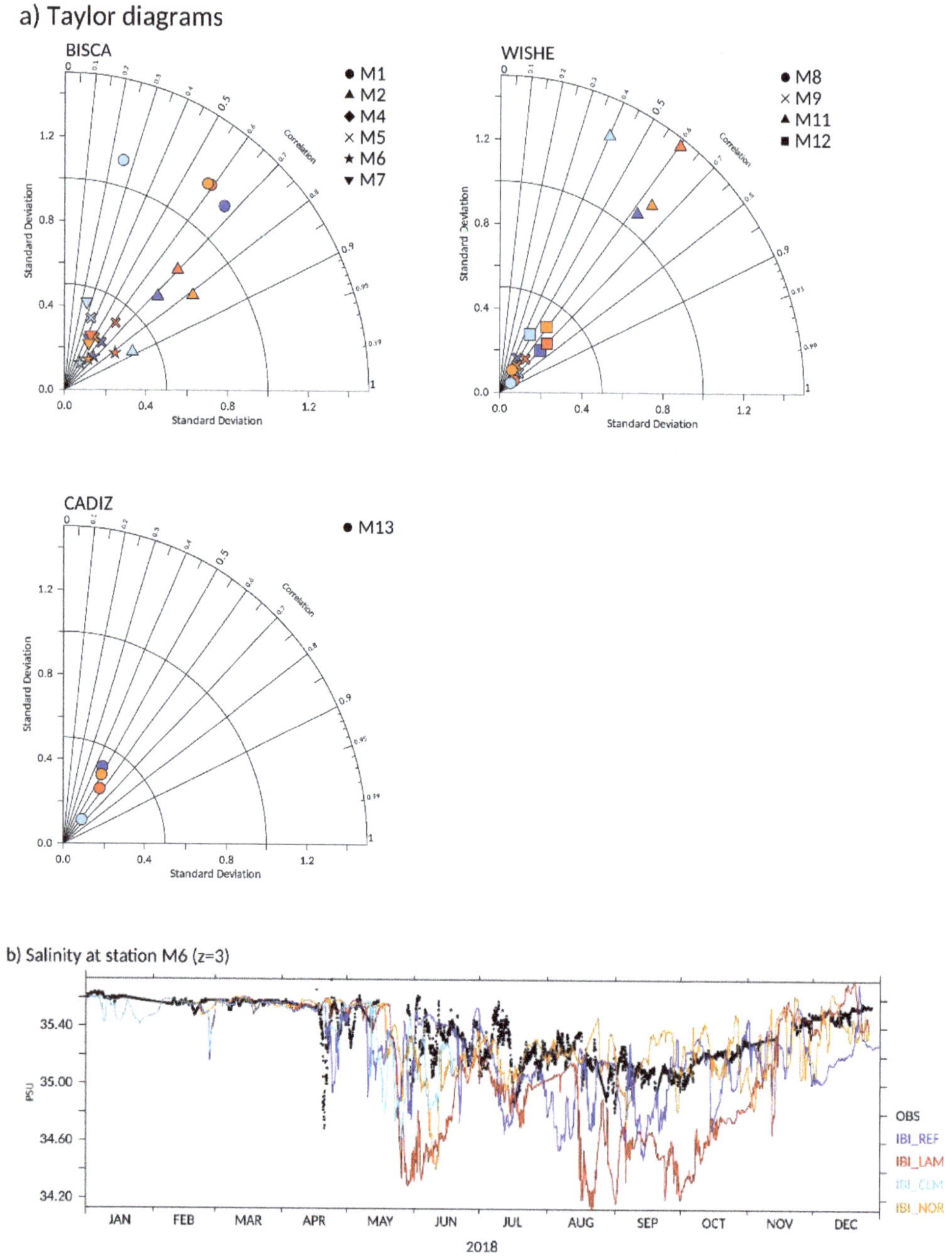

Figure 7. (**a**) Taylor diagrams showing observed-modelled salinity comparisons at moorings in the BISCA, WISHE and CADIZ areas. Each location is represented by a different symbol, whereas colors represent the model scenario simulation (IBI_REF in blue, IBI_CLM in light blue, IBI_LAM in red and IBI_NOR in orange). M3, M4 and M10, because of the important bias between observations and simulations, are not represented here. (**b**) Timeseries of salinity at station M6, at the shelf-break of Santander, Northern Spain, in BISCA. The color code is the same as in panel (**a**).

The RECOPESCA campaign took place from 24 April to 30 May 2018. Even though its temporal coverage is limited to one month, it covers the whole French shelf and allows to estimate the extension of the ROFI (Figure 8). The low-salinity front does not reach the shelf break on the wider part of the shelf (north of 46° N). All the simulations overestimate the offshore extension of the ROFI, but the run forced by a climatology which, as seen from the in-situ timeseries, degrades the variability of salinity, features a smaller bias at the time of the campaign (1.55 PSU at the surface) than the other simulations. We can hypothesize that all the "realistic" river forcing data introduce too much discharged water on the shelf and/or do not manage to evacuate it. Nevertheless, validation of the salinity field is difficult because of the great variability of salinity on the shelf in the BISCA area due to large inputs of fresh water and strong variability.

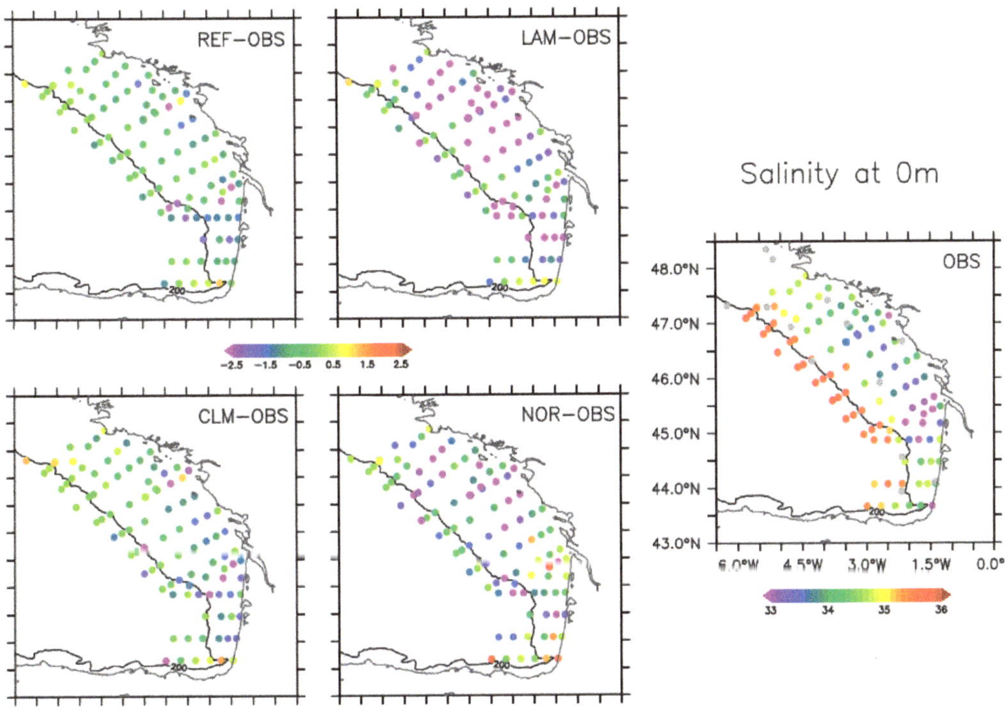

Figure 8. Sea surface salinity (SSS) measured with XBTs from the RECOPESCA campaign for May 2018 in the BISCA area. The right panel shows observed (OBS) salinity, whereas panels on the left show the respective differences between the simulated salinity by each model scenario (i.e., IBI_REF, IBI_LAM, IBI_CLM, IBI_NOR) and the RECOPESCA observations.

The combined input from the extra coastal runoff and LAMBDA (IBI_LAM) creates a significant fresh pool of water on the shelf, particularly in the South, which degrades the solution at the time of the RECOPESCA campaign (Figure 8 and Table 5). However, by removing the extra coastal runoff (IBI_NOR), the solution is improved. The same observation can be made from the moorings' timeseries: the simulation run with LAMBDA but without extra coastal runoff locally improves the solution, meaning that the salinity bias does not come from the LAMBDA forcing itself. In fact, the extra coastal runoff is not needed anymore on the southern shelf of Biscay, as the freshwater discharge is more realistic, even though it is still needed in other areas, where rivers are not parametrized (such as Brittany in the northern BISCA). A full network of coastal salinity observations would be optimal to tune this extra monthly runoff.

Table 5. Mean difference (Bias) and Root Mean Squared Error (RMSE) between observed salinity (from mooring buoys and RECOPESCA XBT profiles) and simulated one (from the IBI_REF, IBI_LAM, IBI_CLM, IBI_NOR model scenarios), over the respective length of the simulations, in the BISCA area. The smallest model bias and RMSE for each dataset are in bold. The mooring data have hourly frequency and RECOPESCA data daily frequency. N is the number of measurements.

Salinity Observations in BISCA			RMSE				Bias (Model Observations)			
			IBI_REF	IBI_LAM	IBI_CLM *	IBI_NOR	IBI_REF	IBI_LAM	IBI_CLM *	IBI_NOR
Mooring (2018)	Depth (m)	N								
M1	5	8314	**1.18**	1.21	1.35	1.54	0.16	**0.04**	0.75	0.97
M2	1	3622	0.96	1.58	**0.64**	1.22	−0.72	−1.36	**−0.52**	−0.94
M3	1	6633	2.68	2.63	**2.47**	3.71	1.28	**1.14**	1.33	2.88
M4	10	2701	**0.78**	1.18	0.91	1.59	**−0.65**	−0.92	−0.69	−1.12
M5	3	6938	0.31	0.49	0.39	**0.28**	−0.11	−0.28	−0.13	**0.00**
M6	3	8160	0.24	0.38	**0.15**	0.19	−0.11	−0.24	−0.04	**−0.03**
M7	3	7022	0.32	**0.30**	0.48	0.38	0.18	**0.11**	0.22	0.29
RECOPESCA (May 2018)	Depth (m)	N								
	0	106	1.69	2.14	**1.55**	1.79	−0.76	−1.24	**−0.52**	−0.79
	5	106	0.99	1.39	**0.74**	1.14	−0.67	−1.03	**−0.39**	−0.71
	10	106	0.47	0.61	**0.46**	0.56	−0.21	−0.32	**−0.03**	−0.21
	20	106	0.38	0.43	**0.28**	0.39	−0.20	−0.25	**−0.10**	−0.19
	50	106	0.11	**0.09**	**0.09**	**0.09**	−0.06	−0.06	−0.05	**−0.04**

* IBI_CLM timeseries are shorter than the others.

3.2.2. WISHE Region

The salinity at the shelf-break of WISHE is validated by comparison against the mooring stations M8, M9 and M12, all situated in the Northern part of the area. Here, salinity is steadier than in the BISCA area and it is mainly controlled by the slope current. However, occasional episodes marked by the advection of lower salinity water masses coming from the shelf (and highly modified by river and coastal freshwater influence) are clearly identified in the mooring sites. As previously commented in the description of the CMEMS salinity product validation, the IBI operational system can capture to some extent such noticeable surface salinity drops occurred in the peripheries of ROFI areas (see the forecast contingency table shown in Figure 5 in Section 3.1). Likewise, the IBI model scenarios generally accurately reproduce these observed freshwater intrusions, with small variations in timing and/or intensity.

As an example, Figure 9 shows the salinity from March to May 2018 at the western shelf-break of Galicia (station M9). A first well-marked event occurred on 20 March 2018 and has been documented by Campuzano [68] and Lorente et al. [17], and a second one less clearly defined occurred around 22 April 2018. This event of low-salinity intrusion at the shelf-break took place all around the Galician shelf (indeed, the same salinity drop event was also observed at the M12 station for the same period, see Figure 5b). All the IBI model simulations capture the events, at both locations, but show different performance characteristics. Thus, IBI_LAM is clearly too fresh during the first SSS drop event, but it is the only model run to replicate the low-salinity minimum observed on 30 April 2018. On the other hand, IBI_REF seems to be the only simulation to accurately capture the salinity drop on 14 April 2018. Finally, IBI_CLM is systematically too salty and IBI_NOR features a false salinity drop on 22 April 2018 at M12, further south (figure not shown).

Snapshots of model salinity fields at the dates when salinity dropped are observed in-situ (figures not shown) and indicate that these recorded events of low salinity are caused by filaments of freshwater, extending from the shelf towards the open ocean. The model simulations, actually all of them, feature these freshwater filaments, but with small spatiotemporal variations (of the order of a day or a few kilometers), meaning that at a given location (such as the mooring station), missing an event can be a matter of a few model grid points. Essentially, the resolution of this dynamical scale is stochastic, and a systematic resolution cannot be expected from a regional configuration such as the IBI one. However, this assessment shows that the IBI model configuration can catch the level of dynamical activity, with all model scenarios reproducing main salinity drop events. Once again, we emphasize that the use of a climatology slightly degrades the variability (timing and intensity) of the salinity.

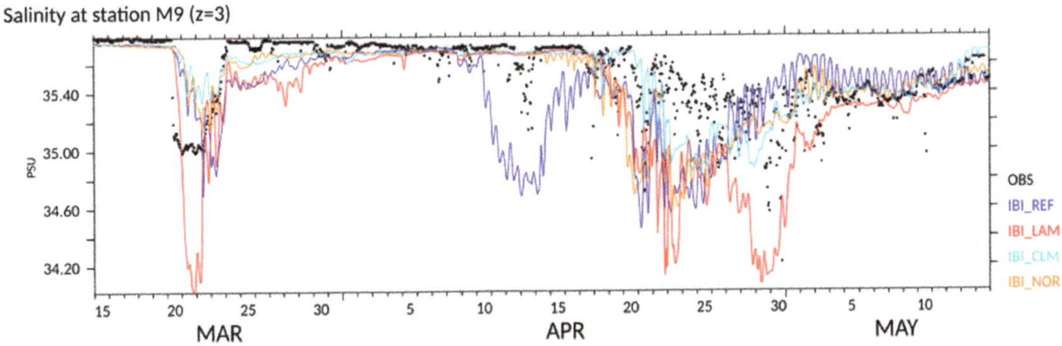

Figure 9. Timeseries of salinity at mooring station M9, at the shelf-break of Northwestern Iberia, from 15 March to 15 May 2018. Observed salinity represented by black dots and the different IBI model scenarios by solid lines (IBI_REF in blue, IBI_CLM in light blue, IBI_LAM in red and IBI_NOR in orange color).

In order to complete the analysis in this area, the weekly salinity profiles obtained from the INTECMAR CTD station records are used, located at same latitude as moorings M10 and M11. These observations are very coastal and taken at the mouth of the Rias Baixas in Galicia. This area is affected by freshwater plumes originating from discharges of several rivers [49], which are not parameterized in the IBI model set-up as proper river sources, but somehow it should be taken into account by means of the extra coastal runoff climatological forcing.

For validating the IBI model scenarios, simulated salinity profiles at the closest model grid point are vertically interpolated into the depth levels of the observations. All the simulations accurately reproduce the variability of salinity with observed data: steady in summer, but with salinity drops in winter and spring, affecting the whole water column in March and April. Table 6 shows statistics for the model validation at the 5 CTD INTECMAR stations. Figure 10 shows temporal evolution of observed and modelled salinities at the INTECMAR station C3 (8.96° W/42.48° N), showing a negative bias for all simulations in winter and the beginning of spring (stronger in November and December for IBI_LAM simulation), and a positive bias in the month of April (stronger for IBI_CLM simulation). Statistically, the simulation without extra coastal runoff has a better variability than the other simulations, with a correlation most of the time better than the other test simulations at both the INTECMAR stations and at the mooring buoy station, and also a smaller bias. It seems that by removing this extra local freshwater input, the simulated variability of salinity is improved. This fact can suggest that the salinity budget in this coastal point is rather controlled by the extension and variability of the western Iberian Buoyant Plume (mainly supplied and controlled by discharges of the Minho, Duero and Mondego rivers) than by the local runoff from the nearby Rias.

There are no timeseries of observed salinity from mooring buoys on the southern part of the WISHE area. However, the IPMA campaign recorded surface salinity transects across the Portuguese shelf, going South from 42° N. These thermo-salinometer measurements (taken from 28 April to 20 May 2018) have been used to validate the different IBI model scenarios in the area. Figure 11 shows how all the simulations have a negative surface salinity bias from 42° N to 39.5° N and at the Tejo river mouth, but inversely are too salty on the southern part of the measured area. As all the model scenarios feature the same pattern of bias in the region, it seems that the river discharge forcing data source is not the main cause of salinity errors on the shelf freshwater budget, with there being other dynamical factors that may explain this model behavior.

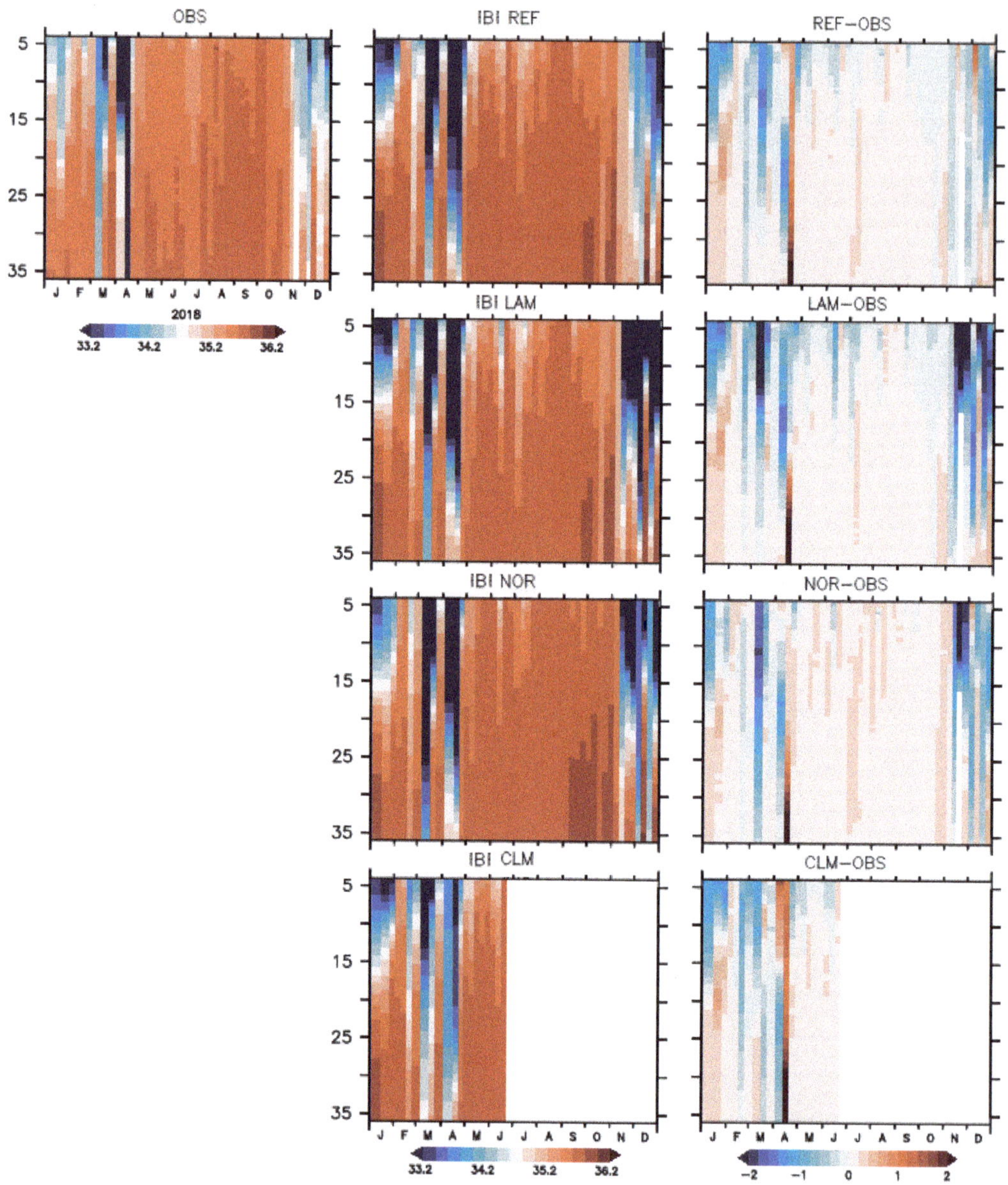

Figure 10. Salinity profiles measured by CTD at the INTECMAR station C3 (8.96° W/42.48° N) for 2018 in WISHE area (**left** panel) in comparison with salinity profiles simulated, at the closest model grid point, by the different model scenarios—IBI_REF, IBI_LAM, IBI_NOR and IBI_CLM (**middle** panel). Differences between modelled and observed salinity profiles (**right** panels). Note that IBI_CLM run is shorter than other simulations.

Table 6. Mean difference (Bias) and Root Mean Squared Error (RMSE) between observed salinity (from mooring buoys, an IPMA campaign with thermo-salinograph and INTECMAR CTD stations) and simulated one (from the IBI_REF, IBI_LAM, IBI_CLM, IBI_NOR model scenarios), over the respective length of the simulations, in the WISHE area. The smallest model bias and RMSE for each dataset are in bold. The mooring data have hourly frequency, the IPMA data was measured every 10 min and INTECMAR measurements at each CTD station have a weekly frequency. N is the number of measurements.

Salinity Observations in WISHE			RMSE				Bias (Model Observations)			
			IBI_REF	IBI_LAM	IBI_CLM *	IBI_NOR	IBI_REF	IBI_LAM	IBI_CLM *	IBI_NOR
Moorings (2018)	Depth (m)	N								
M8	3	8587	0.16	0.13	**0.09**	0.19	0.12	0.10	**0.06**	0.15
M9	3	8676	0.19	0.20	**0.13**	0.17	0.06	0.03	**−0.01**	0.08
M10	3	6981	13.1	**13.0**	17.7	13.48	9.2	**8.9**	14.0	9.62
M11	9	6835	1.43	1.99	1.64	**1.19**	−0.93	−1.35	−0.96	**−0.24**
M12	3	8701	**0.28**	0.34	0.91	0.38	−0.03	−0.10	−0.05	**−0.01**
IPMA	Depth (m)	N								
	surface	1563	0.51	0.73	**0.50**	0.64	−0.14	−0.18	−0.06	**0.01**
INTECMAR		N								
C1		41	0.35	0.45	0.47	**0.32**	0.15	0.25	0.20	**0.02**
C2		38	0.50	0.62	0.78	**0.48**	0.14	0.26	0.17	**0.03**
C3		48	**0.36**	0.51	0.55	0.43	−0.06	0.07	−0.14	**−0.04**
C4		45	0.39	0.53	0.45	**0.38**	0.05	0.14	0.07	**−0.01**
C5		45	**0.48**	0.66	0.59	0.51	0.10	0.21	0.09	**0.04**

* IBI_CLM timeseries are shorter than the others.

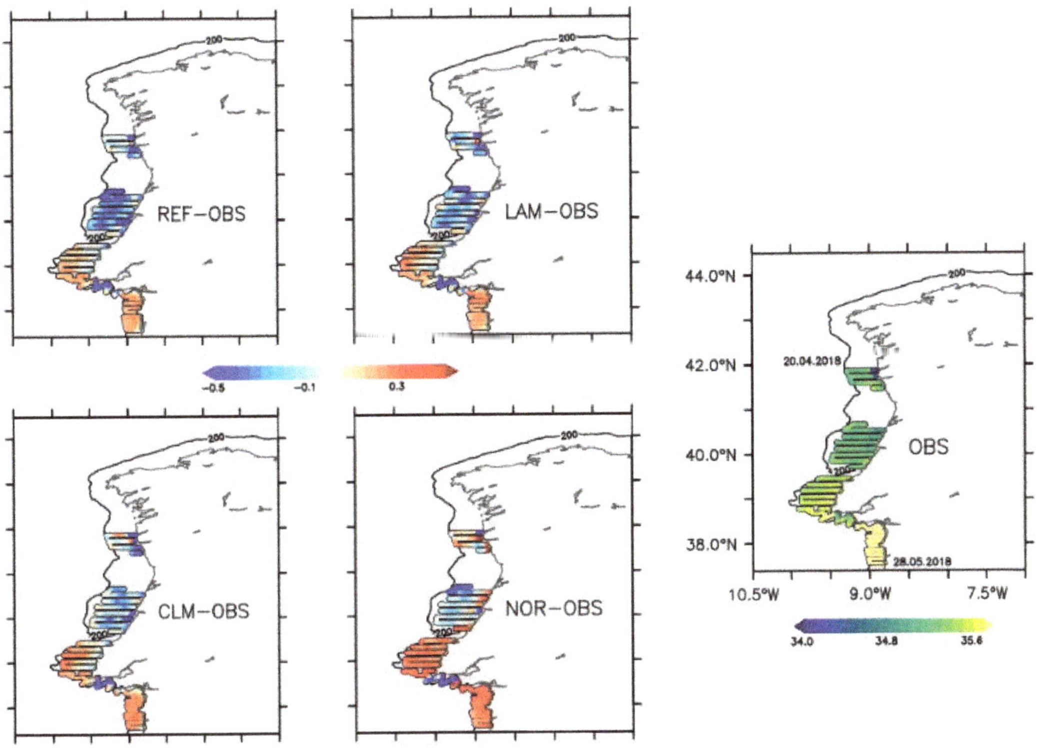

Figure 11. Surface salinity during the IPMA campaign in May 2018 in WISHE area. The **right** panel shows observed (OBS) salinity (measured by means of a thermo-salinometer), whereas panels on the **left** show the respective differences between the simulated salinity (from each model scenario: i.e., IBI_REF, IBI_LAM, IBI_CLM, IBI_NOR) and the IPMA observations.

3.2.3. CADIZ Region

At the northern shelf-break of the Bay of Cádiz, we know from the data at mooring M13 that the salinity is steady around 36.5 PSU, except in summer when some small freshwater intrusions occur (±0.2 PSU, Figure 12b). The simulations have a correct average salinity here (less than 0.2 PSU of bias, Table 7), but they reproduce punctual spurious intrusions, associated with peaks of river discharge. For instance, the strong river discharge of the reference forcing in February (Figure 3) has induced an extension of the ROFI up to the station, which is not realistic and only reproduced by IBI_REF. The same kind of behavior is observed in IBI_LAM in November.

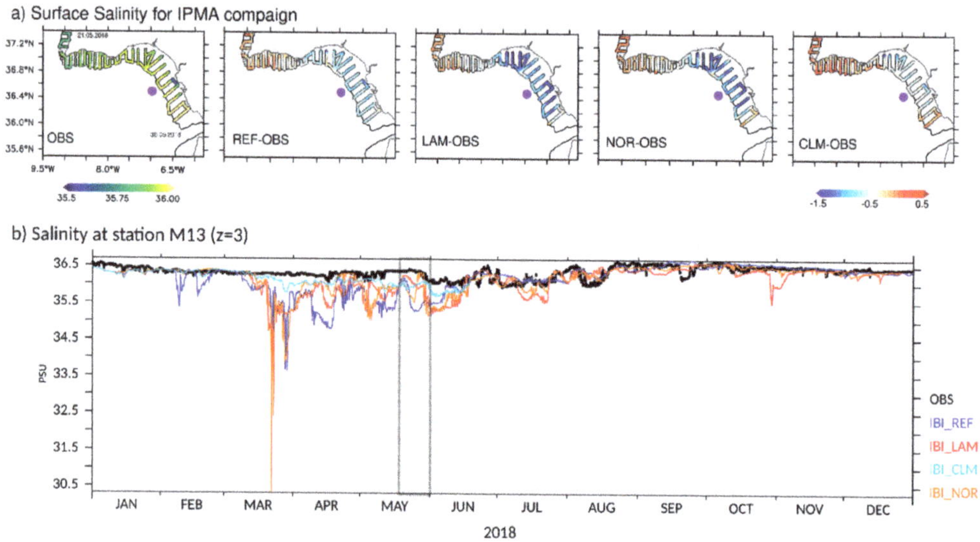

Figure 12. (**a**) Surface salinity during the IPMA campaign in late May 2018 in the CADIZ area. The left panel shows the observed salinity measured (by means of a thermo-salinometer), and the other panels (from left to right) show differences between modelled salinities (from IBI_REF, IBI_LAM, IBI_NOR, IBI_CLM model scenarios) and the observations. The purple dot depicts the location of the mooring buoy M13. (**b**) Timeseries of salinity at station M13, at the shelf-break of the Gulf of Cadiz. Observed salinity is represented by black dots and the different IBI model scenarios by solid lines (IBI_REF in blue, IBI_CLM in light blue, IBI_LAM in red and IBI_NOR in orange color). The grey rectangle highlights the period of the IPMA campaign in the Gulf of Cadiz.

Table 7. Mean difference (Bias) and Root Mean Squared Error (RMSE) between observed salinity (from the M13 mooring buoy and the IPMA campaign) and simulated one (from the IBI_REF, IBI_LAM, IBI_CLM, IBI_NOR model scenarios), over the respective length of the simulations, in the CADIZ area. The smallest model bias and RMSE for each dataset are in bold. The mooring data have hourly frequency and IPMA data was measured every 10 min. N is the number of measurements.

Salinity Observations in CADIZ			RMSE				Bias (Model Observations)			
			IBI_REF	IBI_LAM	IBI_CLM *	IBI_NOR	IBI_REF	IBI_LAM	IBI_CLM *	IBI_NOR
Mooring (2018)	Depth (m)	N								
M13	3	8587	0.45	0.37	**0.21**	0.41	−0.19	−0.20	**−0.15**	−0.17
IPMA (May 2018)	surface	1217	0.59	1.18	**0.42**	1.12	−0.50	−0.87	**−0.28**	−0.71

* IBI_CLM timeseries are shorter than the others.

While the salinity at the M13 shelf-break station features a higher variability from May to August (Figure 12), the model scenario simulations reproduce instabilities over a longer period (March to July) and much fresher waters (up to 5 PSU of bias with respect to the measured salinity). The most significant error occurs in March, where all simulations except the one forced by the climatology reproduce a significative drop of salinity at the shelf-break. The IPMA campaign in CADIZ area took place in late May on the shelf. The comparison against the simulated salinities shows that the bias of salinity observed at the shelf-break mooring for this period of the year extends to the whole shelf, from 8.2° W to the strait of Gibraltar. In this regard, Table 7 shows how the waters simulated by the proposed model scenarios are 0.5 to 0.87 PSU fresher than the ones observed in the IPMA campaign. This bias is especially strong in the simulations forced by LAMBDA, which has the biggest river discharge at this time of the year.

Even though the comparison of IBI model scenarios against surface salinity shelf transects shows a systematic bias in the IBI simulations in May, this assessment cannot be extended to the whole period of study, as there is no salinity measurement available in winter on the shelf to support or invalidate the assumption. In this case, observations from the only mooring buoy available in the region (M13, located at the shelf-break and far away from the coast) show much lower variability than in the mooring buoy cases shown for the other study subregions. Indeed, the existence of salinity drop events, controlled by ROFIs intrusions, are not so evident at this M13 station. The climatological forcing seems to be the most adequate in this CADIZ region, showing lower errors and bias.

4. Discussion

The present study analyzed the response of an operational ocean model regional application to changes in its river and coastal freshwater forcing inputs. The study is focused on the European Atlantic margin, the so-called IBI area, and the study domain (IBBIS) has been divided into three subregions of interest: CADIZ, WISHE and BISCA (see geographical domains in Figure 2). It should be emphasized that the two latter regions (both the Western Iberian Shelf (WISHE) and the Gulf of Biscay (BISCA)) are ROFI areas, where circulation patterns are ruled by significant density differences and the baroclinic frontal features existing between the saltier sea waters and the fresher coastal ones (highly influenced by coastal run-off and river freshwater discharges). It is well-known that differences between existing operational modelled salinity products become particularly evident in ROFI areas (mostly due to the diverse usages that operational forecast services apply to coastal and river freshwater forcing). Thus, a non-optimal representation of coastal and river freshwater signal in the ocean model set-ups can certainly lead to biased simulated ocean patterns, jeopardizing the reliability of ocean forecasts in these ROFI areas.

Simultaneously to this need of operational ocean models to count with an optimal land freshwater forcing, there is a steady effort by the hydrology community to enhance monitoring of rivers and runoff rates. Several multidisciplinary R&D projects and initiatives are working on building new improved freshwater river discharge databases (combining hydrological models and observational data sources) that can be useful to improve ocean model forecasting. With the objective of quantifying the impacts that coastal and river freshwater contributions have on IBI regional ocean model simulations, the new LAMBDA river discharge database (LAMBDA Project [27], Campuzano et al. [55]) was tested as part of the IBI ocean model freshwater forcing.

4.1. How Are the Proposed Ocean Model Scenarios Configured?

Several ocean model simulations were carried out here to evaluate model sensitivity to changes in coastal and river freshwater forcing. The proposed ocean model scenarios were built using the same NEMO model set-up used by the CMEMS IBI-MFC to generate its operational regional forecasts along the year 2018 [56]. This IBI operational model set-up included a freshwater forcing from land composed of two different components: (1) punctual river discharges, imposed as an open boundary condition for the main 33 rivers discharging in the IBI model domain, and (2) an extra monthly climatological runoff applied along all the IBI coastal grid points. This extra coastal freshwater input tends to compensate the identified IBI discharges deficit related to missing rivers and to the use of climatological diminished daily flow rates [69]. The river discharge values originally imposed in this IBI operational model set-up at the 33 rivers result from a combination of different data sources, including: (1) observed inflows from in-situ river discharge stations (daily forcing data applied only at 8 out of the 33 rivers, and only in hindcast and analysis IBI runs), (2) hydrological model estimations (imposed at all the IBI rivers in the forecast horizons, but also applied in IBI hindcast and analysis runs at those rivers where no flow observation is available) and finally, (3) climatological inputs (prepared to be used at any of the 33 IBI rivers as a back-up solution in case of unavailability or failure in the update of any of the aforementioned observed and modelled river data sources). It is pointed out that this combined use of hydrological observations and model outputs to fix the IBI river freshwater forcing was only active until April 2018; afterwards, until the end of the year 2018, the river freshwater input responded to a pure climatological forcing (averaged daily freshwater discharges shown in Figure 3 illustrate this situation).

The four 2018 IBI runs, performed to test model sensitivity to changes in the river and coastal freshwater forcing, were: a control run, the reference (IBI-REF), using a freshwater forcing data identical to the one used in the CMEMS IBI operational forecast service (including both the punctual river contribution and the extra climatological coastal run-off), and a first IBI sensitivity run (defined as IBI-LAM) using the new LAMBDA river discharges at the 33 IBI river inputs and keeping the same extra added climatological coastal run-off. The second model test run (named IBI-NOR) uses the same LAMBDA river discharges, but in this case, no extra coastal run-off contribution is added. Finally, to build the last IBI model scenario (IBI-CLM), a pure climatological river and coastal freshwater forcing was used.

4.2. Is the IBI Model Application (Used in the Scenarios) Suitable to Simulate Salinity Field in the Study Area?

The CMEMS IBI operational forecasts (generated through the NEMO set-up and forced with the above-mentioned river and coastal freshwater forcing) can reproduce salinity fields with an adequate level of accuracy along the whole water column. The assessment of the IBI operational salinity product performed along the study case year (see results in Section 3.1) confirms a level of accuracy analogous to the one provided by Sotillo et al. [69] for longer time periods, with IBI salinity validation metrics (correlation and RMSE) within the ranges provided in the product quality document ([0.4–0.9] and [0.2–0.8 PSU], respectively). The assessment performed illustrates the important role played by Argo observations in the IBI routine validation. With respect to the use of satellite products, the utility of the new SMOS reprocessed maps should be emphasized. This dataset tends to represent salinity gradients and on-shelf small-scale features in the IBI area more realistically than the SMOS operational products (which are currently far from being fit for model validation purposes in coastal IBI waters). Thus, a higher frequency update of this reprocessed SMOS product (i.e., on a quarterly or monthly basis) can positively impact on the routine validation processes of operational models in the IBI area. Nevertheless, the use of these observational data sources as main references limits the operational model salinity assessments to offshore deep waters (where Argo floats' drifts and SMOS products show their highest representativeness), underrepresenting the model validation on coastal and on-shelf waters.

This limitation in the operational model validation can be partially overcome thanks to the availability of in-situ observations at some buoys moored along the shelf-break, on-shelf and coastal IBI waters. The product quality assessment performed comprises comparison of the IBI operational salinity product with these in-situ surface salinity observations, and some of the buoys used in this study (especially the ones moored along the Iberian shelf break and coast: M4–M13) are also operationally used in NARVAL for a routine near-real-time local assessment of IBI products.

According to some validation analyses done with in-situ observations from a buoy moored in NW Spain (the M12 one), Lorente et al. [17] proved that the IBI operational solution was able to capture shelf dynamics nearby ROFI areas, by better representing the horizontal extent and strength of a river freshwater plume. This validation approach is extended here to all the in-situ observations available in 2018 at NW Iberian mooring buoys (the M5–M12 ones). This analysis of IBI model and observed salinity timeseries has allowed identifying extreme salinity drop events (associated with across shelf water movements that extend river plumes and coastal fresher water zones to more offshore locations at the shelf-break and beyond). Thus, 35 remarkable surface salinity drops were recorded, with 24 of them being successfully forecasted by the CMEMS IBI service. On the other hand, 11 of these salinity decreases were missed or underestimated by the IBI model, whereas 22 "false alarm" events (salinity drops simulated by IBI, but not observed in-situ) were identified. This kind of observed and forecasted event categorization, summarized in contingency tables (Figure 5a), is quite helpful to evaluate the IBI model capabilities nearby ROFI areas. It gives a measure of how the IBI model realistically reproduces baroclinic frontal structures and their variability, which may be linked to changes in both local dynamical patterns and freshwater coastal and river discharges.

All these validation results of the IBI operational solution in 2018 confirm the IBI model set-up and the approach followed in it to count with the river and coastal freshwater forcing, as a valid approach to achieve realistic simulations of ocean salinity in the IBI area, and they support the decision of using this IBI NEMO model set-up (analogous to the one used in the CMEMS operational forecast service) to build the model scenarios proposed here.

4.3. Model Scenario Validation: Are Available In-Situ Observations Adequate to Evaluate Model Sensitivity?

The impacts on IBI model salinity related to changes in the river and coastal freshwater forcing are assessed by means of comparing the modelled salinity fields with in-situ salinity observations. To this purpose, a multi-platform observational salinity database (including profiles from CTDs, XBTs and surface observations from operational mooring buoys and a thermo-salinograph campaign) was compiled and used as a reference in the validation of the proposed regional model scenarios. The spatial observational coverage is illustrated in Figure 2. In-situ salinity observations on shelf and along the shelf-break have been prioritized, searching for observational coverage on ROFI areas. This approach showcases that prevailing effects of river plumes and coastal freshwater-induced fronts in such ROFI areas can be analyzed and the response of each model scenario run to simulate them can be assessed.

These IBI model scenario SSS outputs were validated through comparisons with the different in-situ observational data sources available in the study region. The comparison with observed salinity timeseries from moorings, mostly located on-shelf and along the shelf-break, shows that all the IBI simulations globally manage to accurately reproduce the spatial-temporal variability of salinity within the IBBIS domain (statistical metrics are provided for all the stations, and some examples of modelled and observed salinity timeseries at buoys moored in the BISCA, WISH and CADIZ areas are shown as examples in Figures 7, 9 and 12, respectively).

It is important to mention that most of the in-situ buoy stations available for the study are quite offshore, moored close to the shelf-break and far away from nearby coastal discharge areas. At these locations, the seasonal variability of salinity is mostly explained by the general circulation and the mesoscale dynamics. River freshwater discharges do not greatly influence the salinity seasonal cycle at the shelf break, but rather act on the higher (daily) variability of salinity, and coastal freshwater influence seems to be mostly linked to major abrupt freshwater intrusions (such as the ones identified, and discussed, by the validation of the operational IBI salinity performed in Section 3.2).

Moreover, it should be emphasized that local validation of model salinity against in-situ observations at specific moored buoys is very challenging since these abrupt low-salinity water intrusion features, that certainly mark model and observed salinity timeseries at given locations, can be frequently under-represented by the model due to small spatial or temporal feature shifts. Thus, in some of the occasions when the model is unable to reproduce one of these major salinity drops locally measured at a buoy, it is seen how the model is able to simulate the (usually coastal) fresher water mass that causes the recorded salinity drop, but it is not able to precisely locate the baroclinic frontal structure, not reaching (sometimes just by a matter of model grid points) the buoy location. On the other hand, in some other cases when the simulated frontal structure effectively reaches the buoy location, the fresher water mass arrival can be affected by some temporal shift, and then, abrupt salinity drops are simulated earlier or with some delay with respect to the observed event. Keeping this in mind, the model validation performed with in-situ mooring observations has shown that IBI model configuration is able to capture the level of dynamical activity, reproducing all the sensitivity test simulations of the main events with appropriate spatial-temporal accuracy.

4.4. How Different Are the River Flow Estimations Used as Forcing in the Ocean Model Scenarios?

Significant differences in terms of total daily river discharge contributions were identified among the different datasets imposed as freshwater forcing in each sensitivity model run. Table 3 shows how the 2018 daily mean flow (averaged over the whole IBBIS study domain) of the new LAMBDA river forcing reaches up to 8910 m^3s^{-1}, meaning a 70% higher contribution than the river forcing used in the IBI reference control run (5223 m^3s^{-1}) and 128% more than the pure climatological river discharge forcing (3893 m^3s^{-1}). Furthermore, the extra climatological coastal runoff, added to compensate lacks in the IBI freshwater signal, represents in the IBBIS region a yearly contribution of 843 m^3s^{-1}. It is worth noting that in the first months of 2018, the variability and intensity are equivalent in both the reference and the new LAMBDA datasets, with main peaks of freshwater in January and March. However, from April onwards, the reference data, mainly based on climatology, shows a smooth decrease in total river debit, whereas LAMBDA features a minimum in summer too, associated with a reduced variability. It also shows two important peaks in June and November, that do not exist in the IBI (climatological at that time) reference forcing (Figure 3).

4.5. What Are the Major Impacts in Salinity Associated to the Proposed Changes in Freshwater Forcing?

The river forcing plays a significant role in the regional simulation of the IBI sea surface salinity, as it is seen when looking at the modelled salinity fields resulting from the different IBI scenario runs (Figure 6). As expected, due to the higher river discharge of LAMDBA (as mentioned above, overall 70% higher than the reference one), the sensitivity runs based on this forcing (IBI_LAM and IBI_NOR) are fresher than the control one. Major impacts are found during spring and summer in the BISCA subregion, when IBI simulations forced by LAMBDA feature fresher water masses on-shelf, that extend up to open waters. On the contrary, the simulation forced by a climatology (IBI_CLM) is saltier than the other simulations (+0.05 PSU over the IBBIS domain compared to the reference in winter). Indeed, using a climatology instead of realistic daily river discharge significatively changes the surface/subsurface salinity budget, not only at the coast, but on the whole shelf and at the shelf-break. The SSS differences between the various simulations are in general more noticeable along the coast and on shelf waters, and especially obvious on the main IBI ROFI areas. Indeed, small SSS changes identified in offshore deeper waters seem more linked to the expected propagation of differences, when modeling structures after 12 months of free simulation, and there are not such model differences in deep-water areas (outside the Gulf of Biscay) driven by variations in the river and coastal freshwater forcing.

4.6. What Are the Regional Impacts in the Three Areas of Interest?

Each subregion included in the study domain features its proper dynamics and counts with different observational data sources, therefore, the assessment of salinity and frontal structure variability related to the impacts of changes on river discharge inputs has been conducted separately for the three proposed subregions: BISCA, WISHE and CADIZ.

4.7. Impacts in the Gulf of Biscay

The IBI scenario simulations are all coherent in the BISCA region with the local seasonal patterns. The use of climatological river data as forcing degrades the model solution in this region, changing surface/subsurface salinity budgets (not only at the coast, but on the whole shelf and at the shelf-break), and especially in terms of variability (featuring the climatological run in some mooring locations' (e.g., M6) unrealistic variability pattern, especially in wintertime). The model variability is enhanced when daily updated river data is included in the model forcing. This emphasizes the fact that using climatological instead of more realistic, higher frequency (daily) freshwater discharge inputs significantly changes the salinity budget on the shelf. The most important salinity differences between the model scenario runs are found in this BISCA region, and particularly on the French Shelf. In its northern side, discharges linked to the Vilaine and Loire river system impacts on surface and subsurface layers all year-long. LAMBDA freshwater inflows are bigger than the IBI reference ones and model simulations forced with LAMBDA in the region feature on-shelf fresher water masses that extend up to the open ocean (particularly in spring and summer). The combined input from the extra coastal runoff and LAMBDA (IBI_LAM) creates a significant fresh pool of water on the shelf, particularly in the South, which degrades the solution, at least at the RECOPESCA campaign time (Figure 8 and Table 5). However, by removing the extra coastal runoff (IBI_NOR), the solution is improved. This extra coastal runoff is not needed anymore on the southern shelf of Biscay, as the freshwater discharge is more realistic, even though it can be still needed in other zones in the northern BISCA region, such as Brittany, where rivers are not parametrized. In that sense, the BISCA southernmost area, especially along the northern Iberian shelf, arises as the most sensitive zone to the use of an extra coastal runoff (confirmed by the higher salinity simulated in the area by the run without such extra coastal climatological forcing: the IBI_NOR run). In this region, there are no major rivers (at least they are not included in the IBI river forcing as punctual freshwater sources), but many small freshwater inputs whose cumulated flow is not negligible. Then, here, using a monthly coastal runoff contribution added to the daily river freshwater forcing can locally improve simulated coastal salinity. However, this extra climatological coastal freshwater input must be tuned to avoid overestimations of total freshwater inputs that may result in unrealistic simulation of coastal freshwater fronts or across-shelf intrusions. Unfortunately, the lack of a network of coastal salinity observations is a limitation in the area to adequately tune available coastal freshwater inputs.

4.8. Impacts in the Western Iberian Shelf

Salinity in the WISHE zone is steadier than in the BISCA area and, as documented in the literature, it is defined by a coastal fresher water mass, usually limited to the narrow shelf, but occasionally extended offshore. The location of the baroclinic front is mainly controlled by the Iberian poleward slope current and its interactions with the shelf waters, resulting in complex front dynamics with occasional episodes marked by strong across-shelf-break advection of lower salinity water masses coming from coastal areas. These intrusions of waters, highly modified by the river and coastal freshwater influence, are clearly identified in the salinity records at mooring sites and they are captured to some extent by the IBI model application (as previously proven through the analysis of the IBI operational performance during the main 2018 observed salinity drops). The IBI model application is adequate to simulate shelf dynamics in ROFI areas of the WISHE zone and the different IBI model scenarios generally reproduce these observed freshwater intrusions, with small variations in timing and/or intensity (with the IBI-NOR and IBI-CLM runs showing better statistical metrics, depending on the buoy). Along the western Galician coast (monitored through the very coastal periodic INTECMAR CTD stations located in front of the Rias), it is the IBI-NOR run, forced with the new LAMBDA river data but without extra coastal runoff, the model scenario that better performs statistically (showing lower RMSE and bias when compared to the INTECMAR 2018 periodic salinity profiles). Emphasize that present IBI model set-up does not include any riverine inflows along this Galician coast, except the Minho and the major, but southern, regional freshwater contributors: the Douro and Mondego rivers. This would mean that salinity budget at this coastal zone off the Rias is rather controlled by the extension of the western Iberian buoyancy plume (mostly fed by freshwater contribution from these three major rivers) than by the local runoff from the nearby Rias itself. Further south, the model scenario assessment performed with the IPMA SSS campaign along the Portuguese shelf shows how all runs feature a similar salinity pattern. This suggests that, in this zone, variations in the imposed river discharge data may not be the main cause of salinity model errors on the shelf freshwater budget. This is so even nearby the Tejo river mouth, where all runs show a similar pattern of salinity differences, pointing out other dynamical factors as more determinant to explain this consistent model behavior.

4.9. *Impacts in the Gulf of Cadiz*

This consistency between model scenarios is also seen in the Gulf of Cadiz. Here, a similar salinity pattern is identified, with main differences between model runs located on-shelf, and especially on coastal areas close to the two rivers considered by IBI (the Guadalquivir and Guadiana) in the CADIZ zone. The occasional comparison with the SSS IPMA campaign (limited to a 9-day campaign in May 2018) indicates an error pattern marked by too-salty modelled waters in the western shelf, getting fresher as moving eastwards closer to the 2 IBI rivers' mouths. The model run that uses climatological discharges at these two rivers shows lower RMSE and bias values, whereas the 2 runs using LAMBDA inputs (IBI-LAM and IBI-NOR) show higher negative biases. This general anomalously lower model salinity seen in the area for all the runs is directly linked to either the excess of river freshwater contribution imposed at the 2 rivers, especially when using LAMBDA data, or an unrealistic accumulation of freshwater at the river mouths due to the model dynamics. All the nearby on-shelf areas seem affected, as confirmed at the single buoy available in the region (M13, moored offshore at deep waters, but not far from the Guadalquivir mouth).

5. Conclusions and Future Research Directions

In conclusion, the effect of varying imposed river outflows in the IBI operational ocean model system was investigated, showcasing the potential impacts that a new river freshwater model database (such as the LAMBDA one) can have on regional operational forecasts. Some enhancements in model capabilities to better represent salinity and especially baroclinic frontal structures linked to coastal and river freshwater buoyancy plumes have been demonstrated.

Although major impacts were identified on ROFI areas associated with bigger river discharges (i.e., the French shelf in the Gulf of Biscay or the Northwestern Iberian coast), it is found that in some other regions (such as the Portuguese shelf) or in areas with lower riverine freshwater contribution along the study year (such as the Gulf of Cadiz), these impacts related to changes in the imposed river inflows are lower, with other dynamical factors playing a more important role in governing the modelled salinity field.

The CMEMS IBI operational model set-up can reproduce salinity fields with an adequate level of accuracy along the whole water column, including remarkable salinity drop events linked to across shelf water movements that extend river plumes and coastal fresher water zones to more offshore locations at the self-break and beyond. The validation of the 2018 IBI operational solution confirms that the IBI model set-up (and the approach used to include in it the river and coastal fresh water forcing) is valid to achieve realistic simulations of ocean salinity in the IBI area, and they support the decision of using this IBI NEMO model set-up as a base to build the model scenarios proposed here.

The river forcing plays a significant role in the regional simulation of the IBI sea surface salinity, as it is seen when looking at the modelled salinity fields resulting from the different IBI scenario runs. Major impacts are found during spring and summer in the BISCA subregion, when IBI simulations forced by LAMBDA feature fresher water masses on shelf, that extend up to open waters. On the contrary, the simulation forced by a climatology is saltier than the other simulations. The SSS differences between the various simulations are in general more noticeable along the coast and on shelf waters, and especially obvious on the main IBI ROFI areas.

Significant differences in terms of total daily river discharge contributions were identified among the different river flow datasets imposed as freshwater forcing in each sensitivity model run. The use of "realistic" daily updated model-derived river inputs can benefit operational ocean models, especially improving their ability to capture salinity variability, and as demonstrated through the model sensitivity tests performed, their uses can be an option to avoid, or at least to minimize, the use of more static climatological approaches (such as the coastal climatological correction currently applied in the IBI model set-up). Furthermore, some regions, identified here as sensitive to changes in the freshwater forcing data, are of special interest to upgrade their river freshwater contribution, replacing the current freshwater climatological approach. The most noticeable case is the northern Iberian coast, where the current IBI operational model set-up does not count with any river source (with all the regional freshwater contribution coming from the coastal runoff correction applied), despite the existence of numerous small rivers in the area whose cumulated freshwater contribution into the ocean is not negligible (all year round).

A full network of coastal salinity observations would be optimal to evaluate and tune the contribution of freshwater forcing data in ocean simulations. This need of an enhanced in-situ observational coverage should be met for on-shelf and coastal zones (and especially in ROFI areas). Recovered ancillary databases or specific campaigns' data (such as the RECOPESCA or IPMA data used here) can be useful to study cases, but sustained operational monitoring (i.e., the moored buoys) or routine periodic repetition of salinity observations at specific locations (such as the INTECMAR data) make it possible to analyze the salinity variability. Improvements of satellite-based products are also desired. the with the SMOS products available for coastal uses being mostly reprocessed ones (with an intense use of in-situ observations for calibration and filling gap purposes), again, the in-situ component arises as critical. It should be highlighted that enhancement of the river hydrological observation component is also a key issue: without river flow observations, calibration of hydrological models is limited, decreasing the accuracy of freshwater river flow estimations available to force ocean models.

The progressive replacement of static climatological river inputs in ocean models is likely to become a key research line, needed to upgrade operational ocean circulation models, with a broadening of scope, especially from coastal model services, to increase the number of rivers to be considered as part of the land boundary contribution. This would not be only a matter of including freshwater contribution from major rivers in ocean models (currently the most common approach in operational systems), but also to count with relatively minor ones, which can play a significant role in local coastal dynamics, enhancing its influence under extreme weather event conditions. In that sense, Ruiz-Parrado et al. [70] and Sotillo et al. [71] show how a lack of adequate real-time updated river inputs is a major limitation for the performance of ocean circulation models, including the IBI one, during specific storm events.

Future research lines could include the application of these updated river freshwater model estimations into ocean models, not only in hindcast or analysis mode, but also in forecast runs. The objective would be to substitute the present common approach, mostly based on the use of persistence of the last available river discharge value, by near-real-time updated forecasted river estimations. This upgrade, that can certainly enhance ocean forecast skill, may lead to more integrated multi-disciplinary approaches based on combined near-real-time forecast runs of ocean and hydrological models (both models using the same atmospheric forcing to keep consistency).

Author Contributions: Conceptualization, M.G.S.; methodology, K.G., A.M. and M.G.S.; software, K.G., A.M. and M.A.A.-B.; validation, P.L., K.G. and A.M.; resources, F.C., F.S., E.O., A.N. and M.G.S.; writing—original draft preparation, M.G.S., K.G., P.L. and A.M.; writing—review and editing, F.C., F.S., E.O. and A.N.; project administration, M.G.S., F.C. and A.N. All authors have read and agreed to the published version of the manuscript.

Funding: Part of this research work was conducted in the framework of the following 2 projects: The EU Interreg Atlantic Area MyCoast Project EAPA_285/2016 (F.C., A.M. and M.G.S.) and the CMEMS Service Evolution Project LAMBDA (F.C., F.S., E.O. and A.N.). It was also supported by activity from the CMEMS IBI-MFC (K.G., P.L. and A.A.B.).

Institutional Review Board Statement: Not applicable.

Informed Consent Statement: Not applicable.

Data Availability Statement: The following publicly available datasets were analyzed in this study: Model data: The CMEMS IBI MFC operational forecast product can be found in the CMEMS catalogue (IBI-MFC forecast analysis product: https://resources.marine.copernicus.eu/?option=com_csw&view=details&product_id=IBI_ANALYSISFORECAST_PHY_005_001 (accessed on 8 April 2021)); LAMBDA river data can be found in: http://www.cmems-lambda.eu/#data-portal (accessed on 8 April 2021); data from the different ocean model scenarios analyzed in the study can be available on request from the corresponding author. Observational data: The data from in-situ buoys, Argo floats and Recopesca can be found in the CMEMS catalogue (Insitu-TAC Near-Real-Time observational product: https://resources.marine.copernicus.eu/?option=com_csw&view=details&product_id=INSITU_IBI_NRT_OBSERVATIONS_013_033 (accessed on 8 April 2021)). The salinity data from SMOS (the low-resolution level 3 SSS product computed with smoothening spatial window of 50-km radius) can be found in the BEC public ftp catalogue: http://bec.icm.csic.es/bec-ftp-service/ (accessed on 8 April 2021); Finally, the INTECMAR and IPMA observational salinity datasets used in this study are 3rd Party Data, and restrictions are applied to their availability (contact with the institutions owner of these datasets would be required for access permission).

Acknowledgments: The authors are grateful to the following experts (and institutions) for their support to this research: Pedro Montero (INTECMAR) for supporting with the INTECMAR CTD data, Guillaume Charria (Ifremer) for supporting with the RECOPESCA data, Diogo André Reis de Sousa and Paulo Oliveira (both from the Instituto Português do Mar e da Atmosfera) for the IPMA campaign data. Other MyCoast ocean modelers: Juan Taboada (MeteoGalicia), Joao Sobrino (IST), Tomasz Dabrowski (IMI). This study has been conducted using E.U. Copernicus Marine Service Information. The authors especially thank the CMEMS Insitu-TAC (for the moored buoys and Argo profilers) and the CMEMS IBI-MFC (for their IBI operational model products and validation assessments).

Conflicts of Interest: The authors declare no conflict of interest.

References

1. Dai, A.; Trenberth, K.E. Estimates of Freshwater Discharge from Continents: Latitudinal and Seasonal Variations. *J. Hydrometeorol.* **2002**, *3*, 660–687. [CrossRef]
2. Garvine, R.W.; Whitney, M.M. An estuarine box model of freshwater delivery to the coastal ocean for use in climate models. *J. Mar. Res.* **2006**, *64*, 173–194. [CrossRef]
3. Fong, D.A.; Geyer, W.R. The Alongshore Transport of Freshwater in a Surface-Trapped River Plume. *J. Phys. Oceanogr.* **2002**, *32*, 957–972. [CrossRef]
4. Simpson, J.H. Physical processes in the ROFI regime. *J. Mar. Syst.* **1997**, *12*, 3–15. [CrossRef]
5. Otero, P.; Ruiz-Villarreal, M.; Peliz, A. Variability of river plumes off Northwest Iberia in response to wind events. *J. Mar. Syst.* **2008**. [CrossRef]
6. Kourafalou, V.H.; Androulidakis, S.Y. Influence of Mississippi River induced circulation on the Deepwater Horizon oil spill transport. *J. Geophys. Res.* **2013**, *118*, 3823–3842. [CrossRef]
7. Banas, N.S.; MacCready, P.; Hickey, B.M. The Columbia River plume as cross-shelf exporter and along-coast barrier. *Cont. Shelf Res.* **2009**, *29*, 292–301. [CrossRef]
8. Santos, A.M.P.; Chícharo, A.; Dos Santos, A.; Moita, T.; Oliveira, P.B.; Peliz, Á.; Ré, P. Physical–biological interactions in the life history of small pelagic fish in the Western Iberia Upwelling Ecosystem. *Prog. Oceanogr.* **2007**, *74*, 192–209. [CrossRef]
9. Zuo, H.; de Boisséson, E.; Zsoter, E.; Harrigan, S.; de Rosnay, P.; Wetterhall, F.; Prudhomme, C. Benefits of dynamically modelled river discharge input for ocean and coupled atmosphere-land-ocean systems. In Proceedings of the EGU General Assembly, Vienna, Austria, 4–8 May 2020. [CrossRef]

10. Bourdalle-Badie, R.; Treguier, A.M. *Mercator-Ocean Report. A Climatology of Runoff for the Global Ocean-Ice Model ORCA025*; Technical Report MOO-RP-425-365-MER; Mercator Ocean: Toulouse, France, 2006.
11. Dai, A.; Qian, T.; Trenberth, K.E.; Milliman, J.D. Changes in Continental Freshwater Discharge from 1948 to 2004. *J. Clim.* **2008**, *22*, 2773–2792. [CrossRef]
12. Global Runoff Data Centre. Available online: https://www.bafg.de/GRDC/EN/Home/homepage_node.html (accessed on 2 June 2020).
13. Harrigan, S.; Zsoter, E.; Alfieri, L.; Prudhomme, C.; Salamon, P.; Wetterhall, F.; Barnard, C.; Cloke, H.; Pappenberger, F. GloFAS-ERA5 operational global river discharge reanalysis 1979–present. *Earth Syst. Sci. Data Discuss* **2020**. [CrossRef]
14. Kourafalou, V.H.; de Mey, P.; Staneva, J.; Ayoub, N.; Barth, A.; Chao, Y.; Cirano, M.; Fiechter, J.; Herzfeld, M.; Kurapov, A.; et al. Coastal Ocean Forecasting: Science foundation and user benefits. *J. Oper. Oceanogr.* **2015**, *8*, 147–167. [CrossRef]
15. Campuzano, F.; Brito, D.; Juliano, M.; Fernandes, R.; de Pablo, H.; Neves, R. Coupling watersheds, estuaries and regional ocean through numerical modelling for western iberia: A novel methodology. *Ocean Dyn.* **2016**, *66*, 1745–1756. [CrossRef]
16. Zheng, L.; Weisberg, R.H. Modelling the West Florida Coastal Ocean by Downscaling from the Deep Ocean, Across the Continental Shelf and into the Estuaries. *Ocean Model.* **2012**, *48*, 10–29. [CrossRef]
17. Lorente, P.; Sotillo, M.; Amo-Baladrón, A.; Aznar, R.; Levier, B.; Sánchez-Garrido, J.C.; Sammartino, S.; de Pascual-Collar, Á.; Reffray, G.; Toledano, C.; et al. Skill assessment of global, regional, and coastal circulation forecast models: Evaluating the benefits of dynamical downscaling in IBI (Iberia–Biscay–Ireland) surface waters. *Ocean Sci.* **2019**, *15*, 967–996. [CrossRef]
18. Sotillo, M.G.; Cerralbo, P.; Lorente, P.; Grifoll, M.; Espino, M.; Sanchez-Arcilla, A.; Álvarez-Fanjul, E. Coastal ocean forecasting in Spanish ports: The SAMOA operational service. *J. Oper. Oceanogr.* **2019**, *13*, 37–54. [CrossRef]
19. EU Copernicus Marine Environment Monitoring Service (CMEMS). Available online: https://marine.copernicus.eu/ (accessed on 3 September 2020).
20. Sotillo, M.G.; Cailleau, S.; Lorente, P.; Levier, B.; Aznar, R.; Reffray, G.; Amo-Baladrón, A.; Alvarez-Fanjul, E. The MyOcean IBI ocean forecast and reanalysis systems: Operational products and roadmap to the future Copernicus service. *J. Oper. Oceanogr.* **2015**, *8*, 63–79. [CrossRef]
21. Le Traon, P.Y.; Reppucci, A.; Fanjul, E.A.; Aouf, L.; Behrens, A.; Belmonte, M.; Bentamy, A.; Bertino, L.; Brando, V.E.; Kreiner, M.B.; et al. From observation to information and users: The Copernicus Marine service perspective. *Front. Mar. Sci.* **2019**, *6*, 234. [CrossRef]
22. Capet, A.; Fernández, V.; She, J.; Dabrowski, T.; Umgiesser, G.; Staneva, J.; Mészáros, L.; Campuzano, F.; Ursella, L.; Nolan, G.; et al. Operational Modeling Capacity in European Seas—An EuroGOOS Perspective and Recommendations for Improvement. *Front. Mar. Sci.* **2020**, *7*, 129. [CrossRef]
23. MyCoast Project (EU INTERREG Atlantic Area Transnational Cooperation Programme). Available online: http://www.mycoast-project.org/ (accessed on 3 September 2020).
24. Matulka, A.; Lorente, P.; Sotillo, M.G.; Campuzano, F.; Sobrinho, J.; Taboada, J.; Melo, P.; Ferrer, L.; Robert, M.; Dabrowski, T.; et al. MyCOAST Regional & Coastal Ocean Models. In Proceedings of the 1st MyCOAST Regional Workshop Southeastern Bay of Biscay, San Sebastian, Spain, 11–13 November 2019.
25. CMEMS STAC. Copernicus Marine Environment Monitoring Service (CMEMS) Service Evolution Strategy: R&D priorities. Version 3. 2017. Available online: http://marine.copernicus.eu/wp-content/uploads/2017/06/CMEMS-Service_evolution_strategy_RD_priorities_V3-final.pdf (accessed on 3 September 2020).
26. Bronco Project. CMEMS Service Evolution Project. Available online: https://www.mercator-ocean.fr/en/portfolio/bronco-2/ (accessed on 3 September 2020).
27. Lambda Project. CMEMS Service Evolution Project. Available online: http://www.cmems-lambda.eu/ (accessed on 6 April 2021).
28. Pingree, R.D.; Le Cann, B. A shallow meddy (a smeddy) from the secondary Mediterranean salinity maximum. *J. Geophys. Res. Ocean.* **1993**, *98*, 20169–20185. [CrossRef]
29. Jia, Y. Formation of an Azores Current Due to Mediterranean Overflow in a Modeling Study of the North Atlantic. *J. Phys. Oceanogr.* **2000**, *30*, 2342–2358. [CrossRef]
30. Carracedo, L.; Gilcoto, M.; Mercier, H.; Pérez, F. Seasonal dynamics in the Azores–Gibraltar Strait region: A climatologically-based study. *Prog. Oceanogr.* **2014**, *122*, 116–130. [CrossRef]
31. Sánchez, R.F.; Relvas, P. Spring–summer climatological circulation in the upper layer in the region of Cape St. Vincent, Southwest Portugal. *ICES J. Mar. Sci.* **2003**, *60*, 1232–1250. [CrossRef]
32. Folkard, A.M.; Davies, P.A.; Fiúza, A.F.G.; Ambar, I. Remotely sensed sea surface thermal patterns in the Gulf of Cadiz and the Strait of Gibraltar: Variability, correlations, and relationships with the surface wind field. *J. Geophys. Res. Ocean.* **1997**, *102*, 5669–5683. [CrossRef]
33. Lobo, F.J.; Le Roy, P.; Mendes, I.; Sahabi, M.; Chiocci, F.L.; Chivas, A.R. (Eds.) *Continental Shelves of the World: Their Evolution During the Last Glacio-Eustatic Cycle. 9. The Gulf of Cádiz Continental Shelves*; The Geological Society of London: London, UK, 2014. [CrossRef]
34. Peliz, Á.; Santos, A.M.P.; Oliveira, P.B.; Dubert, J. Extreme cross-shelf transport induced by eddy interactions southwest of Iberia in winter 2001. *Geophys. Res. Lett.* **2004**, *31*. [CrossRef]

35. Lobo, F.; Sánchez, R.; González, R.; Dias, J.; Hernández-Molina, F.; Fernández-Salas, L.; del Río, V.D.; Mendes, I. Contrasting styles of the Holocene highstand sedimentation and sediment dispersal systems in the northern shelf of the Gulf of Cadiz. *Cont. Shelf Res.* **2004**, *24*, 461–482. [CrossRef]
36. González, A.F.; Otero, J.; Guerra, A.; Prego, R.; Rocha, F.J.; Dale, A.W. Distribution of common octopus and common squid paralarvae in a wind-driven upwelling area (Ria of Vigo, northwestern Spain). *J. Plankton Res.* **2005**, *27*, 271–277. [CrossRef]
37. Haynes, R.; Barton, E.D.; Pilling, I. Development, Persistence, and Variability of Upwelling Filaments off the Atlantic Coast of the Iberian Peninsula. *J. Geophys. Res.* **1993**, *98*, 22681–22692. [CrossRef]
38. Peliz, A.; Dubert, J.; Santos, A.M.P.; Oliveira, P.B.; Le Cann, B. Winter upper ocean circulation in the Western Iberian Basin—Fronts, Eddies and Poleward Flows: An overview. *Deep Sea Res.* **2005**, *52*, 621–646. [CrossRef]
39. Kouttsikopoulos, C. Physical processes and hydrological structures related to the Bay of Biscay anchovy. *Sci. Mar.* **1996**, *60*, 9–19.
40. Kersalé, M.; Marié, L.; Le Cann, B.; Serpette, A.; Lathuilière, C.; Le Boyer, A.; Rubio, A.; Lazure, P. Poleward along-shore current pulses on the inner shelf of the Bay of Biscay. *Estuar. Coast. Shelf Sci.* **2016**, *179*, 155–171. [CrossRef]
41. Batifoulier, F.; Lazure, P.; Bonneton, P. Poleward coastal jets induced by westerlies in the Bay of Biscay. *J. Geophys. Res. Ocean.* **2012**, *117*. [CrossRef]
42. Solabarrieta, L.; Rubio, A.; Cárdenas, M.; Castanedo, S.; Esnaola, G.; Méndez, F.; Medina, R.; Ferrer, L. Probabilistic relationships between wind and surface water circulation patterns in the SE of Bay of Biscay. *Ocean Dyn.* **2015**, *65*, 1289–1303. [CrossRef]
43. Charria, G.; Theethen, S.; Vandermeirsch, F.; Yelekçi, O.; Audriffren, N. Interannual evolution of (sub)mesoscale dynamics in the Bay of Biscay. *Ocean Sci.* **2017**, *13*, 777–797. [CrossRef]
44. Reverdin, G.; Marié, L.; Lazure, P.; d'Ovidio, F.; Boutin, J.; Testor, P.; Martin, N.; Lourenco, A.; Gaillard, F.; Lavin, A.; et al. Freshwater from the Bay of Biscay shelves in 2009. *J. Mar. Syst.* **2013**, *109–110*, S134–S143. [CrossRef]
45. Castaing, P.; Froidefond, J.; Lazure, P.; Weber, O.; Prud'homme, R.; Jouanneau, J. Relationship between hydrology and seasonal distribution of suspended sediments on the continental shelf of the Bay of Biscay. *Deep Sea Res. Part II Top. Stud. Oceanogr.* **1999**, *46*, 1979–2001. [CrossRef]
46. Valencia, V.; Franco, J.; Borja, A.; Fontan, A.; Borja, A.; Collins, M. (Eds.) *Oceanography and Marine Environment of the Basque Country Hydrography of the Southeastern Bay of Biscay*; Elsevier: Amsterdam, The Netherlands, 2004; Chapter 7; pp. 159–194.
47. Ferrer, L.; Fontán, A.; Mader, J.; Chust, G.; Gonzáles, M.; Valencia, V.; Uriarte, A.; Collins, M.B. Low-salinity plumes in the oceanic region of the Basque Country. *Cont. Shelf Res.* **2009**, *29*, 970–984. [CrossRef]
48. Copernicus Marine In-Situ Tac Data Management Team. Product User Manual for Multiparameter Copernicus In Situ TAC NRT Product (PUM). 2020. Available online: https://archimer.ifremer.fr/doc/00324/43494/ (accessed on 8 April 2021).
49. Venâncio, A.; Montero, P.; Costa, P.; Regueiro, S.; Brands, S.; Taboada, J. An Integrated Perspective of the Operational Forecasting System in Rías Baixas (Galicia, Spain) with Observational Data and End-Users. In *Computational Science—ICCS 2019, Lecture Notes in Computer Science*; Springer: Cham, Switzerland, 2019; Volume 11539, pp. 229–239. ISBN 978-3-030-22746-3.
50. Losada, D.E.; Montero, P.; Brea, D.; Allen-Perkins, S.; Vila, B. Clustering Hydrographic Conditions in Galician Estuaries. In *Computational Science—ICCS 2019, Lecture Notes in Computer Science*; Springer: Cham, Switzerland, 2019; Volume 11539, pp. 346–360. ISBN 978-3-030-22746-3.
51. Szekely, T.; Bezaud, M.; Pouliquen, S.; Reverdin, G.; Charria, G. CORA-IBI, Coriolis Ocean Dataset for Reanalysis for the Ireland-Biscay-Iberia region. *SEANOE* **2017**. [CrossRef]
52. Leblond By, E.; Lazure, P.; Laurans, M.; Rioual, C.; Woerther, P.; Quemener, L.; Berthou, P. RECOPESCA: A new example of participative approach to colletc in-situ environmental and fisheries data. *Jt. Coriolis Mercator Ocean Q. Newsl.* **2010**, *37*, 40–55.
53. Brito, D.; Campuzano, F.J.; Sobrinho, J.; Fernandes, R.; Neves, R. Integrating operational watershed and coastal models for the Iberian Coast: Watershed model implementation—A first approach. *Estuar. Coast. Shelf Sci.* **2015**, *167 Pt A*, 138–146. [CrossRef]
54. Neves, R. The MOHID concept. In *Ocean Modelling for Coastal Management—Case Studies with MOHID*; Mateus, M., Neves, R., Eds.; IST Press: Lisbon, Portugal, 2013; pp. 1–11.
55. Campuzano, F.; Santos, F.; Ramos de Oliveira, A.I.; Simionesei, L.; Fernandes, R.; Brito, D.; Olmedo, E.; Turiel, A.; Alba, M.; Novellino, A.; et al. Framework for improving land boundary conditions in regional ocean products. In Proceedings of the EGU General Assembly, Vienna, Austria, 4–8 May 2020. [CrossRef]
56. Amo, A.; Reffray, G.; Sotillo, M.G.; Aznar, R.; Guihou, K. Product User Manual (PUM) for Atlantic-Iberian Biscay Irish-Ocean Physics Analysis and Forecasting Product: IBI_ANALYSIS_FORECAST_PHYS_005_0012020. Available online: https://resources.marine.copernicus.eu/documents/PUM/CMEMS-IBI-PUM-005-001.pdf (accessed on 3 September 2020).
57. Madec, G. *NEMO Ocean General Circulation Model, Reference Manual, Internal Report*; LODYC/IPSL: Paris, France, 2008.
58. Lellouche, J.-M.; Le Galloudec, O.; Drévillon, M.; Régnier, C.; Greiner, E.; Garric, G.; Ferry, N.; Desportes, C.; Testut, C.-E.; Bricaud, C.; et al. Evalution of global monotoring and forecasting systems at Mercator Océan. *Ocean Sci.* **2013**, *9*, 57–81. [CrossRef]
59. Chune, S.L.; Nouel, L.; Fernandez, E.; Derval, C.; Tressol, M.; Dussurget, R. Product User Manual (PUM) for GLOBAL Ocean Sea Physical Analysis and Forecasting Products GLOBAL_ANALYSIS_FORECAST_PHY_001_024. 2020. Available online: https://resources.marine.copernicus.eu/documents/PUM/CMEMS-GLO-PUM-001-024.pdf (accessed on 27 October 2020).
60. Carrere, L.; Lyard, F.; Cancet, M.; Guillot, A.; Picot, N. FES 2014, a new tidal model—Validation results and perspectives for improvements. In Proceedings of the ESA Living Planet Conference, Prague, Czech Republic, 9–13 May 2016.

61. Aznar, R.; Sotillo, M.G.; Cailleau, S.; Lorente, P.; Levier, B.; Amo-Baladrón, A.; Reffray, G.; Álvarez-Fanjul, E. Strengths and weaknesses of the CMEMS forecasted and reanalyzed solutions for the Iberia–biscay–Ireland (IBI) waters. *J. Mar. Syst.* **2016**, *159*, 1–14. [CrossRef]
62. Lorente, P.; Sotillo, M.G.; Amo-Baladrón, A.; Aznar, R.; Levier, B.; Aouf, L.; Dabrowski, T.; de Pascual, Á.; Dalphinet, G.R.A.; Toledano, C.; et al. The NARVAL Software Toolbox in Support of Ocean Models Skill Assessment at Regional and Coastal Scales. In *Computational Science—ICCS 2019, Lecture Notes in Computer Science*; Springer: Cham, Switzerland, 2019; Volume 11539, pp. 315–328. ISBN 978-3-030-22746-3. [CrossRef]
63. Olmedo, E.; Martínez, J.; Turiel, A.; Ballabrera-Poy, J.; Portabella, M. Debiased non-Bayesian retrieval: A novel approach to SMOS Sea Surface Salinity. *Remote Sens. Environ.* **2017**, *193*, 103–126. [CrossRef]
64. Grodsky, S.; Reverdin, G.; Carton, J.; Coles, V. Year to year salinity changes in the Amazon plume: Contrasting 2011 and 2012 Aquarius/SACD and SMOS satellite data. *Remote Sens. Environ.* **2014**, *140*, 14–22. [CrossRef]
65. Fournier, S.; Lee, T.; Gierach, M. Seasonal and Interannual variations of sea surface salinity associated with Mississippi River plume observed by SMOS and Aquarius. *Remote Sens. Environ.* **2016**, *180*, 431–439. [CrossRef]
66. Olmedo, E.; González-Haro, C.; Hoareau, N.; Umbert, N.; González-Gambau, V.; Martínez, J.; Gabarró, C.; Turiel, A. Nine years of SMOS Sea Surface Salinity global maps at the Barcelona Expert Center. *Earth Syst. Sci. Data Discuss.* **2020**. [CrossRef]
67. Olmedo, E.; González-Gambau, V.; Martínez, J.; González-Haro, C.; Turiel, A.; Portabella, M.; Arias, M.; Sabia, R.; Oliva, R.; Corbella, I. Characterization and Correction of the Latitudinal and Seasonal Bias in BEC SMOS Sea Surface Salinity Maps. In Proceedings of the 2019 IEEE International Geoscience and Remote Sensing Symposium (IGARSS'19), Yokohama, Japan, 28 July–2 August 2019. [CrossRef]
68. Campuzano, F. Coupling Watersheds, Estuaries and Regional Seas through Numerical Modelling for Western Iberia. Ph.D. Thesis, Instituto Superior Técnico, Universidade de Lisboa, Lisboa, Portugal, 2018. Available online: http://www.mohid.com/PublicData/products/Thesis/PhD_Francisco_Campuzano.pdf (accessed on 6 April 2021).
69. Sotillo, M.G.; Levier, B.; Lorente, P.; Guihou, K.; Aznar, R.; Amo, A.; Aouf, L.; Ghantous, M. CMEMS Quality Information Document (QUID) for Atlantic-Iberian Biscay Irish-Ocean Physics Analysis and Forecasting Product:IBI_ANALYSISFORECAST_PHYS_005_001. 2020. Available online: https://resources.marine.copernicus.eu/documents/QUID/CMEMS-IBI-QUID-005-001.pdf (accessed on 8 January 2021).
70. Ruiz-Parrado, I.; Genua-Olmedo, A.; Reyes, E.; Mourre, B.; Rotllán, P.; Lorente, P.; García-Sotillo, M.; Tintoré, J. Coastal ocean variability related to the most extreme Ebro River discharge over the last 15 years. *Section* in Copernicus Marine Service Ocean State Report, Issue 4. *J. Oper. Oceanogr.* **2020**, *13* (Suppl. 1), s160–s165. [CrossRef]
71. Sotillo, M.G.; Mourre, B.; Mestres, M.; Lorente, P.; Aznar, R.; García-León, M.; Liste, M.; Santana, A.; Espino, M.; Álvarez, E. Evaluation of the operational CMEMS and coastal downstream ocean forecasting services during the storm Gloria (January 2020). *Front. Mar. Sci.* **2021**, *8*, 300. [CrossRef]

MDPI
St. Alban-Anlage 66
4052 Basel
Switzerland
Tel. +41 61 683 77 34
Fax +41 61 302 89 18
www.mdpi.com

Journal of Marine Science and Engineering Editorial Office
E-mail: jmse@mdpi.com
www.mdpi.com/journal/jmse

www.ingramcontent.com/pod-product-compliance
Lightning Source LLC
Chambersburg PA
CBHW061837210926
51850CB00006B/913